Methods in Molecular Biology

Series Editor
John M. Walker
School of Life and Medical Sciences
University of Hertfordshire
Hatfield, Hertfordshire, UK

For further volumes:
http://www.springer.com/series/7651

For over 35 years, biological scientists have come to rely on the research protocols and methodologies in the critically acclaimed *Methods in Molecular Biology* series. The series was the first to introduce the step-by-step protocols approach that has become the standard in all biomedical protocol publishing. Each protocol is provided in readily-reproducible step-by-step fashion, opening with an introductory overview, a list of the materials and reagents needed to complete the experiment, and followed by a detailed procedure that is supported with a helpful notes section offering tips and tricks of the trade as well as troubleshooting advice. These hallmark features were introduced by series editor Dr. John Walker and constitute the key ingredient in each and every volume of the *Methods in Molecular Biology* series. Tested and trusted, comprehensive and reliable, all protocols from the series are indexed in PubMed.

Cardiac Tissue Engineering

Methods and Protocols

Second Edition

Edited by

Kareen L. K. Coulombe

School of Engineering, Brown University, Providence, RI, USA

Lauren D. Black III

Department of Biomedical Engineering, Tufts University, Medford, MA, USA

Humana Press

Editors
Kareen L. K. Coulombe
School of Engineering
Brown University
Providence, RI, USA

Lauren D. Black III
Department of Biomedical Engineering
Tufts University
Medford, MA, USA

ISSN 1064-3745 ISSN 1940-6029 (electronic)
Methods in Molecular Biology
ISBN 978-1-0716-2260-5 ISBN 978-1-0716-2261-2 (eBook)
https://doi.org/10.1007/978-1-0716-2261-2

This Humana imprint is published by the registered company Springer Science+Business Media, LLC, part of Springer Nature.
The registered company address is: 1 New York Plaza, New York, NY 10004, U.S.A.

Preface

Due to the limited regenerative potential of the adult human heart, development of alternative treatment options is needed for the variety of conditions that result in the loss, or loss of function, of contractile tissue. Perhaps the prime example of this is myocardial infarction (MI), which results in death of tens of millions of cardiomyocytes. While this loss is significant, many patients receive minimal intervention following MI, and their cardiac health is managed with drugs and changes to diet and exercise with satisfactory results. However, within 5 years, more than half of MI patients will develop heart failure, for which the only successful late-stage treatment is heart transplantation.

In the first edition of this book, we published a wide array of protocols important for cardiac tissue engineering research. Many of these were geared towards in vitro advancement ahead of future clinical work. In the time since the first edition, the field has continued to advance in two key directions. The first is a deeper understanding of human pluripotent stem cell-derived cardiomyocytes in three-dimensional engineered microenvironments and their use in downstream applications. The second is continued movement towards clinical therapies using tissue engineering approaches in the heart in large animal models and human clinical trials.

In this book, the second edition of *Cardiac Tissue Engineering: Methods and Protocols*, an updated collection of state-of-the-art protocols in cardiac tissue engineering is presented. These protocols demonstrate advancements in cell sourcing, assembly, and use of engineered cardiac tissues, imaging and diagnostics, and applications. Platforms developed for broad use to study development and disease in vitro enable genotype to phenotype evaluation, and many are customized for contractility, arrhythmia, or heart repair in vivo using small and large animal models. New animal models, biomaterials, and quantitative analyses are described for broad adoption.

The diversity of research in cardiovascular development and disease has inspired the development of a number of techniques and platforms that can be used to address an array of questions, often leveraging the use of human pluripotent stem cell-derived cardiomyocytes (hiPSC-CMs). Gene editing for high-throughput genetic screening using a custom CRISPR library in cultured cardiomyocytes is described by DeLuca and Bursac. Challenges of protein and mRNA quantification from small samples of hiPSC-CMs are described and overcome by customized methods presented by Entcheva and colleagues to enable high-throughput analyses.

Tissue engineering methods are diverse, enabling selection of an approach based on the study question and design, including considerations for size, throughput, geometry, and endpoint metrics. Methods for forming self-assembled 3D spheroids of hiPSC-CMs and heterotypic myocardial cells are described by Matthys and McDevitt. Fabrication of myocardial scaffolds and slices, integrated with electrical and mechanical stimulation, is presented by Liao and colleagues. For more complex geometries, bioprinting of a 3D ventricle and tips for custom shapes using the freeform reversible embedding of suspended hydrogels (FRESH) method is shared by Feinberg and colleagues. For hydrogel-based elongated engineered heart tissues designed for contractile, structural, and transcriptional studies, the Sniadecki group provides templates and insights for lab-made racks of silicone posts in a standard 24-well dish format.

Fit-for-purpose platforms require alignment of the system's biology with quantitative readouts and has become ever-present as the field of cardiac tissue engineering plunges into drug testing and disease modeling in vitro. Use of microelectrode arrays for high-throughput field potential measurements in 2D plated hiPSC-CMs to assess drug responses is detailed by Wu and colleagues. Micro-heart muscle array technology from the Huebsch Lab enables moderate throughput in pharmacology and pharmacogenomic studies by visual assessment of action potential (AP), calcium transient, and contractility with compatibility for protein and gene analyses. Studies concerned with propagation velocity, or conduction velocity, as related to arrhythmia will benefit from high-speed visual fluorescence imaging of calcium waves and analysis algorithms presented by McCain and colleagues in an aligned 2D cardiac platform. A heterotypic hiPSC-CM and human cardiac fibroblast self-assembled 3D microtissue platform for arrhythmia assessment by quantitative evaluation of all phases of the action potential is presented by Kofron, Choi, and Coulombe. While the focus of much of the cardiac tissue engineering space is on ventricular tissue, an atrial cardiac 3D-engineered tissue model is described by Eschenhagen, Stenzig, et al. using elastomeric posts for auxotonic contractions with applications in atrial-specific studies of drug responses and disease modeling. A cardiac fibrosis model based on the Biowire II platform for contractility assessment from the Radisic group enables local high-fibroblast content to create scar-like tissue adjacent to normal cardiac tissue for studying fibrosis and therapeutics.

As the field of cardiac tissue engineering advances towards clinical heart regeneration, multiple approaches to remuscularize injured hearts with hiPSC/hESC-CMs are moving towards phase I trials. Transepicardial hiPSC-CM delivery in a swine model of acute myocardial infarction is described by Laflamme and colleagues, while defined cellular and culture conditions in collagen-based engineered heart tissue (EHT) is provided by Zimmermann and colleagues for disease modeling, drug screening, and heart repair. Methods for constructing tubular cardiac tissue from multilayered cell sheets are provided by Okano, Sekine et al. for applications in heart failure. Finally, in a critically important analysis of engraftment, Brandt and Mahmoud detail their methods for quantifying cardiomyocyte proliferation and nucleation in repaired hearts via robust histological methods.

Novel therapeutics that enable in situ repair and alternative approaches for regeneration highlight the innovation in the field of cardiac tissue engineering. Using a biomaterials intervention, Christman and colleagues describe injectable extracellular matrix (ECM) scaffolds that have been in use in small and large animal models for cardiac repair and initial safety assessment in a Phase I clinical trial. Use of ECM for encapsulation of cells for echocardiography-directed injection in rodent models is described by Shakya, Brown, and Davis. The impact of a biomaterials-based repair strategy on the monocyte population is described by Suuronen and colleagues using flow cytometry analyses to quantify the levels of major leukocyte subtypes isolated from mouse hearts. Finally, a novel model of patch-based repair is provided by Black and colleagues, where cardiovascular patches are implanted to widen the right ventricular outflow tract in young, rapidly growing porcine hearts to emulate congenital heart defect reconstructive surgery.

Bringing new technologies and therapies to the clinic is a challenging task, but one that is attainable, particularly if we as a field work in collaboration. This second edition of

Cardiac Tissue Engineering: Methods and Protocols aims to be your primary resource for implementing these cutting-edge approaches in your research. With this book, we hope to inspire advancement of cardiotoxicity assessment, drug discovery, and heart repair and regeneration to accelerate heart health around the globe.

Providence, RI, USA ***Kareen L. K. Coulombe***
Medford, MA, USA ***Lauren D. Black III***

Contents

Contributors

Emilio I. Alarcon • *BioEngineering and Therapeutic Solutions (BEaTS), Division of Cardiac Surgery, University of Ottawa Heart Institute, Ottawa, ON, Canada; Department of Biochemistry, Microbiology, and Immunology, University of Ottawa, Ottawa, ON, Canada*

Nadjet Belbachir • *Stanford Cardiovascular Institute, Stanford University, Stanford, CA, USA; Division of Cardiovascular Medicine, Department of Medicine, Stanford University, Stanford, CA, USA*

Lauren D. Black III • *Cellular, Molecular and Developmental Biology Program, Graduate School of Biomedical Sciences, Tufts University, Boston, MA, USA; Department of Biomedical Engineering, Tufts University, Medford, MA, USA*

Rebecca L. Braden • *Sanford Consortium for Regenerative Medicine, Department of Bioengineering, University of California San Diego, La Jolla, CA, USA*

Emma B. Brandt • *Department of Cell and Regenerative Biology, University of Wisconsin-Madison School of Medicine and Public Health, Madison, WI, USA*

Samantha Bremner • *Institute for Stem Cell and Regenerative Medicine, University of Washington, Seattle, WA, USA; Department of Bioengineering, University of Washington, Seattle, WA, USA*

Milton E. Brown • *Wallace H. Coulter Department of Biomedical Engineering, Georgia Institute of Technology and Emory University School of Medicine, Atlanta, GA, USA; Division of Cardiology, Department of Medicine, Emory University School of Medicine, Atlanta, GA, USA*

Nenad Bursac • *Department of Biomedical Engineering, Duke University, Durham, NC, USA*

Bum-Rak Choi • *Cardiovascular Research Center, Cardiovascular Institute, Rhode Island Hospital and Alpert Medical School of Brown University, Providence, RI, USA*

Karen L. Christman • *Sanford Consortium for Regenerative Medicine, Department of Bioengineering, University of California San Diego, La Jolla, CA, USA*

Brian D. Coffin • *Department of Materials Science and Engineering, Carnegie Mellon University, Pittsburgh, PA, USA*

Kareen L. K. Coulombe • *School of Engineering, Center for Biomedical Engineering, Brown University, Providence, RI, USA*

Nathan Cunningham • *Stanford Cardiovascular Institute, Stanford University, Stanford, CA, USA; Division of Cardiovascular Medicine, Department of Medicine, Stanford University, Stanford, CA, USA*

Michael E. Davis • *Wallace H. Coulter Department of Biomedical Engineering, Georgia Institute of Technology and Emory University School of Medicine, Atlanta, GA, USA; Division of Cardiology, Department of Medicine, Emory University School of Medicine, Atlanta, GA, USA; Children's Heart Research and Outcomes (HeRO) Center, Children's Healthcare of Atlanta and Emory University, Atlanta, GA, USA*

Amy L. de Jongh Curry • *Department of Biomedical Engineering, University of Memphis, Memphis, TN, USA*

Sophia DeLuca • *Department of Cell Biology, Duke University, Durham, NC, USA*

Alexandra Eder • *Department of Experimental Pharmacology and Toxicology, University Medical Center Hamburg-Eppendorf (UKE), Hamburg, Germany; DZHK (German Centre for Cardiovascular Research), Hamburg/Kiel/Lübeck, Germany*

Emilia Entcheva • *Department of Biomedical Engineering, School of Engineering and Applied Science, The George Washington University, Washington, DC, USA*

Thomas Eschenhagen • *Department of Experimental Pharmacology and Toxicology, University Medical Center Hamburg-Eppendorf (UKE), Hamburg, Germany; DZHK (German Centre for Cardiovascular Research), Hamburg/Kiel/Lübeck, Germany*

Adam W. Feinberg • *Department of Materials Science and Engineering, Carnegie Mellon University, Pittsburgh, PA, USA; Department of Biomedical Engineering, Carnegie Mellon University, Pittsburgh, PA, USA*

Alex J. Goldstein • *Institute for Stem Cell and Regenerative Medicine, University of Washington, Seattle, WA, USA; Department of Materials Science and Engineering, University of Washington, Seattle, WA, USA; Department of Laboratory Medicine & Pathology, University of Washington, Seattle, WA, USA*

Julie L. Han • *Department of Biomedical Engineering, School of Engineering and Applied Science, The George Washington University, Washington, DC, USA*

Ty Higashi • *Institute for Stem Cell and Regenerative Medicine, University of Washington, Seattle, WA, USA; Department of Mechanical Engineering, University of Washington, Seattle, WA, USA*

Yi Hong • *Department of Bioengineering, University of Texas at Arlington, Arlington, TX, USA*

Andrew R. Hudson • *Department of Biomedical Engineering, Carnegie Mellon University, Pittsburgh, PA, USA*

Nathaniel Huebsch • *Department of Biomedical Engineering, Washington University in St. Louis, St. Louis, MO, USA; NSF Science and Technology Center for Engineering Mechanobiology, Washington University in St. Louis, St. Louis, MO, USA; Center for Cardiovascular Research, Center for Regenerative Medicine, Center for Investigation of Membrane Excitability Diseases, Washington University in St. Louis, St. Louis, MO, USA*

Jervaughn D. Hunter • *Sanford Consortium for Regenerative Medicine, Department of Bioengineering, University of California San Diego, La Jolla, CA, USA*

Todd D. Johnson • *Sanford Consortium for Regenerative Medicine, Department of Bioengineering, University of California San Diego, La Jolla, CA, USA*

Celinda M. Kofron • *School of Engineering, Center for Biomedical Engineering, Brown University, Providence, RI, USA*

Julia Krause • *Department of Experimental Pharmacology and Toxicology, University Medical Center Hamburg-Eppendorf (UKE), Hamburg, Germany; DZHK (German Centre for Cardiovascular Research), Hamburg/Kiel/Lübeck, Germany*

Raymond K. Kudej • *Department of Clinical Sciences, Tufts University School of Veterinary Medicine, North Grafton, MA, USA*

Michael A. Laflamme • *McEwen Stem Cell Institute, University Health Network, Toronto, ON, Canada; Peter Munk Cardiac Centre, University Health Network, Toronto, ON, Canada; Department of Laboratory Medicine and Pathobiology, University of Toronto, Toronto, ON, Canada*

Andrew Lee • *Department of Biomedical Engineering, Carnegie Mellon University, Pittsburgh, PA, USA*

MARTA LEMME • *Department of Experimental Pharmacology and Toxicology, University Medical Center Hamburg-Eppendorf (UKE), Hamburg, Germany; DZHK (German Centre for Cardiovascular Research), Hamburg/Kiel/Lübeck, Germany*

JUN LIAO • *Department of Bioengineering, University of Texas at Arlington, Arlington, TX, USA*

WEIZHEN LI • *Department of Biomedical Engineering, School of Engineering and Applied Science, The George Washington University, Washington, DC, USA*

AHMED I. MAHMOUD • *Department of Cell and Regenerative Biology, University of Wisconsin-Madison School of Medicine and Public Health, Madison, WI, USA; University of Wisconsin-Madison School of Medicine and Public Health, Madison, WI, USA*

INGRA MANNHARDT • *Department of Experimental Pharmacology and Toxicology, University Medical Center Hamburg-Eppendorf (UKE), Hamburg, Germany; DZHK (German Centre for Cardiovascular Research), Hamburg/Kiel/Lübeck, Germany*

ORIANE B. MATTHYS • *UC Berkeley-UC San Francisco Graduate Program in Bioengineering, San Francisco, CA, USA; Gladstone Institutes, San Francisco, CA, USA*

MEGAN L. MCCAIN • *Laboratory for Living Systems Engineering, Department of Biomedical Engineering, USC Viterbi School of Engineering, University of Southern California, Los Angeles, CA, USA; Department of Stem Cell Biology and Regenerative Medicine, Keck School of Medicine of USC, University of Southern California, Los Angeles, CA, USA*

TODD C. MCDEVITT • *Gladstone Institutes, San Francisco, CA, USA; Department of Bioengineering and Therapeutic Sciences, University of California, San Francisco, CA, USA*

SARAH MCLAUGHLIN • *BioEngineering and Therapeutic Solutions (BEaTS), Division of Cardiac Surgery, University of Ottawa Heart Institute, Ottawa, ON, Canada; Department of Cellular and Molecular Medicine, University of Ottawa, Ottawa, ON, Canada*

TIM MEYER • *Institute of Pharmacology and Toxicology, University Medical Center Göttingen, Göttingen, Germany; DZHK (German Center for Cardiovascular Research), Partner Site Göttingen, Göttingen, Germany*

TERUO OKANO • *Institute of Advanced Biomedical Engineering and Science, Tokyo Women's Medical University, Tokyo, Japan; Center for Advanced Medical and Life Science, Tokyo Women's Medical University, Tokyo, Japan; Cell Sheet Tissue Engineering Center (CSTEC), School of Medicine and College of Pharmacy, Department of Pharmaceutics and Pharmaceutical Chemistry, University of Utah, Salt Lake City, UT, USA*

ANDREW P. PETERSEN • *Laboratory for Living Systems Engineering, Department of Biomedical Engineering, USC Viterbi School of Engineering, University of Southern California, Los Angeles, CA, USA*

ELIZABETH C. PORTER • *Cellular, Molecular and Developmental Biology Program, Graduate School of Biomedical Sciences, Tufts University, Boston, MA, USA*

BEIPING QIANG • *McEwen Stem Cell Institute, University Health Network, Toronto, ON, Canada*

MILICA RADISIC • *Institute of Biomedical Engineering, University of Toronto, Toronto, ON, Canada; Department of Chemical Engineering and Applied Chemistry, University of Toronto, Toronto, ON, Canada; Toronto General Research Institute, University Health Network, Toronto, ON, Canada*

ROCCO ROMAGNUOLO • *McEwen Stem Cell Institute, University Health Network, Toronto, ON, Canada*
PIERRE-LUC SATIN • *Institute of Pharmacology and Toxicology, University Medical Center Göttingen, Göttingen, Germany; DZHK (German Center for Cardiovascular Research), Partner Site Göttingen, Göttingen, Germany*
HIDEKAZU SEKINE • *Institute of Advanced Biomedical Engineering and Science, Tokyo Women's Medical University, Tokyo, Japan; Center for Advanced Medical and Life Science, Tokyo Women's Medical University, Tokyo, Japan; Cell Sheet Tissue Engineering Center (CSTEC), School of Medicine and College of Pharmacy, Department of Pharmaceutics and Pharmaceutical Chemistry, University of Utah, Salt Lake City, UT, USA*
MICKEY SHAH • *Department of Biomedical Engineering, The University of Akron, Akron, OH, USA*
PREETY SHAKYA • *Wallace H. Coulter Department of Biomedical Engineering, Georgia Institute of Technology and Emory University School of Medicine, Atlanta, GA, USA*
DANIEL W. SIMMONS • *Department of Biomedical Engineering, Washington University in St. Louis, St. Louis, MO, USA; NSF Science and Technology Center for Engineering Mechanobiology, Washington University in St. Louis, St. Louis, MO, USA*
JACOB SMITH • *Department of Chemical Engineering and Applied Chemistry, University of Toronto, Toronto, ON, Canada*
DAVID SMYTH • *Cardiac Function Laboratory, University of Ottawa Heart Institute, Ottawa, ON, Canada*
NATHAN J. SNIADECKI • *Institute for Stem Cell and Regenerative Medicine, University of Washington, Seattle, WA, USA; Department of Bioengineering, University of Washington, Seattle, WA, USA; Department of Mechanical Engineering, University of Washington, Seattle, WA, USA; Department of Laboratory Medicine & Pathology, University of Washington, Seattle, WA, USA*
JUSTUS STENZIG • *Department of Experimental Pharmacology and Toxicology, University Medical Center Hamburg-Eppendorf (UKE), Hamburg, Germany; DZHK (German Centre for Cardiovascular Research), Hamburg/Kiel/Lübeck, Germany*
WHITNEY L. STOPPEL • *Department of Biomedical Engineering, Tufts University, Medford, MA, USA; Department of Chemical Engineering, University of Florida, Gainesville, FL, USA*
ERIK J. SUURONEN • *BioEngineering and Therapeutic Solutions (BEaTS), Division of Cardiac Surgery, University of Ottawa Heart Institute, Ottawa, ON, Canada; Department of Cellular and Molecular Medicine, University of Ottawa, Ottawa, ON, Canada*
MALTE TIBURCY • *Institute of Pharmacology and Toxicology, University Medical Center Göttingen, Göttingen, Germany; DZHK (German Center for Cardiovascular Research), Partner Site Göttingen, Göttingen, Germany*
BÄRBEL ULMER • *Department of Experimental Pharmacology and Toxicology, University Medical Center Hamburg-Eppendorf (UKE), Hamburg, Germany; DZHK (German Centre for Cardiovascular Research), Hamburg/Kiel/Lübeck, Germany*
BO WANG • *Joint Department of Biomedical Engineering, Marquette University and the Medical College of Wisconsin, Milwaukee, WI, USA*
ERIKA YAN WANG • *Institute of Biomedical Engineering, University of Toronto, Toronto, ON, Canada*
LAKIESHA N. WILLIAMS • *Department of Biomedical Engineering, University of Florida, Gainesville, FL, USA*

JOSEPH C. WU • *Stanford Cardiovascular Institute, Stanford University, Stanford, CA, USA; Division of Cardiovascular Medicine, Department of Medicine, Stanford University, Stanford, CA, USA; Department of Radiology, Stanford University School of Medicine, Stanford, CA, USA*

FANNY WULKAN • *McEwen Stem Cell Institute, University Health Network, Toronto, ON, Canada*

GE ZHANG • *Department of Biomedical Engineering, The University of Akron, Akron, OH, USA*

WOLFRAM-HUBERTUS ZIMMERMANN • *Institute of Pharmacology and Toxicology, University Medical Center Göttingen, Göttingen, Germany; DZHK (German Center for Cardiovascular Research), Partner Site Göttingen, Göttingen, Germany; Cluster of Excellence "Multiscale Bioimaging: From Molecular Machines to Networks of Excitable Cells" (MBExC), University of Göttingen, Göttingen, Germany; Center for Neurodegenerative Diseases (DZNE), Göttingen, Germany; Fraunhofer Institute for Translational Medicine and Pharmacology (ITMP), Göttingen, Germany*

Chapter 1

CRISPR Library Screening in Cultured Cardiomyocytes

Sophia DeLuca and Nenad Bursac

Abstract

CRISPR-Cas9-based screening technologies enable precise, high-throughput genetic and epigenetic manipulation to study mechanisms of development and disease and identify new therapeutic targets. Here, we describe a general protocol for the generation of custom, pooled CRISPR sgRNA libraries for screening in cardiomyocyte cultures. This methodology can address a variety of lab-specific research questions in cardiomyocytes and other cell types, as the genes to be modified can be curated or whole genomes can be investigated. The use of lentiviral sgRNA delivery followed by high-throughput sequencing allows for rapid comparison and identification of candidate genes and epigenetic modifiers, which can be further validated individually or in sub-pooled libraries following screening.

Key words CRISPR, Genetic screen, Cardiomyocyte, Knock-out, High-throughput, Proliferation, Maturation, Survival

1 Introduction

The application of clustered regularly interspersed palindromic repeat (CRISPR) methods to modern genome engineering has been transformative for the field, as the ability to perform precise genetic manipulations at almost any locus of interest has become both expedient and accessible [1]. One important application of CRISPR-Cas9 technology is the ability to perform efficient and high-throughput in vitro genetic screens to address a broad spectrum of scientific questions. There are several types of CRISPR screens that can be performed (Table 1), with gene knock-out (CRISPR-KO) screens being widely used [2]. However, newer techniques for epigenetic screening that involve CRISPR-based gene upregulation or repression through the use of catalytically inactive Cas9 (dCas9) fused to functional domains of chromatin-modifying proteins are also available [3]. These epigenetic activation and inhibition screens offer additional flexibility for the study of regulatory mechanisms in the development and disease and for identification of novel therapeutic targets [4].

Kareen L. K. Coulombe and Lauren D. Black III (eds.), *Cardiac Tissue Engineering: Methods and Protocols*,
Methods in Molecular Biology, vol. 2485, https://doi.org/10.1007/978-1-0716-2261-2_1,

Table 1
A summary of CRISPR screening technologies applicable to pooled CRISPR sgRNA library screening

Function	Type of Cas9	Targeting location	Notes
Knock-out	spCas9	First exon	Cas9 derived from *S. pyogenes* is DNA endonuclease used to induce frame-shift mutations at the targeted locus, resulting in gene inactivation
Activation	dCas9 fused to VP64	Promoter	VP64 is a tetrameric repeat of the herpes simplex virus protein VP16, which induces transcriptional gene activation [23, 24]
Activation	dCas9 fused to p300	Promoter	p300 is a histone acetyltransferase which facilitates gene transcription [25]
Activation	dCas9 fused to PRDM9	Promoter	PRDM9 is a histone methyltransferase used to stabilize gene expression via induction of the activating mark H3K4me3
Inhibition or activation	dCas9 fused to HDAC3	Promoter	HDAC3 is a histone deacetylase associated with both gene activation and repression depending on the targeted locus [26]
Inhibition	dCas9 fused to Dnmt3a	Promoter or CpG islands	Dnmt3a is a DNA methyltransferase used to induce targeted DNA methylation and suppress gene transcription [27, 28]
Inhibition	dCas9 fused to KRAB	Promoter or enhancer	Kruppel-associated box domain (KRAB) recruits a complex responsible for both histone methylation and deacetylation, resulting in heterochromatin formation and repression of gene transcription [29–31]
Inhibition	dCas9 fused to KRAB–MeCP2	Promoter or CpG islands	KRAB in combination with methyl CpG-binding protein 2 (MeCP2) aids in gene silencing via complex formation with histone deacetylases and via direct interaction with transcription factors [32]
Inhibition	dCas9 fused to LSD1	Promoter	LSD1 is a histone demethylase used to repress enhancers by removing H3K4me2 mark from histone, resulting in reduced gene expression due to enhancer inactivation [33]
Mutagenesis	dCas9 fused to AIDx	Gene or regulatory region	AID is activation-induced cytidine deaminase with the ability to generate a wide array of targeted point mutations in high-throughput screens for disease-related variants [34, 35]

In vitro CRISPR screens typically involve lentiviral delivery of a library of single guide RNA sequences (sgRNAs) targeting a subset of genes or the whole genome. As lentivirus stably integrates into the DNA, the sgRNA will be present in the cell's genome and identifiable by high-throughput sequencing [5]. Differential prevalence of sequenced sgRNAs will point to the specific genes involved in the phenotype of interest. Library size and the desired sensitivity of detecting an effect will determine the cell number necessary for screening. Screens for cell survival or proliferation can be performed without selecting the cells for a specific phenotype, with all cells being subjected to sequencing. Screens for cell differentiation, maturation, or a particular phenotypic change will require use of fluorescent phenotypic reporters followed by cell sorting or antibiotic resistance-based selection to separate cells for sequencing. Regardless of cell selection prior to sequencing, the false-positive threshold for detecting sgRNA hits should be set above the DNA replication rate since any baseline DNA synthesis will result in sgRNA amplification.

This protocol describes a generalized method for designing, validating, and screening a custom CRISPR library in cultured cardiomyocytes, making a protocol for cardiomyocyte-specific high-throughput genetic screening accessible to the broad research community. While the use of a particular cardiomyocyte type (e.g., neonatal rat [6], mouse postnatal [7], human pluripotent stem cell-derived [8]) or culture model (monolayer [9] or 3D engineered tissue [10–12]) will require some adjustments, the described protocol will outline custom library design including generation of a gene list, in silico sgRNA design, and library cloning to provide flexibility when addressing lab-specific research questions. If a custom library is not desired, many pre-made pooled sgRNA screening libraries, including those targeting whole genome, are available through various resources such as Addgene.

2 Materials

2.1 Molecular Biology

1. LentiCRISPR V2 Plasmid (Addgene #52961, [2]).
2. Molecular biology grade agarose.
3. Tris-acetate-EDTA (TAE) buffer: 40 mM Tris base, 2 mM EDTA, 20 mM acetic acid, pH 8.5.
4. Sybr Safe DNA Stain (Thermo Fisher).
5. DNA Gel Box and Power Supply.
6. Thermocycler.
7. Phusion High-Fidelity DNA Polymerase (NEB).
8. Deoxynucleotide triphosphates (dNTPs).
9. Restriction enzymes.
10. 100 bp DNA Ladder.
11. 1 kb DNA Ladder.
12. Zymo Gel DNA Extraction Kit (Zymo Research).

13. Zymo Clean and Concentrator Kit (Zymo Research).
14. Razor blades.
15. UV box for viewing DNA gels.
16. UV face shield.
17. T4 DNA Ligase.
18. Endura Electrocompetent *E. coli* (Lucigen).
19. Gene Pulser Xcell™ Total Electroporator (Biorad).
20. 0.1 cm Gap Electroporator Cuvettes.
21. SOC Medium: 20 g/L tryptone, 5 g/L yeast extract, 0.5 g/L NaCl, 20 mM glucose.
22. 10 mL round bottom tubes.
23. Agar plates containing antibiotic.
24. Luria Bertani (LB) broth containing antibiotic.
25. Two-Liter Bacterial Culture Flask.
26. Bacterial Shaking Incubator.
27. Maxi Prep Kit (Qiagen).
28. Centrifuge.
29. Molecular Biology Grade Ethanol.
30. NanoDrop™ 2000 (Thermo Fisher).
31. Custom amplification primers.
32. Custom-pooled sgRNA oligonucleotides.
33. Genomic DNA Isolation Kit.
34. Custom Amplification Primers.

2.2 Cell Culture

1. HEK293T cells (ATCC).
2. Cardiomyocytes of choice (e.g., neonatal rat, neonatal mouse, human pluripotent stem cell-derived).
3. Click-iT™ EdU Alexa Fluor™ 488 Flow Cytometry Assay Kit (Invitrogen).
4. 4′,6-diamidino-2-phenylindole (DAPI) stain.
5. Live/Dead Cell Viability/Cytotoxicity Kit (Invitrogen).
6. 10 cm tissue culture-treated dishes.

3 Methods

3.1 Library Preparation and In Silico sgRNA Sequence Design

1. Generate gene list for CRISPR library using Gene Ontology resources based on specific research interests [13, 14] (*see* **Notes 1** and **2**).

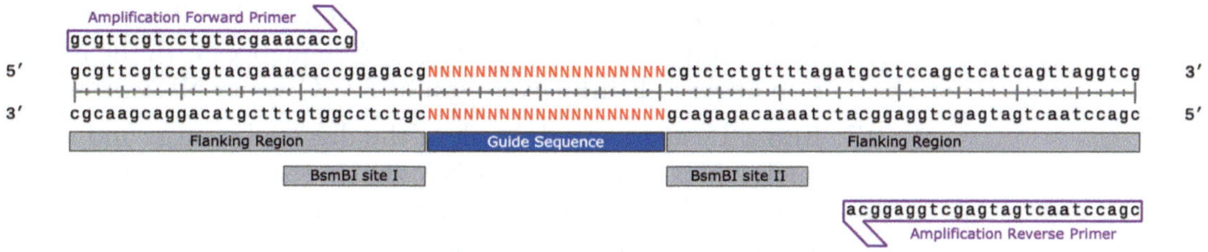

Fig. 1 A sample sgRNA sequence denoted by "*n*" flanked by directional restriction sites and amplification primer-binding sites

2. Design targeting sgRNA sequences using the Broad Institute's GPP sgRNA Designer website (https://portals.broadinstitute.org/gpp/public/analysis-tools/sgrna-design) [15–18] or other source if desired.
3. Sort ranked sgRNA sequences using Excel or Matlab and export to a final list. Choose the five lowest overall ranked sgRNA sequences for each gene, which correspond to sequences with the best predicted on-target activity and lowest predicted off-target activity (*see* **Notes 3** and **4**).

3.2 Sequence Design for sgRNA Oligonucleotide Cloning

1. When designing the universal flanking regions for synthesized sgRNA sequences, it is recommended to include primer-binding sites for oligonucleotide amplification, directional restriction sites for cloning, and sequencing primer sites for high-throughput sequencing (Fig. 1) (*see* **Notes 5** and **6**).
2. Append universal flanking regions containing primer-binding sites and restriction sites as per Fig. 1 to designed sgRNA sequences using Excel or other preferred software.
3. Order pooled sgRNA oligonucleotides containing flanking regions.

3.3 sgRNA Library Plasmid Cloning

1. Upon receipt of pooled oligonucleotides containing sgRNA sequences, PCR-amplify them using designed amplification primers (*see* **Note 7**).
2. Perform at least five 20 μL PCR reactions in parallel with Phusion DNA polymerase and pool the PCR product.
3. Check the PCR by running 5 μL of pooled PCR product on a 1% agarose gel containing Sybr Safe or other preferred DNA stain with a 100 bp DNA ladder.
4. Confirm that a single band of the correct size is present on the gel by viewing the gel over a UV light source while wearing appropriate personal protective equipment such as a UV face shield.
5. Purify the remaining PCR product with a Clean and Concentrator Kit and elute DNA in 10 μL sterile, nuclease-free water.

6. Restriction digest the 1 μg of LentiCRISPR V2 plasmid and 10 μL of the PCR-amplified sgRNA sequences with the restriction enzyme BsmBI overnight at 37 °C as per NEB protocol.
7. Run a 1% agarose gel for all of the digested vectors and oligonucleotides containing sgRNA sequences. Include a 1 kb DNA ladder and undigested vector as a control, making sure to skip at least one lane between the digested and undigested vectors.
8. Cut out bands over a UV light source with a clean razor blade, making sure to use a new blade for each band and to utilize appropriate UV protection.
9. Gel purify the digested oligonucleotides and digested vector with the Gel Purification kit (*see* **Note 8**).
10. Determine purified DNA concentrations with Nanodrop.
11. Ligate overnight at 4 °C with T4 DNA Ligase using a molar ratio of five parts insert to one part vector.
12. Transform ligated DNA in triplicate into electrocompetent *E. coli* in a 0.1 cm gap electroporation cuvette using the following settings on an electroporator: 1800 V, 10 μF, 600 Ω.
13. Immediately after electroporation, quickly wash each cuvette out with 5 mL SOC medium and transfer contents to a sterile 10 mL round bottom culture tube.
14. Incubate the 10 mL round bottom tubes containing the transformed library culture for 1 h at 37 °C and 250 rpm.
15. After 1 h, take 100 μL of the transformed library culture from each of the three tubes and plate 100 μL of at least two dilutions (1:10 and 1:100) in LB broth on agar plates containing the appropriate antibiotic, which will allow for the calculation of transformation efficiency. Culture agar plates overnight in a stationary incubator at 37 °C and count colonies the next morning (*see* **Note 9**).
16. Combine rest of the transformed library culture in the three 10 mL round bottom tubes and transfer to a two-liter culture flask containing 250 mL of LB broth with the appropriate antibiotic (*see* **Note 10**).
17. Incubate the two-liter flask overnight (12–16 h) in a bacterial shaking incubator at 37 °C and 250 rpm.
18. Following overnight incubation, spin the transformed library culture down and isolate plasmid DNA as per maxiprep protocol.
19. Elute DNA in sterile water inside a biological safety cabinet. This DNA will be used to make lentivirus and thus should be kept sterile.
20. Perform high-throughput sequencing on the purified plasmid to ensure that sgRNA representation is maintained in the library following amplification and cloning (*see* **Note 11**). PAUSE POINT.

3.4 Assessment of Baseline Cardiomyocyte Proliferation

1. Before performing the screen, assess baseline rate of DNA synthesis of cardiomyocytes by adding 10 mM EdU to culture medium for 24 h.
2. Fix cells, then stain for EdU per manufacturer protocol and add DAPI stain at 1:1000 concentration. For less than 100% cardiomyocyte purity, also include a cardiomyocyte-specific antibody stain, such as cardiac troponin T or sarcomeric alpha-actinin.
3. Perform flow cytometry to quantify the percent EdU-positive cardiomyocytes as a proportion of total cardiomyocytes (*see* **Note 12**). PAUSE POINT.

3.5 Assessment of Antibiotic Susceptibility

1. To determine the optimal antibiotic concentration for selection of transduced cardiomyocytes, assess puromycin susceptibility of the cardiomyocytes by treating wells with cells with a series of concentrations of puromycin in normal culture medium from 1 to 10 μg/mL.
2. Apply antibiotic selection for 5 days with a media change every 48 h and assess cardiomyocyte viability with the Live/Dead Cell Viability/Cytotoxicity Kit as per manufacturer protocol. PAUSE POINT.

3.6 sgRNA Lentivirus Library Preparation

1. Prepare lentivirus from the purified library plasmid generated in Subheading 3.2 as per Addgene protocol (https://www.addgene.org/protocols/lentivirus-production/) or any other preferred protocol in HEK293T cells [19, 20].
2. Determine functional (infectious) titer of the lentivirus by infecting cardiomyocytes with several concentrations of virus such that concentration can be correlated with transduction efficiency and used for determination of functional titer (*see* **Note 13**).
3. Deliver puromycin (or other appropriate antibiotic) 48 h post transduction at the lowest antibiotic concentration that caused nearly 100% non-transduced cardiomyocyte death by 5 days.
4. Choose wells with concentrations of virus that resulted in roughly 20–40% transduction such that individual cells are easily quantifiable and calculate the percent surviving cells (percent transduction) after 5 days and therefore determine the functional titer of the virus in μL virus per number of cells transduced (*see* **Notes 14** and **15**).
5. Choose the virus concentration that generates a multiplicity of infection (MOI) of 0.2–0.4, which reduces the likelihood that cardiomyocytes will be transduced with more than one viral particle while also minimizing the number of cells necessary to ensure sufficient library coverage.

3.7 Library Delivery and Screening

1. To determine the cell number needed for screening, use the following formula: (Number of sgRNAs in library) * (Library coverage) * (Number of Timepoints)/(MOI).
2. Choose a cell number that ensures library coverage is at least 300-500×, i.e., 300–500 cells will receive each sgRNA from the library.
3. Plate cardiomyocytes at the desired seeding density (*see* **Note 16**). Depending on the experiment, the library lentivirus can be delivered at the time of plating or after several days to weeks.
4. Deliver the titered library lentivirus at an MOI of 0.2–0.4 overnight in normal culture medium.
5. Perform a media change 24 h after lentiviral delivery.
6. Isolate genomic DNA (gDNA) from the initial timepoint at 48 h post-lentivirus delivery, which will provide the initial library representation. It is not necessary to perform FACS at this step (*see* **Notes 17** and **18**).
7. Determine gDNA concentration using Nanodrop or other DNA quantification method. Store the gDNA at −20 °C until the other samples have been collected.
8. At specific timepoints, isolate gDNA from all cells (or selected cells with the desired phenotype) and store at −20 °C. PAUSE POINT.
9. PCR-amplify sgRNA sequences from gDNA using custom amplification primers.
10. Gel-purify amplified sgRNA sequences using the Gel DNA Extraction Kit.
11. Prepare sequencing library and sequence on an Illumina MiSeq or NextSeq platform depending on the size of the library.
12. Determine sgRNA sequence representation relative to the initial timepoint. Select the most overrepresented sgRNA sequences for validation.

3.8 Validation of sgRNA Targets

1. Order individual oligonucleotides containing the sgRNA sequences of interest. Choose at least three sgRNA sequences per gene.
2. Clone individual sgRNA sequences into the vector and make lentivirus as outlined above (*see* **Note 19**).
3. Deliver lentivirus to cultured cardiomyocytes at an MOI of 1 or higher.
4. Assess for desired phenotype at 72 h or later via flow cytometry or microscopy.

4 Notes

1. Use of Gene Ontology resources allows for the generation of a gene list that contains genes annotated to any chosen pathway or phenotype. However, it is also possible to use any other publicly available data set such as the Human Protein Atlas (http://www.proteinatlas.org) [21], which contains tissue-specific expression data.
2. Cost of sgRNA oligonucleotide synthesis increases with the number of oligonucleotides needed, which may factor into the choice of the number of studied genes and final library size. If the gene list is very long, it may be more cost-effective to use a published whole-genome library available on Addgene rather than ordering synthesized oligonucleotides. For numbers of oligonucleotides greater than 100–200, pooled batch synthesis may be more cost-effective, while for small numbers of oligonucleotides individual synthesis may be less costly.
3. Any sgRNA sequences that contain the restriction site used for cloning should be removed.
4. It is recommended to use at least five sgRNA sequences per gene with the assumption that not all sequences will necessarily be functional. It is likely that multiple sgRNAs will be functional if five are chosen. To limit numbers of needed cardiomyocytes and oligonucleotides, it is possible to use as few as three sgRNAs per gene, but less than three is not recommended.
5. If possible, it is recommended to use single restriction sites such as BsmBI that have recognition sites next to the cut site. This allows for two different "sticky" ends to be generated in the vector with the use of only one enzyme, allowing for directional cloning of the insert. It should be further ensured that the cloning strategy does not disrupt the guide scaffold present in the vector. To achieve high bulk cloning efficiency, primers should be carefully designed and validated with several individual sgRNAs before cloning the full library.
6. Amplicon length for NGS should not exceed Illumina sequencing length capacity of 50 or 75 nucleotides, depending on the exact sequencing method chosen.
7. This is a point at which library preparation can easily fail. Primers should be validated in advance and the optimal melting temperature empirically determined using gradient PCR. High-fidelity DNA polymerase, sterile PCR grade water, and sterile filter tips should be used when working with PCR reagents to avoid contamination.

8. It is critical that the bands are the correct sizes and there are no additional, unexpected bands. The digested vector most likely will have two bands, and the correct band is usually the larger band. Confirm the expected band sizes and that all expected bands are present. Restriction enzymes should not be expired and should be kept at the appropriate storage temperature.
9. Recommended library coverage is at least 10,000×. For example, for a library of 100 genes, plasmid should be introduced into at least 10^6 bacteria. Number of transformed bacteria should be calculated as follows: (Total Culture Volume/Volume Plated) * (Number of Colonies Counted). If the efficiency is low, ligation and transformation should be repeated.
10. While the mammalian selection antibiotic for LentiCRISPR V2 is puromycin, the bacterial antibiotic is ampicillin. Ensure use of the correct antibiotic for bacterial selection.
11. If it is necessary to reduce cost, Sanger sequencing could be performed on many individual colonies. While this method will not provide the overall library representation, it will be more cost-effective to validate that the cloning was successful. It will be evident if library representation is not maintained following cloning if a single guide sequence or several guide sequences are predominant. If this is the case, it may be necessary to troubleshoot the library cloning steps starting with guide oligonucleotide amplification.
12. In addition to flow cytometry, EdU incorporation could be quantified using fluorescence microscopy. The assessment of baseline EdU incorporation rate will serve to determine how much sgRNA sequence overrepresentation could be attributed to endogenous cardiomyocyte proliferation rather than a phenotype induced by the library.
13. For PEG precipitated lentivirus, between 0.05 μL and 8 μL virus per million cells has been used successfully, but the concentration should be determined empirically and can vary widely based on the lentivirus purification protocol.
14. LentiCRISPR V2 contains a puromycin resistance gene that is delivered and integrates in DNA along with the sgRNA sequence and Cas9. If there is no antibiotic resistance gene in the vector, fluorescence genes or other markers can be also used. If no markers are present, an ELISA-based titer determination can be performed; however, this method is not preferred as it does not always correspond to the functional titer [22].
15. Low titer lentivirus should not be used for screening. If lentiviral titer is low, it may be necessary to optimize lentivirus production protocols and/or remake the virus.

16. Seeding densities of 130,000/cm^2 for neonatal rat ventricular myocytes (NRVMs) and 210,000/cm^2 for human-induced pluripotent stem cell-derived cardiomyocytes (hiPSC-CMs) are recommended for proliferation-related screens. However, a lower seeding density can also be used if necessary to observe a phenotype of interest.
17. It is essential to isolate gDNA at an initial timepoint following transduction, but before sgRNA sequence representation changes due to the occurrence of the phenotype of interest. Forty-eight hours is a reasonable timepoint for a lentivirally delivered library to allow for viral transduction and genomic integration of the sgRNA sequences to occur without changes in library representation. This timing may be experiment dependent and should be adjusted for alternative methods of library delivery, such as with adeno-associated viruses which take longer to express than lentivirus.
18. The manufacturer-recommended cell number per column and the column capacity should not be exceeded, as the resulting DNA loss will lead to a loss of library coverage.
19. If many targets are being validated, it is possible to make a sub-pooled library and re-screen. For a small number of targets, each target should be validated individually.

Acknowledgments

This work was supported by NIH grants U01HL134764, U01EB028901, HL132389, and a grant from Foundation Leducq to NB and the T32 Developmental and Stem Cell Biology Training Grant 5T32HD040372-18 to SD.

References

1. Ishino Y, Krupovic M, Forterre P (2018) History of CRISPR-Cas from encounter with a mysterious repeated sequence to genome editing technology. J Bacteriol 200(7): e00580–e00517. https://doi.org/10.1128/JB.00580-17
2. Sanjana NE, Shalem O, Zhang F (2014) Improved vectors and genome-wide libraries for CRISPR screening. Nat Methods 11: 783–784. https://doi.org/10.1038/nmeth.3047
3. Schuster A, Erasimus H, Fritah S et al (2019) RNAi/CRISPR screens: from a pool to a valid hit. Trends Biotechnol 37:38–55. https://doi.org/10.1016/j.tibtech.2018.08.002
4. Doench JG (2018) Am I ready for CRISPR? A user's guide to genetic screens. Nat Rev Genet 19:67–80. https://doi.org/10.1038/nrg.2017.97
5. Aregger M, Chandrashekhar M, Tong AHY et al (2019) Pooled lentiviral CRISPR-Cas9 screens for functional genomics in mammalian cells. Methods Mol Biol Clifton NJ 1869: 169–188. https://doi.org/10.1007/978-1-4939-8805-1_15
6. Bian W, Badie N, Himel HD, Bursac N (2014) Robust T-tubulation and maturation of cardiomyocytes using tissue-engineered epicardial mimetics. Biomaterials 35:3819–3828. https://doi.org/10.1016/j.biomaterials.2014.01.045

7. Bassat E, Mutlak YE, Genzelinakh A et al (2017) The extracellular matrix protein agrin promotes heart regeneration in mice. Nature 547:179–184. https://doi.org/10.1038/nature22978
8. Jackman CP, Carlson AL, Bursac N (2016) Dynamic culture yields engineered myocardium with near-adult functional output. Biomaterials 111:66–79. https://doi.org/10.1016/j.biomaterials.2016.09.024
9. Zhang D, Shadrin IY, Lam J et al (2013) Tissue-engineered cardiac patch for advanced functional maturation of human ESC-derived cardiomyocytes. Biomaterials 34:5813–5820. https://doi.org/10.1016/j.biomaterials.2013.04.026
10. Jackman C, Li H, Bursac N (2018) Long-term contractile activity and thyroid hormone supplementation produce engineered rat myocardium with adult-like structure and function. Acta Biomater 78:98–110. https://doi.org/10.1016/j.actbio.2018.08.003
11. Shadrin IY, Allen BW, Qian Y et al (2017) Cardiopatch platform enables maturation and scale-up of human pluripotent stem cell-derived engineered heart tissues. Nat Commun 8:1–15. https://doi.org/10.1038/s41467-017-01946-x
12. Jackman CP, Ganapathi AM, Asfour H et al (2018) Engineered cardiac tissue patch maintains structural and electrical properties after epicardial implantation. Biomaterials 159:48–58. https://doi.org/10.1016/j.biomaterials.2018.01.002
13. Ashburner M, Ball CA, Blake JA et al (2000) Gene ontology: tool for the unification of biology. The gene ontology consortium. Nat Genet 25:25–29. https://doi.org/10.1038/75556
14. The Gene Ontology Consortium (2019) The gene ontology resource: 20 years and still GOing strong. Nucleic Acids Res 47:D330–D338. https://doi.org/10.1093/nar/gky1055
15. Doench JG, Fusi N, Sullender M et al (2016) Optimized sgRNA design to maximize activity and minimize off-target effects of CRISPR-Cas9. Nat Biotechnol 34:184–191. https://doi.org/10.1038/nbt.3437
16. Sanson KR, Hanna RE, Hegde M et al (2018) Optimized libraries for CRISPR-Cas9 genetic screens with multiple modalities. Nat Commun 9:1–15. https://doi.org/10.1038/s41467-018-07901-8
17. Sanson KR, DeWeirdt PC, Sangree AK et al (2019) Optimization of AsCas12a for combinatorial genetic screens in human cells. bioRxiv:747170. https://doi.org/10.1101/747170
18. Kim HK, Min S, Song M et al (2018) Deep learning improves prediction of CRISPR–Cpf1 guide RNA activity. Nat Biotechnol 36:239–241. https://doi.org/10.1038/nbt.4061
19. Tiscornia G, Singer O, Verma IM (2006) Production and purification of lentiviral vectors. Nat Protoc 1:241–245. https://doi.org/10.1038/nprot.2006.37
20. Wang X, McManus M (2009) Lentivirus production. J Vis Exp JoVE (32):1499. https://doi.org/10.3791/1499
21. Uhlén M, Fagerberg L, Hallström BM et al (2015) Proteomics. Tissue-based map of the human proteome. Science 347:1260419. https://doi.org/10.1126/science.1260419
22. Geraerts M, Willems S, Baekelandt V et al (2006) Comparison of lentiviral vector titration methods. BMC Biotechnol 6:34. https://doi.org/10.1186/1472-6750-6-34
23. Perez-Pinera P, Kocak DD, Vockley CM et al (2013) RNA-guided gene activation by CRISPR-Cas9–based transcription factors. Nat Methods 10:973–976. https://doi.org/10.1038/nmeth.2600
24. Seipel K, Georgiev O, Schaffner W (1992) Different activation domains stimulate transcription from remote ('enhancer') and proximal ('promoter') positions. EMBO J 11:4961–4968
25. Hilton IB, D'Ippolito AM, Vockley CM et al (2015) Epigenome editing by a CRISPR-Cas9-based acetyltransferase activates genes from promoters and enhancers. Nat Biotechnol 33:510–517. https://doi.org/10.1038/nbt.3199
26. Kwon DY, Zhao Y-T, Lamonica JM, Zhou Z (2017) Locus-specific histone deacetylation using a synthetic CRISPR-Cas9-based HDAC. Nat Commun 8:15315. https://doi.org/10.1038/ncomms15315
27. Liu XS, Wu H, Ji X et al (2016) Editing DNA methylation in the mammalian genome. Cell 167:233–247.e17. https://doi.org/10.1016/j.cell.2016.08.056
28. Vojta A, Dobrinić P, Tadić V et al (2016) Repurposing the CRISPR-Cas9 system for targeted DNA methylation. Nucleic Acids Res 44:5615–5628. https://doi.org/10.1093/nar/gkw159
29. Thakore PI, D'Ippolito AM, Song L et al (2015) Highly specific epigenome editing by CRISPR-Cas9 repressors for silencing of distal regulatory elements. Nat Methods 12:

1143–1149. https://doi.org/10.1038/nmeth.3630

30. Polstein LR, Perez-Pinera P, Kocak DD et al (2015) Genome-wide specificity of DNA binding, gene regulation, and chromatin remodeling by TALE- and CRISPR/Cas9-based transcriptional activators. Genome Res 25: 1158–1169. https://doi.org/10.1101/gr.179044.114
31. Parsi KM, Hennessy E, Kearns N, Maehr R (2017) Using an inducible CRISPR-dCas9-KRAB effector system to dissect transcriptional regulation in human embryonic stem cells. Methods Mol Biol Clifton NJ 1507:221–233. https://doi.org/10.1007/978-1-4939-6518-2_16
32. Yeo NC, Chavez A, Lance-Byrne A et al (2018) An enhanced CRISPR repressor for targeted mammalian gene regulation. Nat Methods 15: 611–616. https://doi.org/10.1038/s41592-018-0048-5
33. Kearns NA, Pham H, Tabak B et al (2015) Functional annotation of native enhancers with a Cas9-histone demethylase fusion. Nat Methods 12:401–403. https://doi.org/10.1038/nmeth.3325
34. Ma Y, Zhang J, Yin W et al (2016) Targeted AID-mediated mutagenesis (TAM) enables efficient genomic diversification in mammalian cells. Nat Methods 13:1029–1035. https://doi.org/10.1038/nmeth.4027
35. Hess GT, Frésard L, Han K et al (2016) Directed evolution using dCas9-targeted somatic hypermutation in mammalian cells. Nat Methods 13:1036–1042. https://doi.org/10.1038/nmeth.4038

Chapter 2

Protein and mRNA Quantification in Small Samples of Human-Induced Pluripotent Stem Cell-Derived Cardiomyocytes in 96-Well Microplates

Weizhen Li, Julie L. Han, and Emilia Entcheva

Abstract

We describe a method for protein quantification and for mRNA quantification in small sample quantities of human-induced pluripotent stem cell-derived cardiomyocytes (hiPSC-CMs). Demonstrated here is how the capillary-based protein detection system Wes™ by ProteinSimple and the *Power* SYBR™ Green Cells-to-CT™ Kit by Invitrogen can be applied to individual samples in a 96-well microplate format and thus made compatible with high-throughput (HT) cardiomyocyte assays. As an example of the usage, we illustrate that Cx43 protein and GJA1 mRNA levels in hiPSC-CMs are enhanced when the optogenetic actuator, channelrodopsin-2 (ChR2), is genetically expressed in them. Instructions are presented for cell culture and lysate preparations from hiPSC-CMs, along with optimized parameter settings and experimental protocol steps. Strategies to optimize primary antibody concentrations as well as ways for signal normalization are discussed, i.e., antibody multiplexing and total protein assay. The sensitivity of both the Wes and Cells-to-CT kit enables protein and mRNA quantification in a HT format, which is important when dealing with precious small samples. In addition to being able to handle small cardiomyocyte samples, these streamlined and semi-automated processes enable quick mechanistic analysis.

Key words High-throughput, hiPSC-CMs, Wes™ ProteinSimple, Cells-to-CT™, Cx43, ChR2, Optogenetic

1 Introduction

Human stem cell-derived cardiomyocytes are an important driver in personalized medicine through the development of patient-specific high-throughput assays [1–4]. These assays yield a range of functional outputs from viability to metabolic function to arrhythmia predictions using all-optical or other technologies [5–10]. It is essential to be able to perform quick analysis of protein and gene expression with minimal cell material, compatible with the HT format of these functional assays.

Kareen L. K. Coulombe and Lauren D. Black III (eds.), *Cardiac Tissue Engineering: Methods and Protocols*, Methods in Molecular Biology, vol. 2485, https://doi.org/10.1007/978-1-0716-2261-2_2,

Western Blots (WB) have been widely used for protein quantification since their development in the 1970s [11–13]. In traditional WB, protein samples are denatured and separated based on molecular weights by SDS-PAGE (sodium dodecyl sulfate, SDS—polyacrylamide gel electrophoresis, PAGE). The separated protein components are transferred onto either nitrocellulose membrane or PVDF (polyvinylidene fluoride) for target protein immunoprobing and chemiluminescent quantification. While this classic WB method is widely used, it is difficult to detect targets in total protein amounts less than 8 μg per sample. Thus, it cannot be used to detect protein expression levels in small cell collections completed in 96-well format.

Improvements of the technique have been made to increase the detection sensitivity, throughput, and reproducibility. Utilizing sequential lateral flow, automation of the immunoprobing steps is commercially available from iBind (Thermo Fisher). A bead-based microarray assay immobilizes separated protein components onto hundreds of microspheres and achieves high-throughput by analyzing bead collections [14]. Microfluidics-based immunoassays have integrated protein separation and detection in microchips [15–17] which has allowed for the reduction of starting material to a single cell level [18]. Capillary electrophoresis-based approaches have also been pursued, with dispersion or ligand binding to generate signals along a narrow capillary at nano-liter volumes for high sensitivity quantification with also very small starting material [19, 20].

Based on capillary electrophoresis, the Wes™ ProteinSimple platform for protein quantification was commercialized and has demonstrated high sensitivity, wide dynamic range, and good reproducibility [21–23]. Since deployment, it has been widely used in cancer [24, 25] and neuroscience [26, 27], with some recent applications to cardiac research [28–34]. The key advantages of this system over standard WB include the small starting material needed (as low as 0.8 μg per sample), the level of automation and high throughput (runs up to 24 samples concurrently) and faster turnover (3–5 h assay time), all of which make it ideal for applications requiring analysis of many protein samples limited by size. Due to high-cost considerations and the need for high-throughput mechanism insights into drug screening applications with human cells, such a system for protein quantification is particularly valuable for studies with human-induced pluripotent stem cell-derived cardiomyocytes (hiPSC-CMs). Indeed, the Wes has been already applied in several cases recently, including signaling pathway inquiries with hiPSC-CMs [35], cardiac pharmacology and toxicology [36], hiPSC differentiation [37, 38], gene therapy [39], and quantifying ion channel expression to assess maturity [34].

It is often useful and/or more straightforward to perform gene expression analysis as a surrogate or as a complement to protein quantification. The basis of the current "gold standard" in quantifying gene expression is the polymerase chain reaction (PCR) method developed in the 1980s by Mullis [40, 41]. A variety of newer techniques have emerged in this area, including the quantitative reverse-transcription qPCR, application of microarrays and RNAseq, extending recently to single-cell and spatial transcriptomics [42].

Critical developments in the 1990s included the use of fluorescent labeling for kinetic (real-time) qPCR to quantify mRNA [43, 44]. This allows for very sensitive detection of gene expression in small samples. Rigorous quantification of mRNA based on a critical time or threshold cycle (Ct) from the amplification curves was developed [45–47]. To further streamline the process and eliminate the RNA isolation step, the direct "Cells-to-CT" method [48] was made commercially available. In this method, the cell lysate is directly incorporated in the qPCR workflow, which makes the process faster and enables further reduction of sample size needed to run the reactions. This technique has been applied to human iPSC-CMs and human cardiovascular progenitor cells in a limited number of studies [34, 49–51].

Here, we provide a detailed protocol of protein detection using the Wes and mRNA quantification using the Cells-to-CT kit and qPCR in hiPSC-CMs samples in a 96-well format, overview in Fig. 1. Frequently used system settings, sample preparation, antibody linear dynamic range test, antibody-multiplexing dilutions, and recommended qPCR settings are summarized in tables for reference. Optimization of antibody multiplexing is illustrated using several pairs of proteins in hiPSC-CMs—with options to extend the method to other protein targets.

Using this method, we demonstrate the potential influence of ChR2 on connexin43 (Cx43) in hiPSC-CMs. ChR2 is introduced into hiPSC-CMs by adenoviral infection and comparison was done with respect to both non-infected samples and an Ad-eYFP control. In the presence of ChR2, there is an increase of GJA1 at the transcriptional level and enhanced expression of the Cx43 protein in hiPSC-CMs.

2 Materials

2.1 hiPSC-CMs Cell Culture

1. Human-induced pluripotent stem cell-derived cardiomyocytes iCell Cardiomyocytes[2] CMC-100-012-001, derived from a female donor, purchased from Fujifilm Cellular Dynamics International.
2. MyCell hiPSC-CM-1X 01395, derived from a male donor, purchased from Fujifilm Cellular Dynamics International.
3. Cell plating medium and maintenance medium are provided by the manufacturer (Fujifilm Cellular Dynamics International).

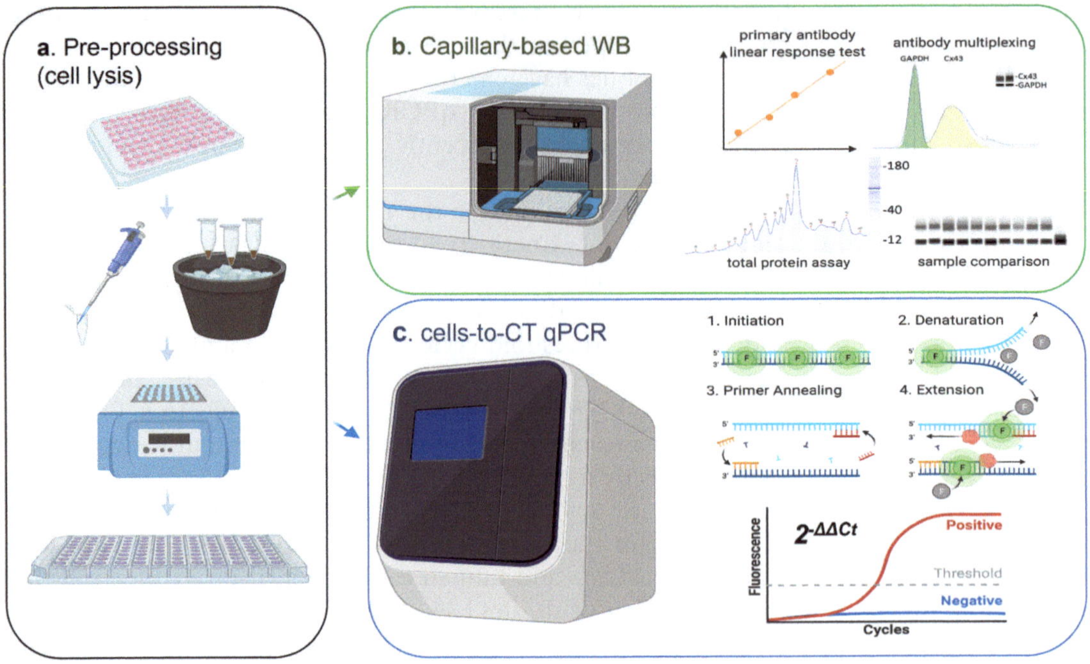

Fig. 1 Small hiPSC-CM sample protein and mRNA quantification overview. (**a**) hiPSC-CMs lysates collected from each of the 96-wells are used directly for protein or mRNA quantification. For Wes quantification, the cell lysate is mixed with reaction reagent and denatured at 75 °C for 10 min. (**b**) Twenty-four samples can be loaded in each Wes microplate and quantified concurrently. After 3–5 h of Wes run, quantification results including target protein linear response, antibody multiplexing or total protein assays can be analyzed in the Compass software, and sample comparison can be displayed using pseudo-gel-bands. (**c**) For qPCR quantification, mRNA in the whole cell lysates from individual wells can be quantified with the Cells-to-CT qPCR approach, involving the steps shown; quantification is by the ΔΔCt method. Figure created with Biorender

4. Fibronectin (Corning).
5. 1× phosphate-buffered saline (PBS).
6. 96-well glass bottom plate (Cellvis).
7. (Optional to illustrate usage by adenoviral infection) Ad-CMV-hChR2 (H134R)-eYFP (Vector Biosystems Inc.).
8. (Optional to illustrate usage by adenoviral infection) Ad-CMV-eYFP as a control (Vector Biosystems Inc.).

2.2 Cell Lysis and Protein Denaturing for WB

1. 1× PBS.
2. Qproteome Mammalian Protein Prep Kit (Qiagen).
3. Sterile 1.5 mL microcentrifuge tubes.
4. Plate shaker, Fig. 2a.
5. Refrigerated high-speed Eppendorf centrifuge (Millipore-Sigma), Fig. 2b.

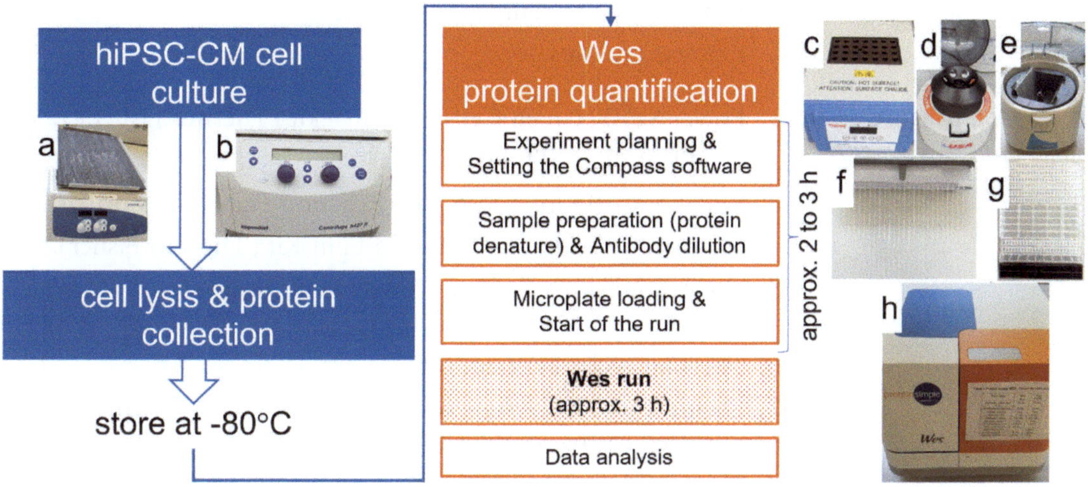

Fig. 2 Detailed workflow and equipment needed for Wes protein quantification. Wes protein analysis workflow can be separated to three main parts: protein collection, experiment preparation, and data analysis. Plate shaker (**a**), a box of ice and refrigerated high-speed centrifuge (**b**) are needed in protein collection. In experiment preparation, the heating block (**c**), microtube centrifuge (**d**), and microplate centrifuge (**e**) are needed. Wes capillary set (**f**) and the prepared microplate (**g**) are inserted in the Wes machine (**h**) for the run

6. Heat block with inserts for microcentrifuge tubes, Fig. 2c.
7. Microcentrifuge/spinner, Fig. 2d.
8. PlateFuge plate centrifuge, Fig. 2e.

2.3 Immunoprobing

1. Kit for running an assay by Protein Simple, containing:
   (a) Jess/Wes 25-capillary cartridge, Fig. 2f.
   (b) Jess/Wes separation module (12–230 kDa) with a pre-filled microplate with split running buffer, Fig. 2g.
   (c) Wash buffer and 10× sample buffer.
   (d) EZ standard pack with lyophilized material in three tubes, as follows: biotinylated ladder, fluorescent standard 5× master mix and dithiothreitol (DTT).
2. Wes system (Protein Simple), Fig. 2h.
3. Primary antibodies tested in this study:
   (a) Cx43 (ab11370), Abcam.
   (b) Kir2.1 (ab65796), Abcam.
   (c) Alpha-tubulin (ab7291), Abcam.
   (d) GAPDH (ab181602), Abcam.
   (e) LDH (H-10) (sc-133,123), Santa Cruz.
4. Secondary antibodies were anti-mouse and anti-rabbit detection modules (Protein Simple), which contained luminol, peroxide, milk-free antibody diluent and streptavidin-HRP in addition to anti-mouse/anti-rabbit secondary HRP antibody.

5. PlateFuge plate centrifuge, Fig. 2e.
6. Pipettes and tips.

2.4 qPCR with the Cells-to-CT™ Kit

1. Power SYBR Green Cells-to-CT™ kit (Thermo Fisher).
2. 1× TE Buffer, pH 8.0.
3. MicroAmp™ Optical 96-well Reaction Plate (Thermo Fisher).
4. MicroAmp™ Optical Adhesive Film Kit (Thermo Fisher).
5. QuantStudio™ 3 Real-Time PCR instrument (Thermo Fisher).
6. Thermal cycler (Eppendorf).
7. QuantStudio™ Design & Analysis Software (Thermo Fisher) was used for to conduct differential expression analysis.
8. Nuclease-free microcentrifuge tubes and pipette tips.

3 Methods

The detailed workflow for protein quantification is illustrated in Fig. 2, with relevant equipment shown.

3.1 Culturing hiPSC-CMs in a 96-Well Plate (50,000 Cells per Well)

Thawing and plating of hiPSC-CMs is done following manufacturer's instructions, adapted by user and summarized as follows:

1. Thaw the commercial plating and maintenance medium at 4 °C overnight; equilibrate to room temperature upon usage.
2. Coat wells (in the 96-well plate) with 100 μL of 50 μg/mL fibronectin diluted in 1× sterile PBS in 37 °C cell culture incubator for at least 2 h. Remove fibronectin solution right before cell plating.
3. Immediately transfer hiPSC-CMs cryovial from liquid nitrogen to 37 °C water bath. Hold the cryovial using floating tube rack and immerse the cryovial in 37 °C water bath for 3 min without submerging the cap (*see* **Note 1**).
4. Move the cryovial into the biosafety cabinet after sterilizing with 70% ethanol.
5. Transfer 1 mL of cell suspension from the cryovial into a 50 mL conical tube. Use 1 mL cell plating medium to rinse the remaining cells in cryovial and add dropwise into the cell suspension at a rate of 1 drop per 5 s and swirl between drops.
6. Gently drop additional 8 mL cell plating medium into the 50 mL conical tube for a total of 10 mL solution, slowly swirl the tube in the process.
7. Gently plate 100 μL cell suspension into each 96-well for a plating density of 50,000 cells per well. Slowly swirl the conical tube while plating as cells will settle over time.

8. Place the plate into 37 °C cell culture incubator after plating, exchange plating media to maintenance medium 4 h after the thaw.
9. Gently replace maintenance medium every 2 to 3 days. Tilt the plate 45 degree and add in medium along the edges. Avoid dropping culture medium directly onto the cell layer (*see* **Note 2**).

3.2 Adenovirus Infection—Optional to Illustrate Usage

1. Start adenovirus infection 5 days after cell plating.
2. Thaw adenovirus from −80 °C to 4 °C on ice. Predilute the virus in sterile 1× PBS.
3. Conduct infection in maintenance media at multiplicity of infection (MOI 50). For calculations, follow the detailed published protocols [52, 53].
4. After 2 h of incubation at (37 °C, 5% CO_2), exchange viral medium with normal maintenance medium.
5. Collect cell lysates 48 h after infection.

3.3 Protein Collection from a 96-Well Microplate

1. Prepare QProteome Mammalian Protein lysis buffer and store on ice. Conduct all subsequent steps on ice.
2. Place the 96-well plate on ice. Aspirate cell culture medium and wash with cold 1× PBS. Completely aspirate the 1× PBS and add 10 μL protein lysis buffer in each well.
3. After adding the lysis buffer, incubate on ice with shake for 5 min.
4. Scrape each well and collect complete lysate into epitubes (*see* **Note 3**).
5. Centrifuge all samples at 447 × *g* for 30 min at 4 °C and separate the supernatant (lysate) from the pellet.
6. Transfer the samples to −80 °C freezer for long-term storage. Upon usage, thaw samples on ice.

3.4 Protein Quantification Using Wes™

An overview of the process for protein quantification using the Wes is illustrated in Fig. 2. In the Wes workflow, the cell lysates and reagents, including the primary antibodies and secondary horseradish peroxidase (HRP) antibodies, are loaded into small compartments in a special assay microplate (25 samples). The microplate is placed onto the system and interfaces a cassette with 25 thin capillaries. Size-based separation of proteins occurs as the proteins migrate in the separation matrix within the capillaries and are immobilized to the walls using proprietary photo-activated chemistry. Within 3–5 h, the protein components in each sample are separated based on molecular weight, immobilized in the capillaries and detected using chemiluminescence. System readout is in the form of digitized "electropherograms," which can be converted to virtual or "pseudo" blot lanes automatically. Target peaks are quantified as area under the curve.

3.4.1 Experiment Planning and Setting the Compass Software

It is crucial to carefully plan the sample layout and to enter all system settings in the software before running an assay.

1. Each 10 μL cell lysate, obtained from a 96-well plate sample, allows for several Wes protein quantification runs, depending on protein target expression levels. For lower-expressing proteins, for example, Kir2.1 in hiPSC-CMs, undiluted (or "neat") samples are preferred and cell lysates from one 96-well may be enough for up to two runs. For abundantly expressed proteins like LDH, Cx43, and GAPDH, samples can be diluted up to four times (*see* **Note 4**).
2. Based on the primary antibody's dilution factor and testing well number, one can calculate the primary antibody volume needed. Prepare for one more well to account for pipette errors.
3. Create a new file in the Compass software based on the type of assay. Choose "Wes size" for typical immunoassay, including antibody multiplexing, or choose "Wes total protein" if testing samples include total protein detection. There are some differences between the two programs, for example, the total protein assay has 30 min of biotin labeling time instead of antibody diluent time in the immunoassay. Frequently used system settings are summarized in Table 1.

Table 1
Frequently used system setting

Parameters	Wes size run	Total protein run
Separation matrix load time	200 s	200 s
Stacking matrix load time	15 s	18 s
Sample load time	9 s	9 s
Separation time	31 min	31 min
Separation voltage	375 volts	375 volts
Standards exposure	4 s	4 s
EE immobilization time	200 s	200 s
Antibody diluent time	30 min	/
Biotin labeling time	/	30 min
Primary antibody time	30 min	30 min
Secondary antibody time	30 min	30 min

4. Enter the sample layout and primary antibodies attributes in the software. Adjust the running parameters. Save the file and print out the layout for referencing on the experimental day.

3.4.2 Sample and Primary Antibody Preparation for Protein Quantification

1. Thaw samples from −80 °C freezer on ice.
2. Preset temperature of the heat block to 75 °C.
3. Start the Wes and the Compass software. Initiate the instrument self-test and save the result. If failure occurs, try again and report the problem if it persists.
4. Open an EZ standard pack and prepare the reagents following the manufacturer's instructions.
5. Dilute the 10× sample buffer to 0.1× buffer to dilute the cell lysates as needed. For analysis of low-abundance proteins, the cell lysates should be used "neat" (undiluted).
6. Mix cell lysates with the 5× fluorescent master mix 1:4 (lysate: fluorescent mix).
7. Denature all samples at 75 °C for 10 min (or at 90 °C for 5 min) on heat block. Vortex and spin the tubes before and after the heating to ensure that the cell lysate and the fluorescent master mix are well homogenized.
8. Place the denatured samples back on ice and spin them in a centrifuge for 5 min at 699 × *g* at room temperature before sample loading.
9. Dilute the primary antibodies with Antibody Diluent2, based on intended final concentration. Working concentrations of each antibody will need optimization as necessary concentrations are antibody dependent. Store in the pre-labeled microcentrifuge tubes.

3.4.3 Microplate Loading and Start of a Wes Run

For reagent loading into the microplates, please *see* Fig. 3 and details below.

1. Take the sealed microplate from the Jess/Wes separation module. To minimize the microwell exposure, peel the Wes microplate foil row by row when pipetting.
2. Loading row A: pipet 5 μL biotinylated ladder in the first well of row A. Pipet 3 μL of each prepared protein sample into the remaining 24 wells of the microplate (*see* **Note 5**).
3. For the biotinylated ladder column (default is the first column), pipet 10 μL of Antibody Diluent2 in row B and C, 10 μL of Streptavidin-HRP in row D.
4. Loading rows B, C, and D with antibody diluent2, primary and secondary antibodies:

   4a. For running an immunoassay, for each of the 24 sample wells, pipette 10 μL of Antibody Diluent2 in row B, 10 μL

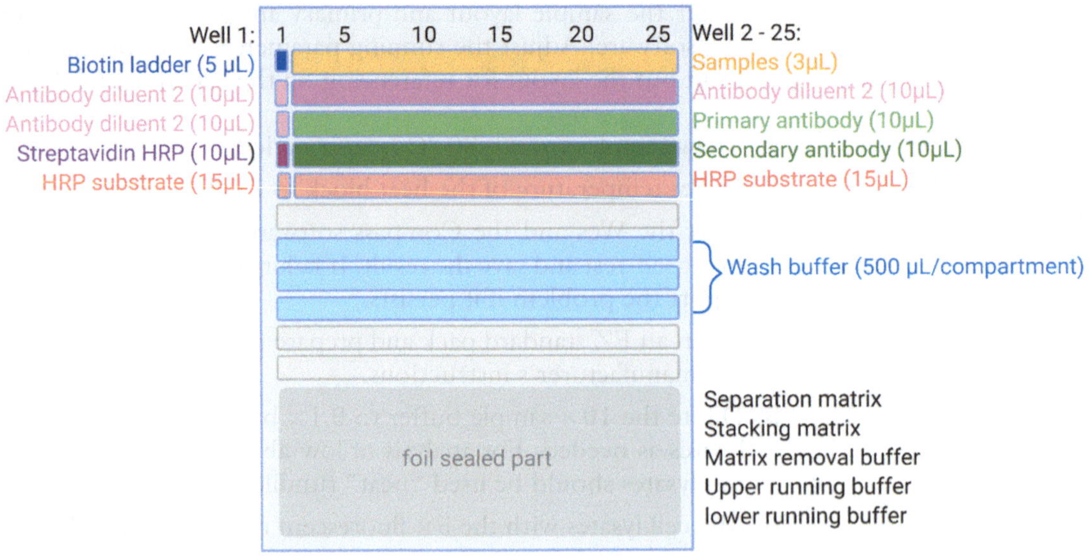

Fig. 3 Microplate loading with reagents. The layout of a Wes microplate includes 6 rows of wells (25 wells per row), 5 rows of wash buffer troughs, and bottom manufactured part. The first 5 rows of oval wells and the first 3 rows of wash buffer troughs are usually loaded as indicated in the figure. Figure created with Biorender

primary antibody in row C, and 10 μL of the corresponding secondary antibody in row D.

4b. For running a total protein assay, for each of the 24 sample wells, pipette 10 μL of Biotinylation reagent in row B, 10 μL of Antibody Diluent2 in row C, and 10 μL of total protein SA-HRP in row D.

5. Loading row E: Mix 200 μL of Luminol with 200 μL of Peroxide and fill row E with 15 μL of Luminol-Peroxide mixture in each well.

6. Cover the microplate with its lid and centrifuge at 2500 rpm for 5 min (using the PlateFuge) at room temperature to eliminate all bubbles; use an unopened microplate for balance.

7. Fill the first three rows of the bigger compartments on the microplate with 500 μL (each) of Wash Buffer.

8. Peel off the assay plate bottom foil. Carefully pop all bubbles in the Separation Matrix wells using 5 μL pipette tips.

9. Open the Wes's door and insert a newly opened capillary cartridge into cartridge holder. The interior light will change from orange to blue.

10. Place the assay plate on the plate holder and make sure the plate is attached with the edge of the tray.

11. Start the Wes run and assign the file save directory. Watch for 10 min to make sure the sample has loaded normally (fluorescent dots travel down evenly).

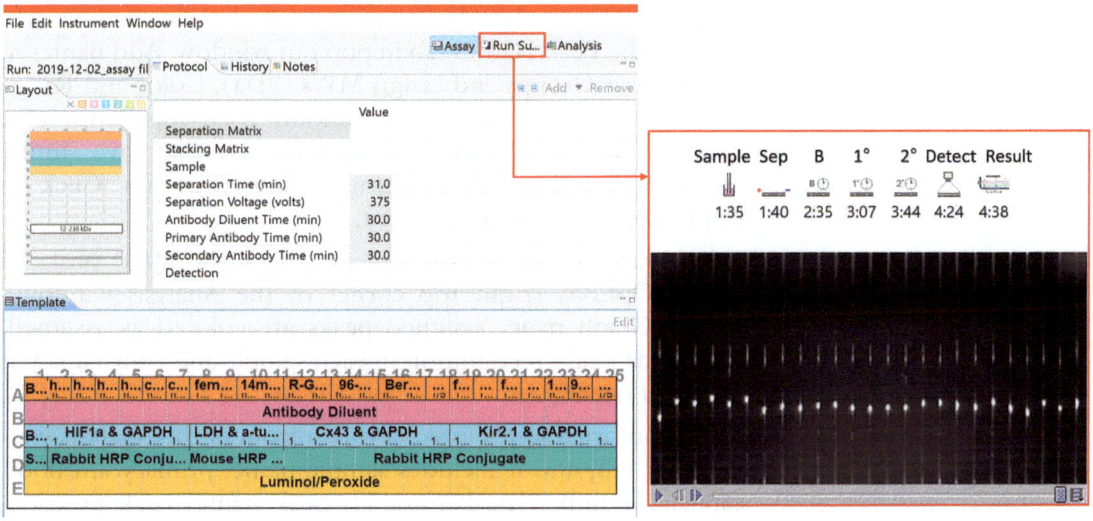

Fig. 4 Compass—data acquisition and analysis software. After creating a new run, the experimental settings, including testing protocol, sample layout, and antibody dilutions are saved as Assay in Compass Data File (.cbz). Once a new run is started, the experimental timeline can be seen in the Run summary. The progression of the sample loading can be monitored in real-time in the Run summary, which is stored as a video after the experiment. Quantification result and signal analysis are available in Analysis

12. The machine will indicate the end-time point for the run (typically takes around 3 h), Fig. 4, inset. After the test, take out and discard the microplate and capillary cartridge.

3.4.4 Data Analysis of Wes Data

1. Check the fluorescent standards assignment in the biotinylated ladder and each sample by clicking on Standard → Single View in the Compass Analysis Graph pane. Standard peaks for 12–230 kDa kit are 1, 29, 230 kDa. 1 kDa standard is the highest peak. The position of the 29 kDa and 230 kDa peaks should be similar for the ladder and the samples. If needed, correct peak position by right-clicking on peak and selecting "Not a Standard" or "Force Standard."
2. If the 1 kDa peaks vary a lot among samples and standard, exclude that peak to increase the accuracy of the protein size assignment. To exclude the 1 kDa standard peak for protein size fit, click Analysis in Edit tab. In the pop out window, under Standard tab, unclick the box for 1 MW (kDa) fit. Click Apply to save the adjusted setting.
3. Confirm biotinylated ladder peaks assignment by clicking on Samples → Single View of first capillary in Compass Analysis Graph pane. Peaks in 12–230 kDa kit are 12, 40, 66, 116, 180, 230 kDa.
4. Make sure the Sample Baseline Corrected and Fit Baseline Corrected are checked in the View Menu pull down options.

5. Name the detected peaks by going to Analysis in the Edit tab, click on the Peak Names tab in pop out window. Add names in the Analysis Groups and assign MWs (kDa), Color and Range (%) for each peak. 10% is commonly used for Range (%), decrease the percentage if two detected peaks have similar molecular weights. Apply setting to desired groups. Click on Apply after the peak assignment.
6. To confirm the peaks assignment, check the fitted peaks in Graph Options (right top corner of the Analysis window). Under Graph pane, assigned peaks are colored as assigned. Check the peaks recognition in each sample and adjust analysis setting if needed.
7. Sample information is listed at the bottom of the Analysis view in Peaks tab, and it includes sample name, primary antibody, capillary number, peak number, peak name, peak position, molecular weight (MW(kDa)), peak height, peak area, percentage area (% peak area/area under the curve), peak width, signal-to-noise ratio (S/N), averaged baseline.
8. For peaks analysis, go to the Capillaries tab at the bottom of Analysis view. The peaks areas are listed with peak names and capillary number. Choose the Area display, copy the sheet by clicking Ctrl + A and Ctrl + C. Paste the data into an Excel file for analysis.
9. Virtual blotting results ("lanes") available in the Compass Lane pane. Add sample, primary antibody, secondary antibody, their attributes, and capillary No. information from Gels Options. Adjust display contrast with the Slider if needed. Copy or screenshot the Lane view.
10. The Compass software can be downloaded for free for further examination and plotting of the data offline.

3.5 Primary Antibody Dilution Optimization

With consistent plating and culturing conditions, the hiPSC-CMs samples collected from 96-wells yield similar protein concentration, therefore the protein amount in each Wes sample loading will be similar. When using an antibody for the first time, optimal primary antibody dilution should be tested. The primary antibody needs to be at a saturating concentration to make sure the HRP signal change is proportional to protein expression difference. Oversaturation will increase the signal background and cause nonspecific detection.

If a target protein has been quantified in traditional WB, a good starting point for the referencing primary antibody dilution in Wes will be to use 20 times higher antibody concentration than in traditional WB. If a target protein could be clearly identified at the expected molecular weight with referencing primary antibody dilution and system peak height/baseline ratio ≥ 3, further tests of

Table 2
Sample preparation for antibody linear dynamic range test

Microcentrifuge sample	Cell lysis	0.1× Sample buffer	Fluorescent master mix
Neat	4 μL	0 μL	1 μL
0.75	3 μL	1 μL	1 μL
0.5	2 μL	2 μL	1 μL
0.25	1 μL	3 μL	1 μL

signal linear response under this dilution could be conducted. Otherwise, increase or decrease the primary antibody dilution according to the signal.

Proper controls should be included to distinguish signal from noise and to validate detection specificity, which includes:

- Neat sample with only Antibody Diluent2 in the microwell of "primary antibody."
- 0.1× sample buffer in the microwell of "sample" with testing antibody.
- Positive and negative controls, i.e., samples confirmed to have abundant and no target protein expression.

For quantifying a protein expression level, confirmation is needed that the measurements are in the linear dynamic range. Prepare four sample attributes using one sample and varying dilutions: neat, 0.75, 0.5, and 0.25 (actual sample attributes are 0.8, 0.6, 0.4, and 0.2 because samples need to be mixed with the Fluorescent Master Mix, as described earlier).

A sample collected from a single 96-well is enough for one linearity test. Repeat the following steps with multiple samples in the same test or in multiple tests to confirm the linearity.

1. For sample preparation, label four microcentrifuge tubes as "neat," "0.75," "0.5," and "0.25." Mix cell lysis, 0.1× sample buffer, and Fluorescent Master Mix as described in Table 2.
2. Other steps are the same as previously described for running an immunoassay by Wes.
3. After the immunoassay, in the Compass Analysis Graph pane, select one sample and click the View Selected on top left corner. Check the All exposures in the Graph options pull down menu. 1 s exposure usually has the highest peak value. If the antibody is not over-saturated, the 1 s, 2 s, and 4 s peaks should be close to each other. Consider decreasing the antibody dilution if the peaks from different exposure times are far apart.

4. Export the peak areas as previously described (*see* **Note 6**). Plot the signal over sample attributes as a dot plot.
5. Add linear regression trendline to the dot plot. Acquire R-squared value of the linear regression. R-squared value above 0.9 suggests the tested antibody dilution has linear response in the range of detection.

The optimized primary antibody dilution is acquired when the high R-squared value of linearity test is consistent, and no over-saturation is observed.

3.6 Normalized Protein Expression by Antibody Multiplexing and Total Protein Linear Response Test

Signal normalization can be achieved by antibody multiplexing or total protein quantification. Both methods have merits. Antibody multiplexing with a loading control protein requires less material, two or more tests could be conducted from a 96-well hiPSC-CMs collection. Total protein quantification could be used when no proper loading control is available.

For antibody multiplexing, the primary antibody of a target protein and the loading control need to be mixed in a proper ratio. Tested target protein and loading control combinations are summarized in Table 3; also *see* **Note 7**. Example results are shown in Fig. 5 for antibody multiplexing using Cx43 and GAPDH (loading control), including the electropherogram with peaks (with areas quantified on the bottom) and the Compass-generated lane view.

For simplicity, start making new antibody combinations by choosing two primary antibodies from the same host species. First, optimize the primary antibody dilution of each target. Make sure the multiplexed antibodies have clean baseline and are sufficiently different in molecular weight. Second, vary the antibody

Table 3
Wes antibody multiplexing dilution examples for use in hiPSC-CMs

Target description	Target protein (antibody)	Loading control	Dilution factor
LDH-A: Lactose dehydrogenase-A, linked to metabolic state	LDH (H-10) (Santa Cruz sc-133123)	Alpha tubulin (Abcam ab7291)	1:300 LDH 1:50 alpha tubulin
Cx43: Connexin43, gap junctional protein found in ventricular heart tissue	Cx43 (Abcam ab11370)	GAPDH (Abcam ab181602)	1:25 Cx43 1:2000 GAPDH
Kir2.1: Ion channel (inward rectifier) responsible for the resting membrane potential	Kir2.1 (Abcam ab65796) (Alomone APC-159)	GAPDH (Abcam ab181602)	1:10 Kir2.1 1: 4000 GAPDH

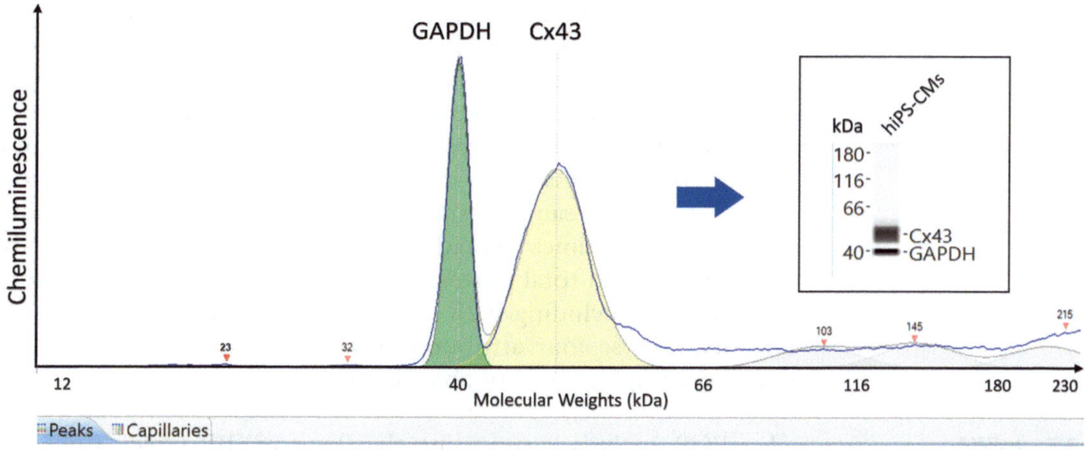

Sample	Primary	Cap	Peak	Name	Position	MW (kDa)	Height	Area	% Area	Width	S/N	Baseline
hiPS-CMs	Cx43 & GAPDH	12	3	GAPDH	378	40	724551.5	6653170	33.1	8.6	10000.9	1000.8
hiPS-CMs	Cx43 & GAPDH	12	4	Cx43	415	50	468053.1	13448959	66.9	27.0	2214.7	1104.9

Fig. 5 Example result of Cx43 and GAPDH multiplexing in human iPSC-CMs. Electropherogram and its conversion to a virtual lane view result of Cx43 and GAPDH multiplexing, using a 96-well collected human iPSC-CMs sample. Information including peak position, molecular weights, height, area, percentage area, width, signal-to-noise ratio and baseline are quantified in the Compass software

dilution to make the detected signals appear at a comparable level. In most cases, the expression levels of two targets are distinct and can differ by order of magnitude. Increase the antibody concentration of the less abundant target protein to balance the signal level. Or further dilute the abundantly expressed protein antibody. Rerun the signal linear response test to make sure the diluted concentration is saturated.

If varying the antibody dilutions can't adjust the signal to a comparable level, alternative solutions are provided by Protein Simple:

- Choose different species for the abundantly expressed protein. Also mix secondary antibodies and ensure they reach saturation. The company provides 20× secondary antibody formulation to ensure saturation of the secondary antibodies.
- Apply an HRP-conjugated secondary with lower HRP load. For example, Simple Western charge secondary antibodies (Anti-Rabbit HRP PN 040-656, Anti-Mouse HRP PN 040-655).
- Mix unlabeled secondary with the HRP-conjugated secondary. Be aware of the stability of secondary antibody mixtures.
- Use HRP-conjugated primary antibody as direct detection. It limits the amplification effect induced by indirect detection, but it is less specific comparing with indirect detection.

Follow Protein Simple's instruction on how to apply these methods.

For signal normalization by total protein assay, total protein linear detection range needs to be identified and included in each immunoassay. Use neat, 0.75, 0.5, and 0.25 sample attributes as four detection points, conduct total protein assay as previously described. Plot the total protein signal over sample attributes and do the linear regression to determine the total protein detection range. If titrated samples show good linearity (R-squared value >0.9), the four-times diluted sample could be used for antibody optimization and total protein assay could be used for signal normalization. If including 0.25 sample attribute influences the signal linearity, choose four attributes above 0.25 and rerun the linear detection range tests until finalizing the detection limit.

3.7 mRNA Quantification in hiPSC-CMs Using Cells-to-CT™ Kit and qPCR

1. qPCR analysis is run using the Power SYBR-Green Cells-to-CT™ kit (Thermo Fisher) and following the manufacturer's instructions.
2. Prepare DNaseI and Lysis solution 1:100 and use 50 μL of it for each well.
3. Aspirate the maintenance medium from the hiPSC-CMs. Wash cells with cold 1× PBS.
4. After aspirating the PBS, add 50 μL of DNaseI and Lysis Solution to each well. Pipet up and down with the pipette set to 30 μL (so to not introduce bubbles) five times.
5. Incubate plate at room temperature (19–25 °C) for 5 min.
6. Add 5 μL of stop solution to each well, mix five times, and incubate at room temperature for an additional 2 min.
7. Store samples at −80 °C as needed or proceed to the next step.
8. Program thermal cycler as indicated in Table 4.
9. Prepare a Reverse Transcriptase Master mix of 25 μL RT buffer, 2.5 μL Reverse Transcriptase, and 12.5 μL nuclease-free water per lysate.
10. Add 10 μL of lysate per 40 μL Reverse Transcriptase Master Mix and mix.
11. Run thermal cycler as indicated.

Table 4
Thermal cycler settings for Reverse Transcriptase

	Stage	Reps	Temp	Time
Reverse transcription	1	1	37 °C	60 min
RT inactivation	2	1	95 °C	5 min
Hold	3	1	4 °C	Indefinite

Table 5
Real-Time PCR cycling parameters

	Stage	Reps (cycles)	Temp	Time
Enzyme activation (hold)	1	1	95 °C	10 min
PCR (cycle)	2	40	95 °C 60 °C	15 s 1 min
Dissociation curve	3	(Use default settings)		

Table 6
PCR cocktail conditions

Component	20uL PCRs (Each reaction)
Power SYBR green PCR master mix	10 μL
Forward and reverse PCR primers	250 nM final concentration of each PCR primer
Nuclease-free water	Variable
cDNA	4 μL

12. After reverse transcriptase, store samples at −20 °C or proceed to the next step.
13. Program the real-time PCR instrument as indicated in Table 5. Reps (cycles) determines the level of amplification.
14. Prepare PCR cocktail as indicated in Table 6. To detect GJA1 mRNA, use Fw_GGTGGTACTCAACAGCCTTATT and Rv_ACCAACATGCACCTCTCTTATC primers and to detect GAPDH, use Fw_GGTGTGAACCATGAGAAGTATGA and Rv_GAGTCCTTCCACGATACCAAAG primers, respectively.
15. Dilute forward and reverse primers in nuclease-free water as needed. Add PCR cocktail and cDNA to each well of the MicroAmp™ Optical 96-well Reaction plate. Scale volumes as needed to account for three technical replicates and the various targets. Per one 96-well of lysate collected, about 12 different gene targets can be tested.
16. Seal MicroAmp™ Optical 96-well Reaction plate with MicroAmp™ Optical Adhesive Film and seal all edges with plastic MicroAmp® Adhesive Film Applicator to form a tight seal between the plate and adhesive film.
17. Run plate on QuantStudio™ 3 Real-Time PCR instrument under the parameters listed on Table 5.
18. Use QuantStudio™ Design & Analysis Software to extract Ct values and normalize Cx43 gene expression quantification to housekeeping gene GAPDH using standard ΔΔCt method [45–47].

3.8 Example Application

We illustrate the described methods for quantification of protein and mRNA within small cell samples (50,000 cells) in a 96-well microplate format by examining the effects of optogenetic transduction (ChR2) on the transcript and protein levels of gap junctional protein Cx43.

Connexin 43 is the main gap junctional protein in ventricular cardiomyocytes with a critical role in their structural and functional maturity [2, 54]. A previous optogenetics study, using HEK293T cells (with minimal Cx43 expression) suggested possible augmentation of Cx43 expression in those cells upon genetic expression of Channelrodopsin-2 (ChR2), a light-gated ion channel [55]. Using the methods described here, we probe for effects of ChR2 on Cx43/GJA1 in human iPSC-CMs from individual wells in a 96-well microplate.

With optimized Cx43 and GAPDH multiplexing in the Wes assay, the Cx43 protein expression levels are compared in non-infected, Ad-CMV-eYFP-infected, and Ad-CMV-hChR2 (H134R)-eYFP-infected hiPSC-CMs using two different hiPSC-CM lines, iCell Cardiomyocytes^2 CMC-100-012-001 (female differentiated) and MyCell hiPSC-CM-1X 01395 (male differentiated). In addition, GJA1 normalized to GAPDH mRNA levels were quantified in the respective conditions using the Cells-to-Ct kit and qPCR.

The results suggest highest Cx43 protein expression levels (after normalization by GAPDH) in ChR2-transduced hiPSC-CMs when compared to Ad-CMV-eYFP-infected and non-infected hiPSC-CM cell lines (Fig. 6a, b). At the transcriptional level, elevated GJA1 is also seen in ChR2-treated hiPSC-CMs compared to the eYFP-expressing cells but not compared to the untreated (Fig. 6c, d). The results indicate that the presence of ChR2 may enhance Cx43 expression in hiPSC-CMs though no statistical significance was reached at the transcription level. Although the mechanism is not clear and warrants further investigation, the findings suggest that an optogenetic perturbation may have an overall positive effect on hiPSC-CMs coupling.

4 Notes

1. Timing the water bath time is critical for cell viability when handling the iPSC-CMs.
2. Keeping the integrity of the cell monolayer is important for a successful protein collection. Proper dilution of fibronectin, sufficient fibronectin coating time, gentle operation during hiPSC-CMs thaw and culture could help improve cell attachment and viability.

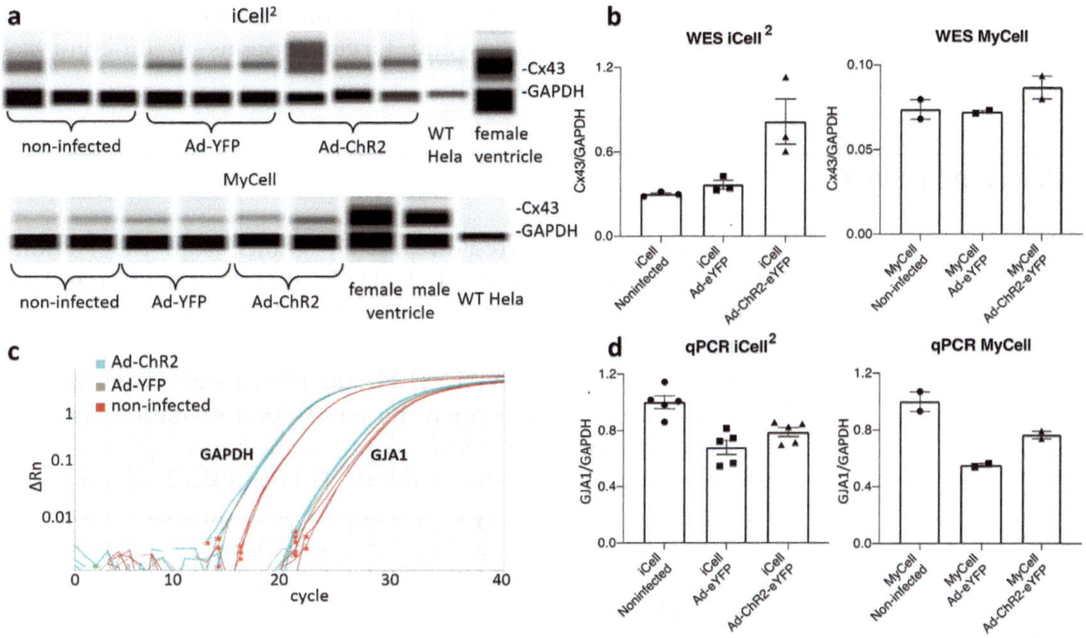

Fig. 6 Quantification of protein (Cx43) and mRNA (GJA1) after ChR2 expression in female iCell² and male MyCell hiPSC-CMs. (**a**, **b**) Protein quantification (Cx43 and GAPDH) in iCell² (**a**) and MyCell iPSC-CMs from individual 96-wells (control non-infected samples, Ad-eYFP-expressing cardiomyocytes, and Ad-ChR2-eYFP-expressing cardiomyocytes); human ventricular tissues as positive control and wild-type HeLa as negative control. Panel (**a**) shows the pseudo-bands automatically generated in the Compass software. Panel (**b**) displays the normalized Cx43/GAPDH values automatically calculated from areas under the curve in the electropherograms. Data are presented as $n = 3$ biological replicates in iCell² and $n = 2$ biological replicates in MyCell. (**c**, **d**) mRNA quantification (GJA1 and GAPDH) in iCell² (**a**) and MyCell iPSC-CMs from individual 96-wells. Panel (**c**) shows example PCR amplification curves from one run with iCell² samples. Panel (**d**) shows the normalized values GJA1/GAPDH calculated using the ΔΔCt method. Data are presented as $n = 5$ biological replicates with $n = 3$ technical replicates for iCell² and $n = 2$ biological replicates with $n = 3$ technical replicates for MyCell. Data are presented as mean +/− S.E.M

3. In the cell lysing process, timing is important—try to operate as quickly as possible to avoid sample degradation. In addition, all operations should be conducted on ice.
4. To minimize the difference in migration speed due to protein concentration, it is preferable to use samples with the same dilution factor in a particular run.
5. Minimize bubble formation by pointing pipette tip to the bottom of each well while pipetting and do it gently.
6. Default exposure is "High Dynamic Range 4.0"; different exposure times yield similar results.
7. Testing is required for each new antibody being considered. Antibodies working in standard WB may not work with Wes, even for the same samples. Different lot numbers of the same product number antibody also can produce variable results and

need to be re-tested. ProteinSimple maintains a database of Wes-tested antibodies and target proteins in different cell types that can be useful.

Acknowledgments

This work was supported in part by grants from the National Institutes of Health (R01-HL144157, R21-EB026152) and the National Science Foundation (EFMA 1830941, CBET1705645, and PFI 1827535).

Author Contributions: WZL, JLH and EE conceived the project. WZL designed and performed the Wes system validation and optimization, antibody multiplexing optimization, total protein linear response test and wrote the first draft. JLH and WZL designed and conducted proof of concept experiments, including qPCR and Wes quantification of Cx43 in non-infected, Ad-eYFP-infected, and Ad-CMV-hChR2 (H134R)-EYFP-infected hiPSC-CMs. WZL, JLH, and EE edited the draft.

References

1. Magdy T, Schuldt AJT, Wu JC, Bernstein D, Burridge PW (2018) Human induced pluripotent stem cell (hipsc)-derived cells to assess drug cardiotoxicity: opportunities and problems. Annu Rev Pharmacol Toxicol 58: 83–103. https://doi.org/10.1146/annurev-pharmtox-010617-053110
2. Gintant G, Burridge P, Gepstein L, Harding S, Herron T, Hong C, Jalife J, Wu JC (2019) Use of human induced pluripotent stem cell-derived cardiomyocytes in preclinical cancer drug cardiotoxicity testing: a scientific statement from the american heart association. Circ Res 125(10):e75–e92. https://doi.org/10.1161/res.0000000000000291
3. Entcheva E, Kay MW (2020) Cardiac optogenetics: a decade of enlightenment. Nat Rev Cardiol 18(5):349–367. https://doi.org/10.1038/s41569-020-00478-0
4. Zhang H, Cohen AE (2017) Optogenetic approaches to drug discovery in neuroscience and beyond. Trends Biotechnol 35(7): 625–639. https://doi.org/10.1016/j.tibtech.2017.04.002
5. Klimas A, Ortiz G, Boggess SC, Miller EW, Entcheva E (2020) Multimodal on-axis platform for all-optical electrophysiology with near-infrared probes in human stem-cell-derived cardiomyocytes. Prog Biophys Mol Biol 154:62–70. https://doi.org/10.1016/j.pbiomolbio.2019.02.004
6. Klimas A, Ambrosi CM, Yu J, Williams JC, Bien H, Entcheva E (2016) Optodyce as an automated system for high-throughput all-optical dynamic cardiac electrophysiology. Nat Commun 7:11542. https://doi.org/10.1038/ncomms11542
7. Sharma A, Burridge PW, McKeithan WL, Serrano R, Shukla P, Sayed N, Churko JM, Kitani T, Wu H, Holmstrom A, Matsa E, Zhang Y, Kumar A, Fan AC, Del Alamo JC, Wu SM, Moslehi JJ, Mercola M, Wu JC (2017) High-throughput screening of tyrosine kinase inhibitor cardiotoxicity with human induced pluripotent stem cells. Sci Transl Med 9(377). https://doi.org/10.1126/scitranslmed.aaf2584
8. McKeithan WL, Savchenko A, Yu MS, Cerignoli F, Bruyneel AAN, Price JH, Colas AR, Miller EW, Cashman JR, Mercola M (2017) An automated platform for assessment of congenital and drug-induced arrhythmia with hipsc-derived cardiomyocytes. Front Physiol 8:1–12. https://doi.org/10.3389/fphys.2017.00766
9. Cerignoli F, Charlot D, Whittaker R, Ingermanson R, Gehalot P, Savchenko A, Gallacher DJ, Towart R, Price JH, McDonough PM, Mercola M (2012) High throughput measurement of ca2+ dynamics for drug risk assessment in human stem cell-derived cardiomyocytes by kinetic image cytometry. J

Pharmacol Toxicol Methods 66(3):246–256. https://doi.org/10.1016/j.vascn.2012.08.167

10. Dempsey GT, Chaudhary KW, Atwater N, Nguyen C, Brown BS, McNeish JD, Cohen AE, Kralj JM (2016) Cardiotoxicity screening with simultaneous optogenetic pacing, voltage imaging and calcium imaging. J Pharmacol Toxicol Methods 81:240–250. https://doi.org/10.1016/j.vascn.2016.05.003
11. Laemmli UK (1970) Cleavage of structural proteins during the assembly of the head of bacteriophage t4. Nature 227(5259):680–685. https://doi.org/10.1038/227680a0
12. Towbin H, Staehelin T, Gordon J (1979) Electrophoretic transfer of proteins from polyacrylamide gels to nitrocellulose sheets: procedure and some applications. Proc Natl Acad Sci U S A 76(9):4350–4354. https://doi.org/10.1073/pnas.76.9.4350
13. Burnette WN (1981) "Western blotting": Electrophoretic transfer of proteins from sodium dodecyl sulfate—polyacrylamide gels to unmodified nitrocellulose and radiographic detection with antibody and radioiodinated protein A. Anal Biochem 112(2):195–203. https://doi.org/10.1016/0003-2697(81)90281-5
14. Treindl F, Ruprecht B, Beiter Y, Schultz S, Döttinger A, Staebler A, Joos TO, Kling S, Poetz O, Fehm T, Neubauer H, Kuster B, Templin MF (2016) A bead-based western for high-throughput cellular signal transduction analyses. Nat Commun 7:12852. https://doi.org/10.1038/ncomms12852
15. Hughes AJ, Herr AE (2012) Microfluidic western blotting. Proc Natl Acad Sci U S A 109(52):21450–21455. https://doi.org/10.1073/pnas.1207754110
16. Hu B, Li J, Mou L, Liu Y, Deng J, Qian W, Sun J, Cha R, Jiang X (2017) An automated and portable microfluidic chemiluminescence immunoassay for quantitative detection of biomarkers. Lab Chip 17(13):2225–2234. https://doi.org/10.1039/c7lc00249a
17. Lin Q, Wu J, Fang X, Kong J (2020) Washing-free centrifugal microchip fluorescence immunoassay for rapid and point-of-care detection of protein. Anal Chim Acta 1118:18–25. https://doi.org/10.1016/j.aca.2020.04.031
18. Hughes AJ, Spelke DP, Xu Z, Kang CC, Schaffer DV, Herr AE (2014) Single-cell western blotting. Nat Methods 11(7):749–755. https://doi.org/10.1038/nmeth.2992
19. Bietlot HP, Girard M (1997) Analysis of recombinant human erythropoietin in drug formulations by high-performance capillary electrophoresis. J Chromatogr A 759(1):177–184. https://doi.org/10.1016/S0021-9673(96)00767-4
20. O'Neill RA, Bhamidipati A, Bi X, Deb-Basu D, Cahill L, Ferrante J, Gentalen E, Glazer M, Gossett J, Hacker K, Kirby C, Knittle J, Loder R, Mastroieni C, MacLaren M, Mills T, Nguyen U, Parker N, Rice A, Roach D, Suich D, Voehringer D, Voss K, Yang J, Yang T, Vander Horn PB (2006) Isoelectric focusing technology quantifies protein signaling in 25 cells. Proc Natl Acad Sci 103(44):16153–16158. https://doi.org/10.1073/pnas.0607973103
21. Anderson GJ, Cipolla CM, Kennedy RT (2011) Western blotting using capillary electrophoresis. Anal Chem 83(4):1350–1355. https://doi.org/10.1021/ac102671n
22. Harris VM (2015) Protein detection by simple western™ analysis. Methods Mol Biol 1312:465–468. https://doi.org/10.1007/978-1-4939-2694-7_47
23. Beekman C, Janson AA, Baghat A, van Deutekom JC, Datson NA (2018) Use of capillary western immunoassay (wes) for quantification of dystrophin levels in skeletal muscle of healthy controls and individuals with Becker and duchenne muscular dystrophy. PLoS One 13(4):e0195850. https://doi.org/10.1371/journal.pone.0195850
24. Farnaby W, Koegl M, Roy MJ, Whitworth C, Diers E, Trainor N, Zollman D, Steurer S, Karolyi-Oezguer J, Riedmueller C, Gmaschitz T, Wachter J, Dank C, Galant M, Sharps B, Rumpel K, Traxler E, Gerstberger T, Schnitzer R, Petermann O, Greb P, Weinstabl H, Bader G, Zoephel A, Weiss-Puxbaum A, Ehrenhöfer-Wölfer K, Wöhrle S, Boehmelt G, Rinnenthal J, Arnhof H, Wiechens N, Wu MY, Owen-Hughes T, Ettmayer P, Pearson M, McConnell DB, Ciulli A (2019) Baf complex vulnerabilities in cancer demonstrated via structure-based protac design. Nat Chem Biol 15(7):672–680. https://doi.org/10.1038/s41589-019-0294-6
25. Park HJ, Kim MJ, Rothenberger C, Kumar A, Sampson EM, Ding D, Han C, White K, Boyd K, Manohar S, Kim YH, Ticsa MS, Gomez AS, Caicedo I, Bose U, Linser PJ, Miyakawa T, Tanokura M, Foster TC, Salvi R, Someya S (2019) Gsta4 mediates reduction of cisplatin ototoxicity in female mice. Nat Commun 10(1):4150. https://doi.org/10.1038/s41467-019-12073-0
26. Schneider JS, Aras R, Williams CK, Koprich JB, Brotchie JM, Singh V (2019) Gm1 ganglioside

modifies α-synuclein toxicity and is neuroprotective in a rat α-synuclein model of parkinson's disease. Sci Rep 9(1):8362. https://doi.org/10.1038/s41598-019-42847-x

27. Alterman JF, Godinho B, Hassler MR, Ferguson CM, Echeverria D, Sapp E, Haraszti RA, Coles AH, Conroy F, Miller R, Roux L, Yan P, Knox EG, Turanov AA, King RM, Gernoux G, Mueller C, Gray-Edwards HL, Moser RP, Bishop NC, Jaber SM, Gounis MJ, Sena-Esteves M, Pai AA, DiFiglia M, Aronin N, Khvorova A (2019) A divalent sirna chemical scaffold for potent and sustained modulation of gene expression throughout the central nervous system. Nat Biotechnol 37(8):884–894. https://doi.org/10.1038/s41587-019-0205-0
28. Pandey R, Yang Y, Jackson L, Ahmed RP (2016) Micrornas regulating meis1 expression and inducing cardiomyocyte proliferation. Cardiovasc Regen Med 3
29. Lu G, Li J, Zhai Y, Li Q, Xie D, Zhang J, Xiao Y, Gao X (2019) Spironolactone suppresses aldosterone-induced kv1.5 expression by attenuating mineralocorticoid receptor-nox1/2/4-mediated ros generation in neonatal rat atrial myocytes. Biochem Biophys Res Commun 520(2):379–384. https://doi.org/10.1016/j.bbrc.2019.10.039
30. Tabebordbar M, Zhu K, Cheng JKW, Chew WL, Widrick JJ, Yan WX, Maesner C, Wu EY, Xiao R, Ran FA, Cong L, Zhang F, Vandenberghe LH, Church GM, Wagers AJ (2016) In vivo gene editing in dystrophic mouse muscle and muscle stem cells. Science 351(6271):407–411. https://doi.org/10.1126/science.aad5177
31. Lue Y, Gao C, Swerdloff R, Hoang J, Avetisyan R, Jia Y, Rao M, Ren S, Atienza V, Yu J, Zhang Y, Chen M, Song Y, Wang Y, Wang C (2018) Humanin analog enhances the protective effect of dexrazoxane against doxorubicin-induced cardiotoxicity. Am J Physiol Heart Circ Physiol 315(3):H634–h643. https://doi.org/10.1152/ajpheart.00155.2018
32. Schiattarella GG, Altamirano F, Tong D, French KM, Villalobos E, Kim SY, Luo X, Jiang N, May HI, Wang ZV, Hill TM, Mammen PPA, Huang J, Lee DI, Hahn VS, Sharma K, Kass DA, Lavandero S, Gillette TG, Hill JA (2019) Nitrosative stress drives heart failure with preserved ejection fraction. Nature 568(7752):351–356. https://doi.org/10.1038/s41586-019-1100-z
33. Beauverger P, Ozoux ML, Bégis G, Glénat V, Briand V, Philippo MC, Daveu C, Tavares G, Roy S, Corbier A, Briand P, Dorchies O, Bauchet AL, Nicolai E, Duclos O, Tamarelle D, Pruniaux MP, Muslin AJ, Janiak P (2020) Reversion of cardiac dysfunction by a novel orally available calcium/calmodulin-dependent protein kinase ii inhibitor, ra306, in a genetic model of dilated cardiomyopathy. Cardiovasc Res 116(2):329–338. https://doi.org/10.1093/cvr/cvz097
34. Li W, Han JL, Entcheva E (2020) Syncytium cell growth increases kir2.1 contribution in human ipsc-cardiomyocytes. American journal of physiology-heart and circulatory. Physiology 319(5):H1112–H1122. https://doi.org/10.1152/ajpheart.00148.2020
35. Guo L, Eldridge S, Furniss M, Mussio J, Davis M (2015) Use of human induced pluripotent stem cell-derived cardiomyocytes (hipsc-cms) to monitor compound effects on cardiac myocyte signaling pathways. Curr Protoc Chem Biol 7(3):141–185. https://doi.org/10.1002/9780470559277.ch150035
36. Guo L, Eldridge S, Furniss M, Mussio J, Davis M (2018) Role of mcl-1 in regulation of cell death in human induced pluripotent stem cell-derived cardiomyocytes in vitro. Toxicol Appl Pharmacol 360:88–98. https://doi.org/10.1016/j.taap.2018.09.041
37. Hernandez I, Baio JM, Tsay E, Martinez AF, Fuentes TI, Bailey LL, Hasaniya NW, Kearns-Jonker M (2018) Short-term hypoxia improves early cardiac progenitor cell function in vitro. Am J Stem Cells 7(1):1–17
38. Zhong Q, Laco F, Liao MC, Woo TL, Oh SKW, Chai CLL (2018) Influencing the fate of cardiac and neural stem cell differentiation using small molecule inhibitors of alk5. Stem Cells Transl Med 7(10):709–720. https://doi.org/10.1002/sctm.17-0246
39. Bezzerides VJ, Caballero A, Wang S, Ai Y, Hylind RJ, Lu F, Heims-Waldron DA, Chambers KD, Zhang D, Abrams DJ, Pu WT (2019) Gene therapy for catecholaminergic polymorphic ventricular tachycardia by inhibition of ca (2+)/calmodulin-dependent kinase ii. Circulation 140(5):405–419. https://doi.org/10.1161/circulationaha.118.038514
40. Saiki RK, Scharf S, Faloona F, Mullis KB, Horn GT, Erlich HA, Arnheim N (1985) Enzymatic amplification of beta-globin genomic sequences and restriction site analysis for diagnosis of sickle cell anemia. Science 230(4732):1350–1354. https://doi.org/10.1126/science.2999980
41. Mullis KB (1990) The unusual origin of the polymerase chain reaction. Sci Am 262(4):56–61, 64–55. https://doi.org/10.1038/scientificamerican0490-56

42. Xia C, Fan J, Emanuel G, Hao J, Zhuang X (2019) Spatial transcriptome profiling by merfish reveals subcellular rna compartmentalization and cell cycle-dependent gene expression. Proc Natl Acad Sci U S A 116(39): 19490–19499. https://doi.org/10.1073/pnas.1912459116
43. Higuchi R, Fockler C, Dollinger G, Watson R (1993) Kinetic pcr analysis: real-time monitoring of DNA amplification reactions. Bio/Technology 11(9):1026–1030. https://doi.org/10.1038/nbt0993-1026
44. Heid CA, Stevens J, Livak KJ, Williams PM (1996) Real time quantitative pcr. Genome Res 6(10):986–994. https://doi.org/10.1101/gr.6.10.986
45. Schmittgen TD, Livak KJ (2008) Analyzing real-time pcr data by the comparative c (t) method. Nat Protoc 3(6):1101–1108. https://doi.org/10.1038/nprot.2008.73
46. Livak KJ, Schmittgen TD (2001) Analysis of relative gene expression data using real-time quantitative pcr and the 2(−delta delta c(t)) method. Methods 25(4):402–408. https://doi.org/10.1006/meth.2001.1262
47. Nolan T, Hands RE, Bustin SA (2006) Quantification of mrna using real-time rt-pcr. Nat Protoc 1(3):1559–1582. https://doi.org/10.1038/nprot.2006.236
48. Van Peer G, Mestdagh P, Vandesompele J (2012) Accurate rt-qpcr gene expression analysis on cell culture lysates. Sci Rep 2:222. https://doi.org/10.1038/srep00222
49. Babiarz JE, Ravon M, Sridhar S, Ravindran P, Swanson B, Bitter H, Weiser T, Chiao E, Certa U, Kolaja KL (2012) Determination of the human cardiomyocyte mrna and mirna differentiation network by fine-scale profiling. Stem Cells Dev 21(11):1956–1965. https://doi.org/10.1089/scd.2011.0357
50. Zhang Y, Cao N, Huang Y, Spencer CI, Fu JD, Yu C, Liu K, Nie B, Xu T, Li K, Xu S, Bruneau BG, Srivastava D, Ding S (2016) Expandable cardiovascular progenitor cells reprogrammed from fibroblasts. Cell Stem Cell 18(3): 368–381. https://doi.org/10.1016/j.stem.2016.02.001
51. Conway M, Xu T, Kirkpatrick A, Ripp S, Sayler G, Close D (2020) Real-time tracking of stem cell viability, proliferation, and differentiation with autonomous bioluminescence imaging. BMC Biol 18(1):79. https://doi.org/10.1186/s12915-020-00815-2
52. Ambrosi CM, Entcheva E (2014) Optogenetic control of cardiomyocytes via viral delivery. Methods Mol Biol 1181:215–228. https://doi.org/10.1007/978-1-4939-1047-2_19
53. Yu J, Entcheva E (2016) Inscribing optical excitability to non-excitable cardiac cells: viral delivery of optogenetic tools in primary cardiac fibroblasts. Methods Mol Biol 1408:303–317. https://doi.org/10.1007/978-1-4939-3512-3_21
54. Kucera JP, Rohr S, Rudy Y (2002) Localization of sodium channels in intercalated disks modulates cardiac conduction. CircRes 91(12): 1176–1182
55. Jia Z, Valiunas V, Lu Z, Bien H, Liu H, Wang HZ, Rosati B, Brink PR, Cohen IS, Entcheva E (2011) Stimulating cardiac muscle by light: cardiac optogenetics by cell delivery. Circ Arrhythm Electrophysiol 4(5):753–760. https://doi.org/10.1161/CIRCEP.111.964247

Chapter 3

Self-Assembled Heterotypic Cardiac Spheroids from Human Pluripotent Stem Cells

Oriane B. Matthys and Todd C. McDevitt

Abstract

Engineered cardiac tissue models aim to recapitulate the multicellular composition of the native myocardium by incorporating multiple tissue-relevant cell populations. Here, we describe the process of generating self-assembled cardiac microtissue spheroids comprised of heterotypic cardiac cell types. The absence of exogenous extracellular matrix (ECM) or scaffolding makes microtissue assembly dependent upon intercellular adhesion interactions over cell–ECM interactions, analogous to early development. Therefore, this approach creates a 3D platform to study how multicellular heterotypic interactions impact tissue structure, function, and phenotype.

Key words Tissue engineering, Scaffold-free, Stem cells, Heterotypic interactions, Cardiomyocytes, Cardiac fibroblasts, Spheroids, Self-assembly, Cardiac differentiation

1 Introduction

Engineered cardiac tissue constructs have become increasingly complex as more tissue-specific cell types are incorporated in order to better mimic the multicellular composition of native heart tissue. Non-parenchymal cardiac cell populations, such as cardiac fibroblasts, endothelial cells, and macrophages, work cooperatively with cardiomyocytes to maintain native tissue homeostasis and function [1]. Furthermore, non-myocytes have also been shown to modulate cardiomyocyte phenotype and cardiac tissue function in in vitro models [2, 3].

Self-assembled cardiac microtissues provide a robust platform to study the direct interactions of heterogeneous cell types in 3D. Since exogenous matrices or scaffold materials are not required for the engineering of self-assembled tissues, changes in tissue function or cell phenotype resulting from different pairings of cardiac cell types can be attributed to altered heterotypic interactions or paracrine signaling mechanisms. Furthermore, self-assembly methods

Kareen L. K. Coulombe and Lauren D. Black III (eds.), *Cardiac Tissue Engineering: Methods and Protocols*, Methods in Molecular Biology, vol. 2485, https://doi.org/10.1007/978-1-0716-2261-2_3,

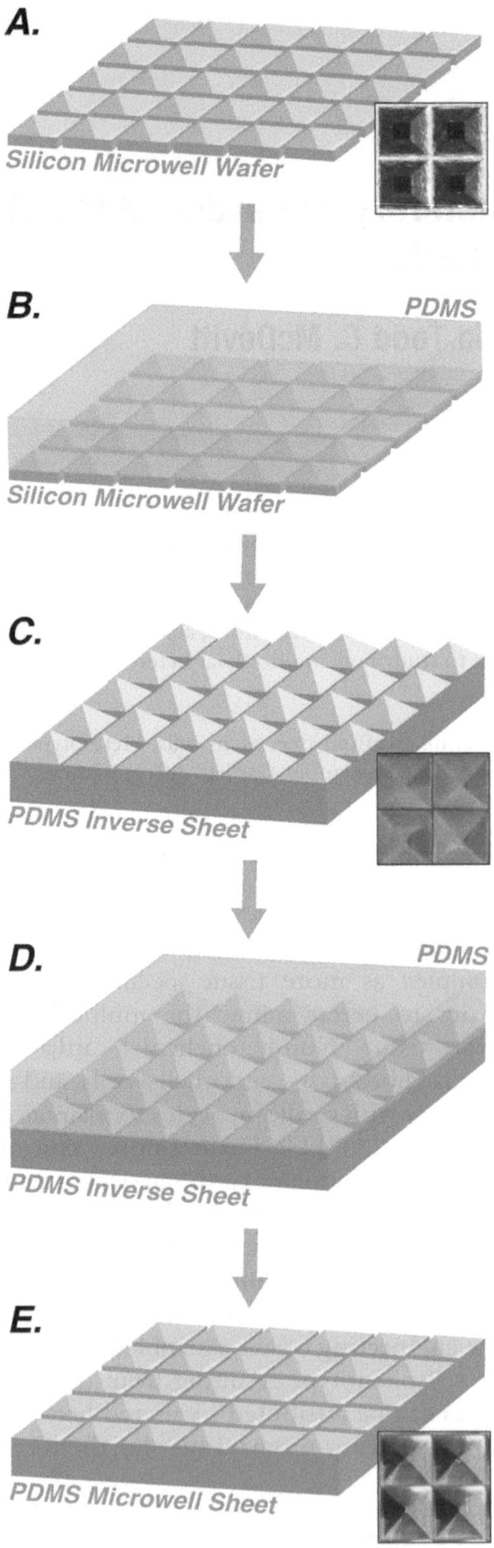

Fig. 1 Replica molding to fabricate sheets of PDMS microwell molds. (**a**) A chemically etched silicon wafer of inverted pyramidal microwells is silanized. (**b**) PDMS is poured on top of the silicon wafer and baked to cure.

of tissue engineering provide precise control over the different cell types and proportions that are seeded into the microtissues [4]. Therefore, self-assembled cardiac microtissues facilitate the study of cardiac intercellular connectivity, both physical and electrical, and ultimately enable determination of the specific contributions of various non-myocyte populations to engineered cardiac tissue structure, function, and phenotype.

Here, we describe the process of forming self-assembled cardiac microtissue spheroids comprised of human-induced pluripotent stem cell-derived cardiomyocytes (hiPSC-CMs) and primary human cardiac fibroblasts (CFs) (Fig. 4) though this method can broadly generate cardiac microtissues of additional heterotypic compositions (i.e., CMs mixed with hiPSC-derived endothelial cells; Fig. 6). We detail the replica molding process that allows us to fabricate arrays of polydimethylsiloxane (PDMS) inverse molds (Figs. 1 and 2), from which we cast the final agarose microwells that the cardiac cell mixtures are seeded into Fig. 3. Overall, our platform of defined, scaffold-free, heterotypic cardiac microtissues can be used to directly study the mechanisms of intercellular interactions in 3D or provide a microscale substrate to interrogate pharmacological effects or disease modeling of human cardiac tissue.

2 Materials

1. Silicon wafer: chemically etched 400 × 400 μm inverted pyramidal microwells.
2. Vacuum chambers.
3. Weigh boats.
4. Aluminum foil.
5. Trichloro(1*H*,1*H*,2*H*,2*H*-perfluorooctyl)silane.
6. Polydimethylsiloxane (PDMS): SYLGARD 184 Silicone Elastomer Kit.
7. Razor blade or scalpel.
8. Steel hole punch: C.S. Osborne Industrial Tools, Arch Punch.
9. Agarose.
10. Glass storage container.
11. Forceps.
12. 24-well plates.

Fig. 1 (continued) (**c**) PDMS layer is removed, creating a sheet of inverse microwells (pyramids). (**d**) PDMS is poured onto the sheet of PDMS inverse microwells (pyramids) and baked to cure. (**e**) PDMS layer is removed, resulting in a sheet of PDMS inverted pyramidal microwells

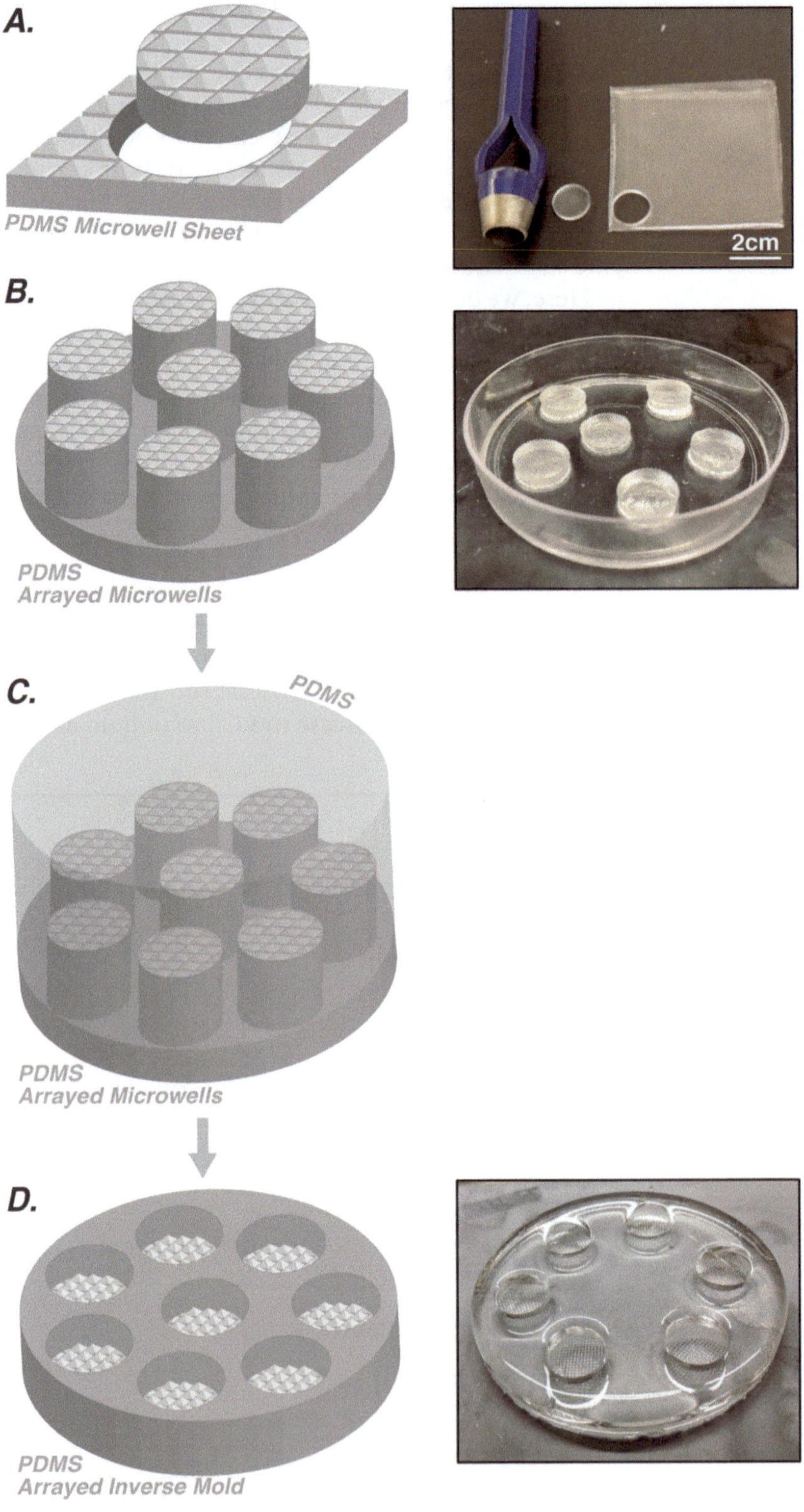

Fig. 2 Replica molding to create arrayed circular inverse molds. (**a**) Circles the size of 24-wells (~15 mm) are punched out of the PDMS inverted pyramidal microwell sheets. (**b**) ~8 circular-punched microwell molds are affixed in a base layer of PDMS in an arrayed pattern and silanized. (**c**) PDMS is poured on top of arrayed circular-punched microwells and baked to cure. (**d**) Top PDMS layer is removed, resulting in a PDMS mold of arrayed circular inverse molds the size of 24-wells

13. Conical tubes.
14. Wide-bore P1000 pipette tips.
15. 10 cm petri dishes.
16. Incubator-grade rotary orbital shaker.

3 Methods

3.1 Replica Molding to Fabricate Sheets of PDMS Microwell Molds

1. Place microfabricated silicon wafer (Fig. 1a; *see* **Note 1**) in a vacuum chamber that is contained within a chemical fume hood. Line a weigh boat with aluminum foil, then place a few drops of silane into the lined weigh boat, and set in vacuum chamber beside wafer. Leave under vacuum overnight (*see* **Notes 2**, **3**, and **4**).
2. Prepare PDMS: weigh base and curing agents and mix thoroughly at a 10:1 ratio of base to curing agent (*see* **Note 5**).
3. De-gas mixed PDMS solution by placing under vacuum until most bubbles have disappeared (~30 min). Can be performed in a vacuum chamber set up on a benchtop.
4. Pour de-gassed PDMS onto silane-treated silicon wafer (*see* **Note 6**; Fig. 1b). De-gas PDMS again by placing under vacuum until all bubbles have disappeared, or at least raised up to the top surface, away from the microwell interface (*see* **Notes 7** and **8**). Place in 60 °C oven for a minimum of 1 h, or until cured (no longer viscous or sticky to the touch).
5. Using a razor blade or scalpel, free PDMS from the edges of the container, being careful not to score or scratch the silicon. Peel the PDMS layer away from the silicon wafer, working slowly so as to not tear the PDMS. This process results in a PDMS sheet of pyramids (i.e., inverse microwells; Fig. 1c).
6. Perform silane surface treatment on the sheet of PDMS inverse microwells as described in Subheading 3.1, **step 1**. Place mixed PDMS under vacuum until most of the bubbles have disappeared and then pour on top of PDMS inverse sheet (*see* **Note 9**; Fig. 1d). De-gas and cure as described in Subheading 3.1, **steps 3** and **4**, respectively. Carefully peel top PDMS layer off of the bottom inverse layer as described in Subheading 3.1, **step 5**. This results in a PDMS sheet of inverted pyramidal microwells (Fig. 1e).

3.2 Repeat Replica Molding to Create Arrayed Circular Inverse Molds

1. Punch the PDMS microwell sheet into circles sized for standard 24-well plates (*see* **Note 10**; Fig. 2a).
2. To create an array of circular-punched microwell molds that fit into a 24-well plate, pour a 6–8 mm-thick layer of de-gassed PDMS into a 10 cm plate and arrange ~8 cut microwell molds

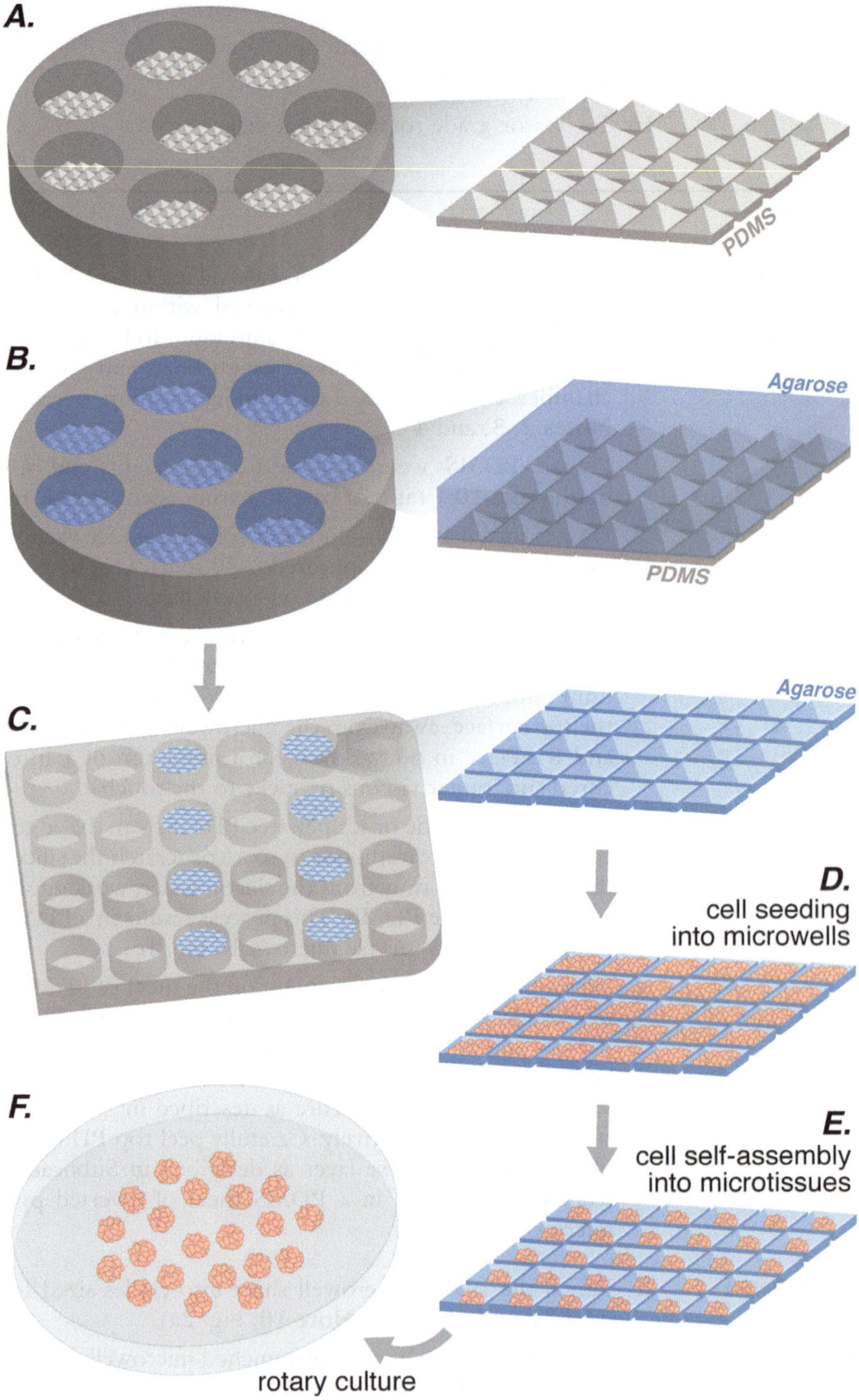

Fig. 3 Casting agarose microwell molds off of PDMS-arrayed circular inverse molds. (**a**) PDMS-arrayed circular inverse mold is autoclaved and (**b**) melted agarose is added to fill the circular wells. (**c**) Once solidified,

into the de-gassed PDMS with the microwell surface facing upward, as illustrated in Fig. 2b. Allow to cure in 60 °C oven for a minimum of 1 h.

3. Silanize the PDMS array of microwell molds as described in Subheading 3.1, **step 1**. Pour de-gassed PDMS into the 10 cm dish holding the arrayed molds (*see* **Note 11**; Fig. 2c). De-gas and cure as described in Subheading 3.1, **steps 3** and **4**. Carefully peel top PDMS layer off of bottom layer. This results in the final PDMS mold, containing an array of circular wells with inverse microwells (i.e., extruded pyramids) covering the bottom surface (Figs. 2d and 3a).
4. Autoclave PDMS-arrayed circular inverse mold before use (*see* **Note 12**).

3.3 Casting Agarose Microwell Molds Off of PDMS-Arrayed Circular Inverse Molds

1. Prepare 3% agarose: weigh 3 g of agarose and add to glass storage container. Pour in 100 mL of Dulbecco's Modified Eagle Medium (DMEM). Microwave and stir to dissolve agarose in DMEM. Autoclave.
2. Microwave 3% agarose until liquid (*see* **Note 13**).
3. In a biosafety cabinet, slowly pipette (using stripette or wide-bore P1000 pipette tip) melted agarose into the circular wells of the PDMS-arrayed circular inverse mold (Fig. 3b), being careful not to introduce bubbles (*see* **Note 14**). Let the agarose cool until solid (~5–10 min).
4. Remove agarose microwell molds by carefully bending the PDMS-arrayed circular inverse mold to release the agarose microwell molds. Make sure to release the agarose microwell molds onto a sterile surface, such as an empty 10 cm or 15 cm round tissue culture plate.
5. Using sterile forceps, carefully place the agarose microwell molds into a 24-well plate, ensuring that the microwell surface is facing up (Fig. 3c). Using the forceps, gently push the agarose microwell mold down into the well until it lays flat along bottom of plate (*see* **Note 15**).
6. Add 0.5 mL of media or PBS into the wells that contain agarose microwell molds. Centrifuge the plate at 2000 × *g* for 5 min in order to force the agarose molds flat against the bottom of the plate and to remove air bubbles from the microwells of the molds (*see* **Note 16**).

Fig. 3 (continued) agarose microwell molds are removed from the arrayed PDMS mold and transferred into wells of a 24-well plate, with the microwell surface facing upward. (**d**) Cardiac cells are seeded into the agarose microwell molds and incubated overnight to allow self-assembly into spheroidal microtissues (**e**). (**f**) Spheroidal microtissues are transferred from microwells to 10 cm petri dishes and maintained on orbital rotary culture

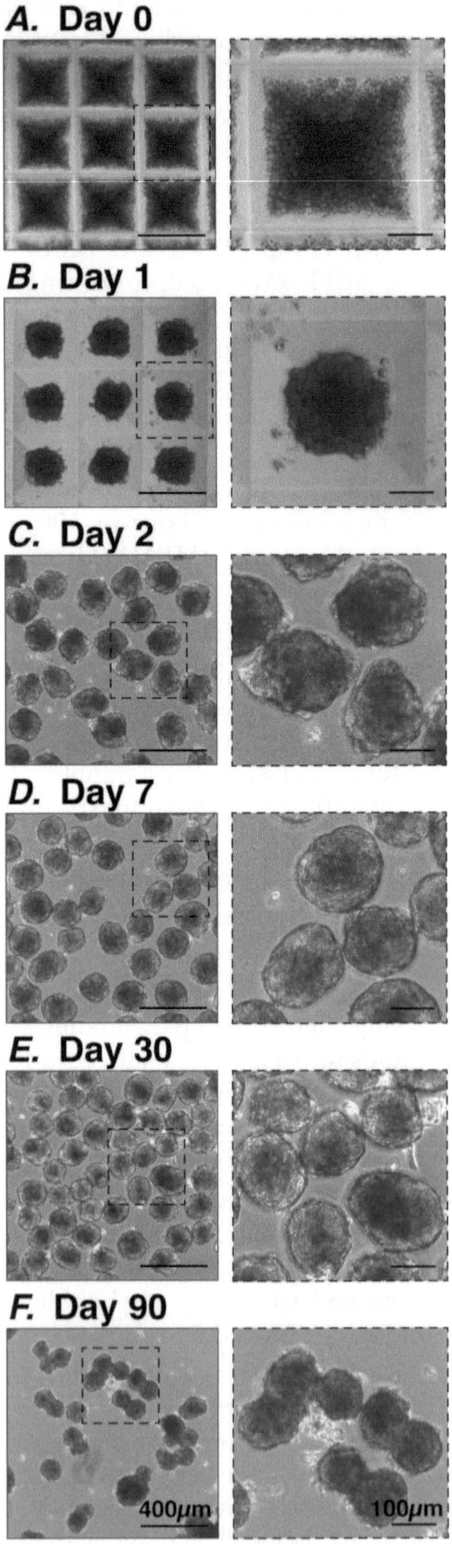

Fig. 4 Self-assembly and long-term culture of heterotypic cardiac microtissue spheroids. (**a**) Heterotypic mixture of cells (cardiomyocytes and cardiac fibroblasts) seeded into agarose microwell molds immediately

7. Agarose microwell molds can be cast and set into 24-well plates up to 24 h before cell seeding. Store in incubator with 0.5–1 mL PBS or media in the wells in order to prevent agarose from drying out.

3.4 Cell Seeding into Agarose Microwell Molds

1. In a biosafety cabinet, rinse adherent cultures of cardiomyocytes (CMs) (*see* **Note 17**) and cardiac fibroblasts (CFs) (*see* **Note 18**) with PBS. Incubate cells in 0.25% Trypsin at 37 °C for 5–10 min, or until cells have lifted off of the plate in a single-cell suspension. Collect CMs and CFs in a conical tube and quench 1:1 volumetrically with 20% FBS in DMEM. Centrifuge at 200 × *g* for 3 min. Aspirate supernatant and resuspend in cardiac maintenance medium (RPMI1640 plus B27 supplement with insulin) and 10 μM ROCK inhibitor (ROCKi).
2. Count CMs and CFs using a hemocytometer. To make 2000-cell microtissues, seed 2×10^6 total cells per 24-well sized agarose microwell mold (*see* **Note 10**). For heterotypic cardiac microtissues at a ratio of 3:1 CMs:CFs, mix 1.5×10^6 CMs with 0.5×10^6 CFs in a conical tube (*see* **Note 19**). Centrifuge conical tube with cell mixture at 200 × *g* for 3 min. Aspirate and resuspend in cardiac maintenance media supplemented with 10 μM ROCKi at a concentration of 1 mL media per agarose microwell mold (*see* **Note 20**).
3. Pipette 1 mL of cardiac cell mixture (2×10^6 total cells) into each agarose microwell mold. Centrifuge at 200 × *g* for 3 min but reduce the acceleration and brake settings (acceleration = 0, brake = 3). Check by microscopy that cells have settled into the microwells (Figs. 3d and 4a).
4. Transfer to incubator, being careful not to jostle the plate. Leave in incubator overnight without disturbing (18–24 h).

3.5 Removing Microtissues from Agarose Microwell Molds

1. After 18–24 h, check that the cells have self-assembled into microtissues in the agarose microwell molds. This will be evident by a sharp spheroidal boundary indicating successful tissue assembly (*see* **Note 21**; Figs. 3e and 4b).
2. Using P1000 wide-bore pipette tips, gently pipette up and down over the agarose microwell molds to dislodge the microtissues from their microwells. Transfer microtissues to conical

Fig. 4 (continued) after centrifugation (day 0). (**b**) The next day (day 1), cells have self-assembled into spheroidal microtissues, as determined by sharp circular borders seen while microtissues are still in the agarose microwell molds. (**c**) Spheroidal microtissues after removal from agarose microwell molds (day 2). Microtissues maintain their spheroidal shape throughout long-term culture but gradually compact in size, as seen after 7 days (**d**), 30 days (**e**), and 90 days (**f**) of rotary culture tubes. Add 1 mL PBS to the agarose microwell molds and gently pipette up and down to collect remaining microtissues. Repeat until all of the microtissues have been collected in the conical tubes, checking by microscope between rinses.

3. Allow the microtissues to settle to the bottom of the conical tube via gravitational force (~5 min; *see* **Note 22**). Carefully aspirate the excess supernatant, taking care not to aspirate the pellet of microtissues.
4. Add fresh cardiac maintenance media to the microtissues in the conical tubes and transfer tissues to 10 cm petri dish (*see* **Note 23**), combining microtissues from 2 to 4 agarose microwell molds per 10 cm dish in a final volume of 10 mL media per 10 cm plate (*see* **Notes 24** and **25**; Figs. 3f and 4c).

3.6 Rotary Suspension Culture

1. Place 10 cm dish on a rotary orbital shaker in an incubator (*see* **Note 26**) rotating at 65 rpm (*see* **Note 27**).
2. Re-feed microtissues every 3 days with cardiac maintenance medium (*see* **Note 28**). To feed, use a stripette to transfer the 10 mL of media with microtissues to a 15 mL conical tube. Allow the microtissues to settle to the bottom of the conical tube (via gravitational force; ~5 min). Carefully aspirate the spent media and add fresh cardiac maintenance media. Transfer the microtissues and fresh media back to the 10 cm petri dish (can use the same dish or a new one) using a stripette or wide-bore P1000 pipette tip.
3. Microtissues can be maintained on the rotary for months (Fig. 4c–f).

4 Notes

1. The chemically etched silicon wafer contains 400 × 400 μm inverted pyramidal microwells.
2. Silane surface treatment prevents PDMS from adhering to the silicon wafer as well as to itself [5].
3. We recommend dedicating a vacuum chamber solely for silanization because the chamber also gets coated during the surface treatment.
4. Silane wafer can be used indefinitely for replica molding, but silanization should be repeated every ~5 uses.
5. Make sure to mix PDMS base and curing agent well; the curing agent is cytotoxic on its own, but becomes safe for use with cells when thoroughly mixed with base.

6. The desired PDMS thickness for this step is ~6–8 mm. In a 10 cm plate, 50 g of mixed PDMS results in ~6 mm thickness and 75 g of PDMS yields ~9 mm thickness.
7. For bubbles to completely disappear in the PDMS once it has been poured onto the silicon wafer of microwells usually takes a minimum of 2 h under vacuum, but PDMS can be left under vacuum overnight.
8. Bubbles at the PDMS-silicon wafer interface will result in defects at the microwell surface that will persist throughout replica molding and result in tissue assembly impurities.
9. For this step, you need a thicker layer of PDMS (~12 mm) because this layer will be cut and adhered in another layer of PDMS. You want the microwell surface of thicker, cut PDMS to be sitting >5 mm above the adherent PDMS layer.
10. To punch the PDMS microwell sheet into circles that fit in a 24-well plate, we use a 15 mm steel hole punch (C.S. Osborne Industrial Tools, Arch Punch; Fig. 2a). Cutting at this size produces ~1000 microwells (and therefore 1000 microtissues) per 24-well mold. The microwell sheet can also be punched in a diameter that fits additional well-plate sizes; a punched circle sized for a 6-well plate contains ~5000 microwells.
11. Pour enough PDMS to ensure that the tops of the circular-punched microwell molds are covered >2 mm (~8–10 mm thick altogether).
12. The PDMS inverse molds can be reused indefinitely. To clean, follow a series of wash steps performed in a sonicating bath: 20 min in 90% EtOH, 20 min in 70% EtOH, 20 min in DI water. Then autoclave.
13. In order to keep the agarose sterile, loosen the lid while microwaving, but do not remove. Each batch of agarose can be microwaved ~3 times before a fresh batch should be made and autoclaved.
14. When pipetting agarose into the circular wells of the PDMS-arrayed inverse mold, make sure to fill up the wells exactly to the top, such that the agarose creates a flat surface. Any concavity or convexity will prevent the agarose microwell mold from sitting flat against the bottom of the 24-well plate, which will create slanted microwells (Fig. 5).
15. Be careful not to touch the forceps to the top microwell surface of the agarose microwell mold. Forceps can easily scratch the agarose, which mars the microwells and creates defects that the cells will settle into. Only touch the sides of the agarose microwell molds when guiding them into the plate.
16. If the thickness of the agarose microwell molds is not uniform, when the agarose molds are forced to sit flat on the bottom of

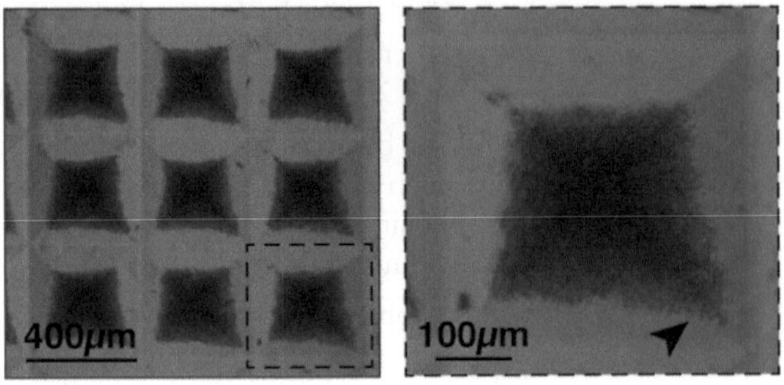

Fig. 5 Troubleshooting slanted agarose microwell molds. Cells that are seeded into slanted microwell molds display a tail effect (arrow) as opposed to the desired equilateral square, as in Fig. 4a

the place, the top microwell surface will sit at a slant. In these instances, the cells cannot fill up the entirety of the microwell when seeded, which alters the self-assembly and size of the microtissues (Fig. 5).

17. Stem cell-derived cardiomyocytes [6, 7] were lactate-purified [8] in order to achieve higher purity of CMs. Using purified populations of cells ensures that the desired ratio of mixed cell types is accurate.
18. Commercially available primary human cardiac fibroblasts were used in these studies. hiPSC-derived cardiac fibroblasts can also be used [9, 10].
19. Additional cell types and/or combinations of cell types can be incorporated into the cardiac microtissues, for example by mixing endothelial cells in with cardiomyocytes (Fig. 6c). However, in order for tissues to self-assemble, the starting CM population need to contain or be mixed with >10% non-myocytes [2]. Therefore, microtissues can be made from heterogeneous cardiac differentiations, which usually contain ~60–80% CMs (Fig. 6a) or from enriched CM differentiations that still contain a small non-myocyte fraction (~85–95% CMs; Fig. 6b).
20. For example, if seeding four agarose microwell molds-worth of heterotypic cardiac microtissues, combine 6×10^6 CMs and 2×10^6 CFs in a conical tube, centrifuge, and resuspend in 4 mL media.
21. Microtissues that do not self-assemble after 18 h lack distinct spherical borders in the microwells (Fig. 7).
22. In the case that microtissues do not settle down to the bottom of the conical tubes on their own, it is possible to gently centrifuge at ~$80 \times g$ for 1–3 min (acceleration = 0, brake = 3). If microtissues are centrifuged too fast or for too long, they will agglomerate.

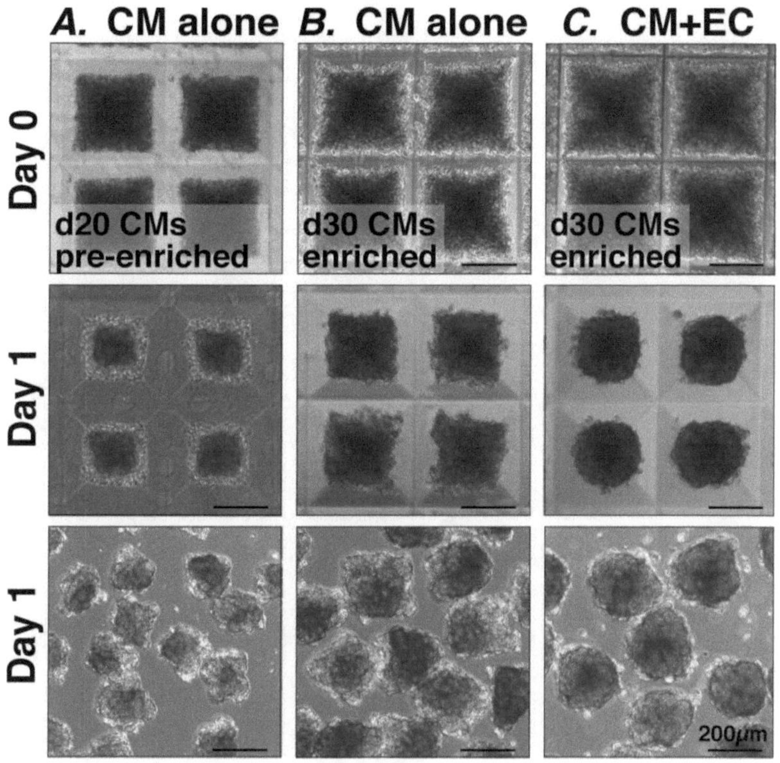

Fig. 6 Additional cardiac microtissue constituents. Cardiac microtissues can be made from cells solely derived in cardiomyocyte differentiations, such as with day 20 hiPSC-CMs that have not been enriched for CMs and therefore typically contain anywhere from 60 to 80% CMs (**a**) or with day 30 hiPSC-CMs that have been enriched via lactate purification methods [8] and contain ~85–95% CMs (**b**). Enriched CMs can also be mixed with other types of non-myocytes, such as hiPSC-derived endothelial cells (ECs) (**c**)

23. Non-tissue culture treated petri dishes, as opposed to tissue culture polystyrene (TCPS) plates, are utilized in order to minimize cell or tissue attachment to the plates. Ultra-low adherent (ULA) plates can be used as well.

24. Smaller sized plates (i.e., ULA 6-well plates) can be used in place of 10 cm petri dishes; however, the scaled-down number of microtissues maintained in each well will need to be optimized.

25. When too few microtissues are cultured in a 10 cm petri dish, the microtissues may begin to agglomerate (Fig. 8a) or start to dissociate and eventually fall apart (Fig. 8b) [4].

26. Using an incubator-grade rotary orbital shaker (we use the Benchmark Scientific Orbi-Shaker CO_2 with a 0.75″ orbital offset distance) will increase the lifetime of the shaker in the humid incubator environment.

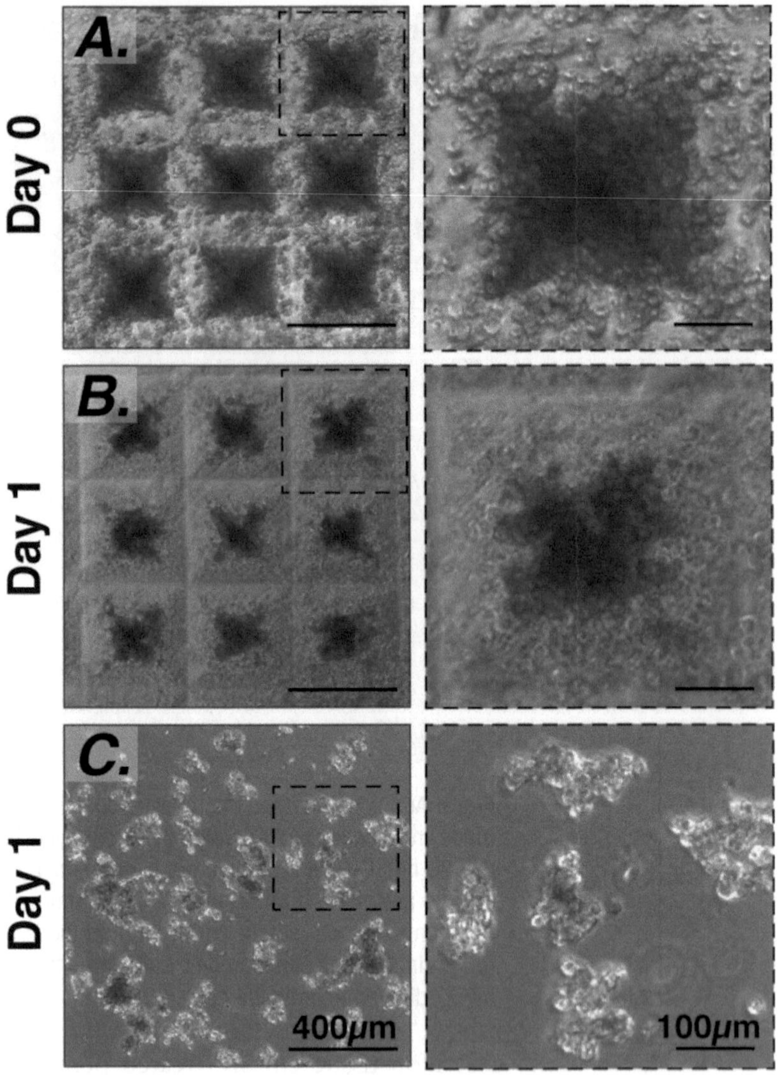

Fig. 7 Troubleshooting poor tissue self-assembly. (**a**) Cardiac cells largely fill up agarose microwell molds after seeding and centrifugation (day 0), but the next day (**b**) cells do not display distinct spherical borders in the microwells; (**c**) as a result, the cells will not result in spheroidal microtissues when removed from the microwells

27. Adjusting the rotation speed of the rotary orbital shaker changes the hydrodynamic forces in the suspension culture, which may alter tissue parameters such as agglomeration, size, or cell survival [4, 11].
28. Other types of microtissues may need to be fed more or less frequently, depending on the metabolic activity of the cultures.

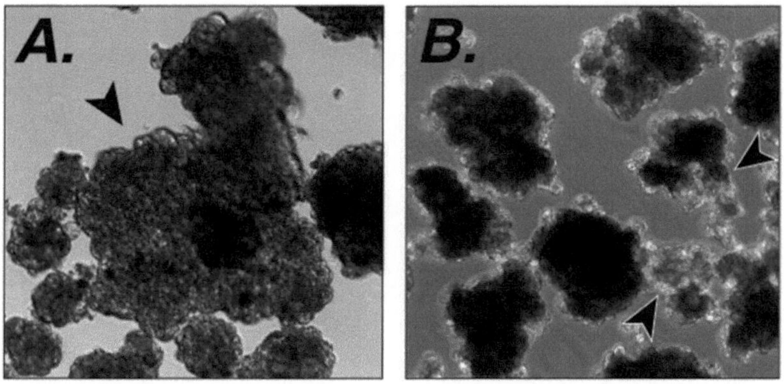

Fig. 8 Troubleshooting rotary tissue culture problems. (**a**) Microtissues that are maintained on rotary culture at too low of a density may start to fuse/agglomerate (arrow). (**b**) Microtissues that are dying start to lose their distinct borders and shed cells or clumps of cells (arrows)

References

1. Pinto AR et al (2016) Revisiting cardiac cellular composition. Circ Res 118(3):400–409
2. Hookway TA et al (2019) Phenotypic variation between stromal cells differentially impacts engineered cardiac tissue function. Tissue Eng Part A 25(9-10):773–785
3. Giacomelli E et al (2017) Three-dimensional cardiac microtissues composed of cardiomyocytes and endothelial cells co-differentiated from human pluripotent stem cells. Development 144(6):1008–1017
4. Hookway TA et al (2016) Aggregate formation and suspension culture of human pluripotent stem cells and differentiated progeny. Methods 101:11–20
5. Ratner BD, Hoffman AS (2013) Physicochemical surface modification of materials used in medicine. Biomaterials Science: An Introduction to Materials: Third Edition:259–275
6. Lian X et al (2012) Robust cardiomyocyte differentiation from human pluripotent stem cells via temporal modulation of canonical Wnt signaling. Proc Natl Acad Sci U S A 109(27): E1848–E1857
7. Lian X et al (2013) Directed cardiomyocyte differentiation from human pluripotent stem cells by modulating Wnt/beta-catenin signaling under fully defined conditions. Nat Protoc 8(1):162–175
8. Tohyama S et al (2013) Distinct metabolic flow enables large-scale purification of mouse and human pluripotent stem cell-derived cardiomyocytes. Cell Stem Cell 12(1):127–137
9. Zhang J et al (2019) Functional cardiac fibroblasts derived from human pluripotent stem cells via second heart field progenitors. Nat Commun 10(1):2238
10. Zhang H et al (2019) Generation of quiescent cardiac fibroblasts from human induced pluripotent stem cells for in vitro modeling of cardiac fibrosis. Circ Res 125(5):552–566
11. Kinney MA, Sargent CY, McDevitt TC (2011) The multiparametric effects of hydrodynamic environments on stem cell culture. Tissue Eng Part B Rev 17(4):249–262

Chapter 4

Acellular Myocardial Scaffolds and Slices Fabrication, and Method for Applying Mechanical and Electrical Simulation to Tissue Construct

Bo Wang, Mickey Shah, Lakiesha N. Williams, Amy L. de Jongh Curry, Yi Hong, Ge Zhang, and Jun Liao

Abstract

Cardiac tissue engineering/regeneration using decellularized myocardium has attracted great research attention due to its potential benefit to myocardial infarction (MI) treatment. Here, we described an optimal decellularization protocol to generate 3D porcine myocardial scaffolds with well-preserved cardiomyocyte lacunae, myocardial slices as a biomimetic cell culture and delivery platform, and a multi-stimulation bioreactor that is able to provide coordinated mechanical and electrical stimulations for facilitating cardiac construct development.

Key words Cardiac tissue engineering/regeneration, Acellular myocardial scaffolds, Acellular myocardial slices, Decellularization, Mechanical simulation, Electrical simulation, Bioreactor

1 Introduction

Myocardial infarction (MI) and heart failure are major causes of mortality worldwide [1]. Currently, the only successful treatment for end-stage heart failure is whole-heart transplantation, which is unfortunately limited by the persistent shortage of suitable heart donors. Newer strategies, including cellular transplantation, intramyocardial gene transfer, and cardiac tissue engineering (TE), have come to the forefront as alternative therapeutic approaches [2–4].

The purpose of cardiac tissue engineering is to develop functional cardiac tissue through integrating cellular components within scaffolds that serve as a structural guide [5–8]. Two major types of scaffold materials have been commonly used for cardiac tissue engineering: synthetic biodegradable material and tissue-derived acellular scaffolds [9–14]. The use of synthetic biodegradable polymers still faces challenges, including inflammatory

Kareen L. K. Coulombe and Lauren D. Black III (eds.), *Cardiac Tissue Engineering: Methods and Protocols*,
Methods in Molecular Biology, vol. 2485, https://doi.org/10.1007/978-1-0716-2261-2_4,

response, mismatched material properties, nonpliability, and difficulty in controlling the degradation rates [13, 15, 16]. Acellular scaffolds, which are derived from native tissues or organs via decellularization, are able to preserve the extracellular matrix (ECM) compositions, overall ultrastructure, shape compatibility, ECM mechanical integrity, and bioactive molecules that benefit cell–ECM adhesion, cell–cell interaction, and de novo ECM formation [17–23].

In practice, it is important to determine the optimal decellularization protocol that can mostly remove cells, cell debris, chromosome fragments, and xenogeneic antigens in order to diminish immunogenicity while at the same time preserve the needed structural and mechanical integrity of the native tissue ECM, which is important for target tissue functionalities [23–26]. For the applications of myocardial ECM scaffolds, Ott et al. decellularized whole rat heart and were able to keep the intact chamber geometry, perfusable vasculature, and competent acellular valves [27]. Badylak et al. and Taylor et al. attempted scaled-up research on whole porcine heart [21, 28]. Yet, many challenges still exist in whole-heart regeneration, such as preservation of myocardial ECM structure, reseeding homogeneity and thoroughness across the ventricle wall thickness, feasibility of reviving the existing vasculature network, and functional integration [21, 22, 29, 30]. Thus, our group has undertaken an effort to harness the potential of decellularized porcine myocardium as a TE scaffold material and focus on tissue-level application [31–33].

Heart walls are constructed of cardiac muscles that consist of cardiomyocytes, which are connected via gap junctions and structurally organized by highly vascularized ECM [34, 35]. As visually demonstrated by our previous diffusion tension MRI study [36], the heart muscles have a highly organized, multilayered helical structure (Fig. 1a, b). The compelling structural beauty of heart muscle fibers hints at the uniqueness and importance of heart muscle ECM. Indeed, the intriguing myocardial ECM network does play key roles in maintaining structural integrity, tethering cardiomyocytes, mediating contraction/relaxation of muscle fibers, and preventing excessive stretching [37–39]. As shown in Fig. 1c, d, removal of the heart muscle fibers (red staining) from the myocardial collagenous network (blue staining) will leave an ECM network that possesses a three-dimensional (3D) morphology and structural anisotropy. Hence, it is understandable that determining how to preserve the 3D ultrastructure of myocardial ECM represents a real challenge in myocardium decellularization via current available decellularization means, which are often beneficial in certain aspects but disruptive at some levels or to certain components [21–23].

In addition, acellular myocardial ECM closely mimic the natural microenvironment of cardiac cells, which make it an optimal cell

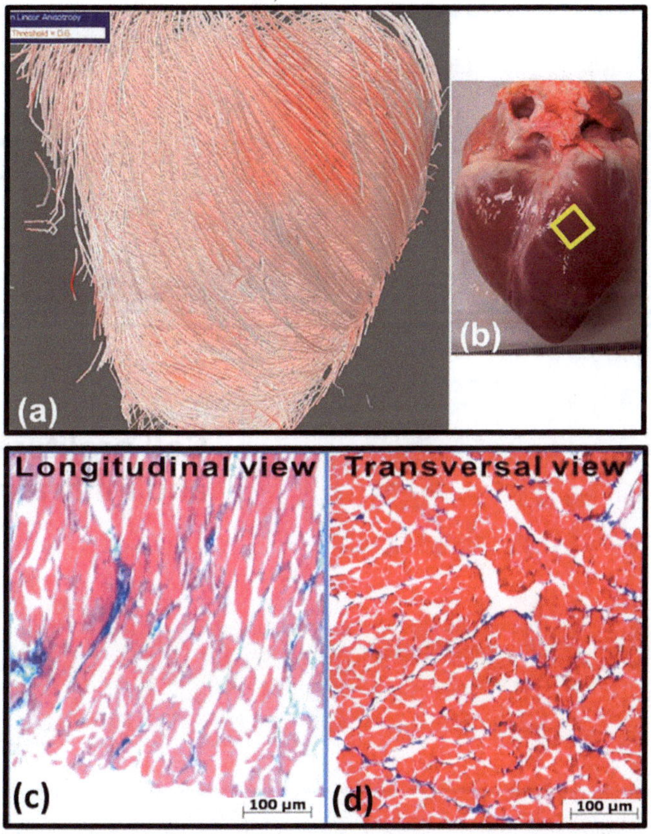

Fig. 1 3D multilayered helical structure of heart muscle fibers. (**a**) Heart muscles have well-organized multilayered helical architecture, which is mediated by 3D myocardial ECM (Diffusion Tensor MRI image by Zhang and Liao, 2010) [36]; (**b**) porcine heart used for DT-MRI imaging. Mason's trichrome staining of longitudinal-section (**c**) and cross-section (**d**) of the native myocardium (red: cardiomyocytes; blue: collagen). Figures reproduced with permission [31, 36]

culture substrate for cardiac applications. However, seeded cells cannot maintain high viability and homogeneous distribution in full-thickness acellular myocardial scaffold because cells in the center of thick scaffold will have insufficient access to oxygen and nutrients. To overcome this challenge, our group has explored the strategy of using a thin layer of acellular myocardial slice as a platform for cell culture and delivery [40–42]. Our results demonstrated that acellular myocardial slice with the thickness of ~300 μm promote cell attachment, growth, homogeneous distribution, and vascular differentiation of stem cells in vitro (Fig. 2).

In this chapter, we introduce an optimal decellularization protocol to generate 3D porcine acellular myocardial scaffolds in which 3D cardiomyocyte lacunae, ECM networks, vasculature templates, and mechanical anisotropy can be well preserved [31, 32]. We also

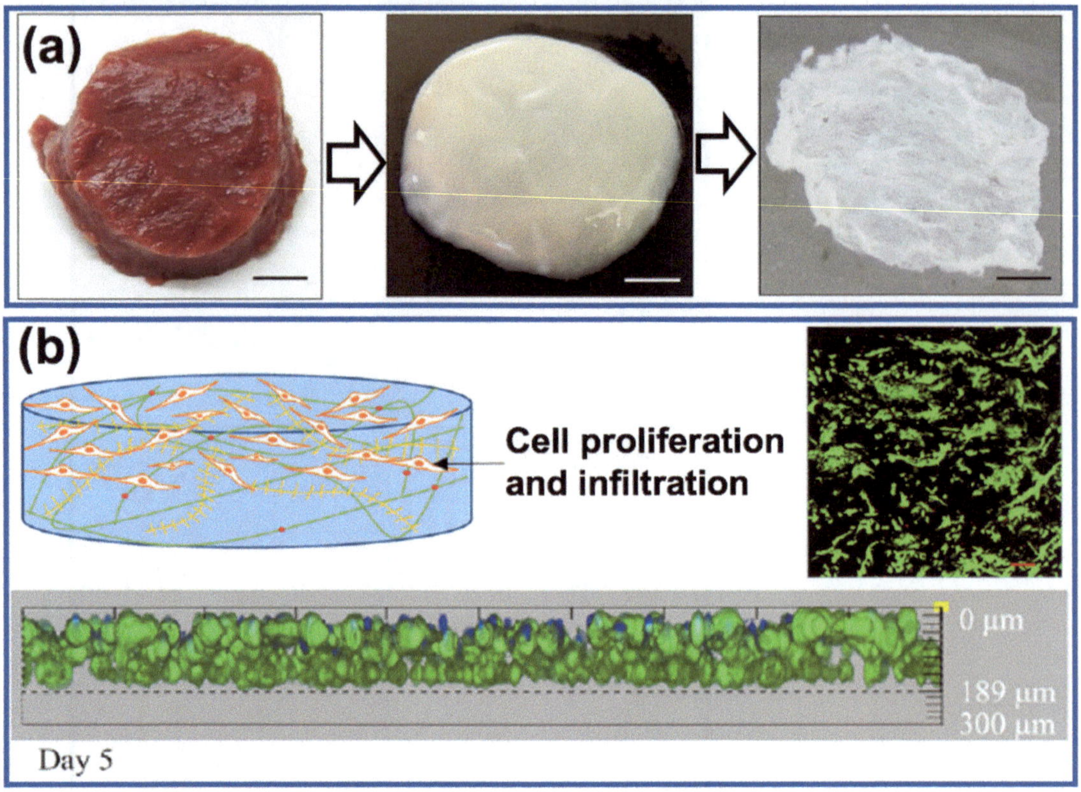

Fig. 2 Acellular porcine myocardial slices. (**a**) Fabrication of the acellular porcine myocardial slices: Left—native myocardial tissue, middle—decellularized myocardial scaffolds, right—acellular myocardial slice at 300 μm thickness. (**b**) Infiltration of pig adipose-derived stem cells (ASCs) on acellular myocardial slice. Image on right: Live and dead cells on acellular myocardial slice after 1 day culture. Image at bottom: Migration distance of pig ASCs seeded on acellular myocardial slice after 5 days culture. Scale bars in (**a**) = 4 mm; Scale bar in (**b**) live and dead cell image = 100 μm

include the detailed experimental protocol for create acellular myocardial slices [40–42]. To further improve the effectiveness and efficiency of cell differentiation and tissue remodeling of the reseeded acellular myocardial scaffolds and slices, we also describe a bioreactor conditioning protocol that is able to apply combined mechanical and electrical stimulations to tissue constructs fabricated with the acellular myocardial scaffolds and slices [33].

2 Materials

2.1 Decellularization Stock Solution

1. 0.1 M phenylmethylsulfonyl fluoride (PMSF): 0.174 g PMSF (Sigma) dissolved in 10 mL of 1-propanol (Sigma).
2. DNase (5 mg/mL): 50 mg of DNase (Sigma) dissolved in 10 mL 1 × PBS.

3. RNase A (5 mg/mL): 50 mg of Ribonuclease A from bovine pancreas (RNase) (Sigma) dissolved in 10 mL 1 × PBS.
4. 100 × antibiotic-antimycotic solution (ABAM) (Life Technologies).
5. 1 × trypsin solution (Sigma).

All the above solutions were stored at −20 °C.

6. 1 × phosphate-buffered saline (PBS) pH 7.4 (Life Technologies) stored at 4 °C.
7. 1% sodium dodecyl sulfate (SDS) solution: 1 g UltraPure™ sodium dodecyl sulfate (SDS) (Life Technologies) dissolved in 100 mL 1 × PBS stored at room temperature.

2.2 Porcine Myocardium

1. Fresh porcine hearts were obtained from juvenile pigs (~6 months old) from a local slaughter house.
2. The porcine hearts were transported to the laboratory in 1 × PBS on ice.
3. A myocardium square (20 × 20 × ~3 mm) was dissected from the middle region of the anterior left ventricular wall of the porcine heart (Fig. 1b) (*see* **Note 1**).
4. All the heart samples were kept in 1 × PBS solution at −80 °C for preservation (*see* **Note 2**).

2.3 Cell Culture Medium, Differentiation Medium, and Complete Medium

1. Mesenchymal stem cell (MSC) medium: Low-glucose Dulbecco's modified Eagle's medium (L-DMEM) with 10% fetal bovine serum (FBS), 1% mesenchymal stem cell growth supplement (Sciencell), 100 U/mL penicillin, and 100 μg/mL streptomycin (Invitrogen).
2. Differentiation medium: L-DMEM, 10% FBS, 3 μmol/L 5-azacytidine (MP Biomedicals), 100 U/mL penicillin, and 100 μg/mL streptomycin (*see* **Note 3**).
3. Complete medium: L-DMEM, 10% FBS, 1% cardiac myocyte growth supplement (Sciencell), 100 U/mL penicillin, and 100 μg/mL streptomycin.
4. All the above media were stored at 4 °C.

3 Methods

1. Myocardium Decellularization Procedure Four pieces of native myocardium samples are thawed at room temperature and washed four times with 100 mL distilled water for 10 min in a 100 mL Simax glass media storage bottle on an orbital shaker (Belly Dancer, Stovall).

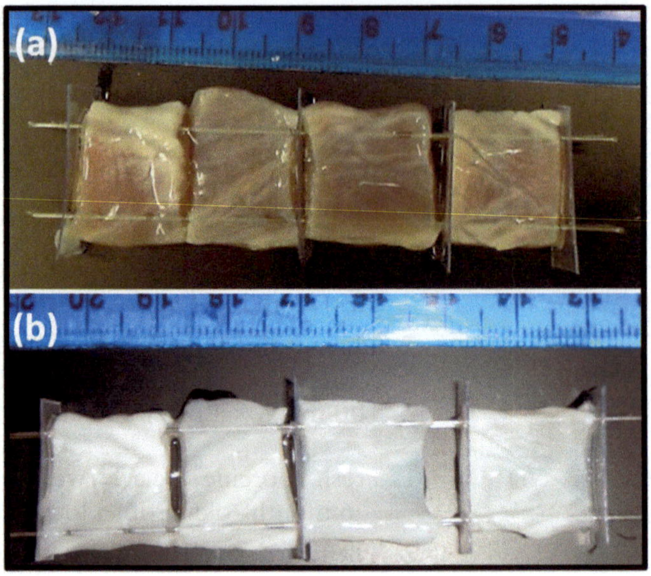

Fig. 3 The frame-pin supporting system. (**a**) Sample morphology after 3-day decellularization; (**b**) sample morphology after 2.5-week decellularization. Figures reproduced with permission [32]

2. A frame-pin supporting system is prepared to better maintain tissue macrogeometry during decellularization (Fig. 3). Briefly, four corners of the myocardium sample are perforated with four 27G × 31/2″ BD Quincke spinal needles and then mounted onto customized rectangular plastic frames (*see* **Note 4**).
3. **After being mounted onto the** frame-pin system**, the myocardium samples are** immersed inside 100 mL decellularization working solution in a 100 mL Simax glass media storage bottle **sealed with cap.**
4. Preparing decellularization working solution: For 50 mL working solution, add 41.8 mL 1 × PBS, 2 mL DNase, 200 μL RNase, 5 mL 1% SDS, 500 μL 0.1 M PMSF, 500 μL ABAM, and 5 μL trypsin solution to make a final solution of 0.1% (SDS), 0.01% trypsin, 1 mM PMSF, 0.2 mg/mL DNase, and 20 μg/mL RNase A.
5. The myocardium patches are then decellularized with agitated decellularization working solution on an orbital shaker (Belly Dancer, Stovall) at 30 revolutions per minute at room temperature.
6. A 10-min ultrasonic treatment (50 HZ, Branson) is applied each day, and the decellularization solution is changed every day to avoid contamination and tissue deterioration.
7. The completeness of myocardium decellularization can be determined when the myocardium patches became a bright

white color, a typical color of collagenous materials; the whole procedure lasts approximately 2.5 weeks (*see* **Notes** 7 and **8**).

8. After decellularization, all the myocardial scaffolds are removed from the frame-pin system and washed four times with 100 mL distilled water for 10 min and then washed four times with 100 mL PBS for 10 min, each time in a 100 mL Simax glass media storage bottle on an orbital shaker.

3.1 Preparation and Handling Tips for Acellular Myocardial Slices

1. The obtained acellular myocardial scaffold is trimmed into a square shape with the size of 2 cm (L) × 2 cm (W) × 1 cm (H).
2. The square sample is washed with 0.01% Triton X-100 for 1 h and then rinsed with 1× phosphate-buffered saline (PBS) for 3 days.
3. The rinsed sample is embedded in Tissue Tek OCT (Fisher Scientific, PA, USA) and snap-frozen on in liquid nitrogen.
4. The embedded sample is cryosectioned into acellular myocardial slices of 300 μm thickness (Fig. 2).
5. The acellular myocardial slices are sterilized prior to cell seeding using absolute ethanol for 45 min followed by three washes with sterilized DI water for 15 min each.

3.2 Preparation of the Multi-stimulation Bioreactor

1. The bioreactor used in this study consists of one tissue culture chamber in which all the structural elements are machined from polysulfone that provides excellent thermal and chemical stability.
2. Inside the tissue culture chamber, a maximum of four pieces of tissue construct (20 mm × 20 mm × ~3 mm) can be mounted between a fixed clamp and a movable clamp (Fig. 4c).
3. Ti-corn blue sutures (#0) are used for connecting the sample with two clamps.
4. The cover of the tissue culture chamber is fabricated with ¼ inch-thick, clear polycarbonate (Small Parts). A hole (2 cm in diameter) is cut on the cover and sealed with PIFE membrane with a 0.2 μm pore size (Millipore) to enable air exchange (Fig. 4c).
5. Linear movement is applied by the movable clamp that is driven by an Xslide assembly and a stepper motor (Velmex) (Fig. 4c) (*see* **Notes 5** and **6**).
6. Electrodes are made from Teflon-coated silver wire (75 μm diameter, A-M Systems). The end part (2 cm) of the Teflon insulation is stripped off, and the naked wires are inserted into the two opposite edges of the tissue construct.
7. The frequency and amplitude of the cyclic stretch are controlled by a customized LabView program (version 2010,

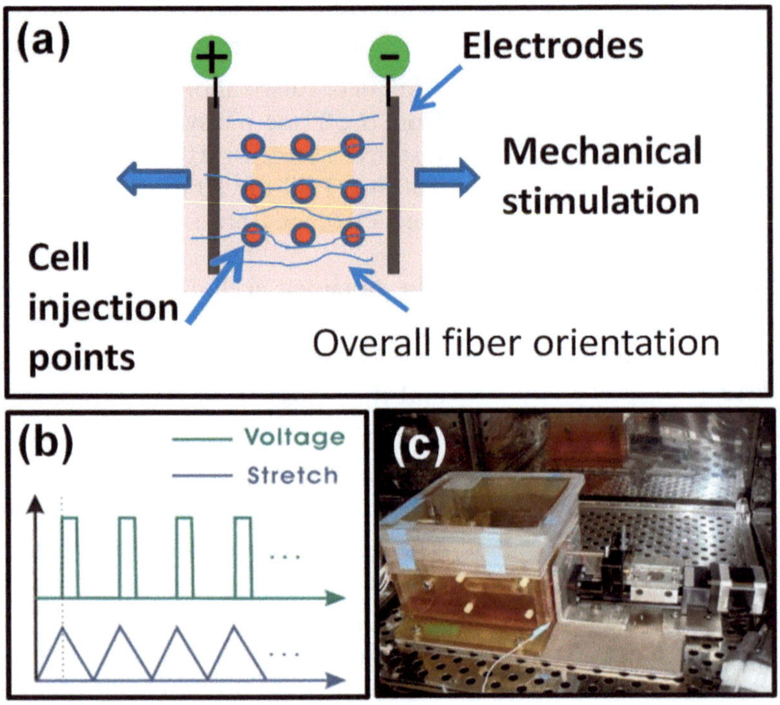

Fig. 4 Design a multi-stimulation bioreactor. (**a**) Schematic illustration of the acellular myocardial scaffold subjected to cell injection, mechanical, and electrical stimulations; (**b**) wave forms of the applied mechanical stretch and electrical pulses; (**c**) the multi-stimulation bioreactor placed in the incubator. Figures reproduced with permission [33]

National Instruments). To simulate similar experiences of the myocardial tissue, electrical pulses are applied at the initial stage of each unloading cycle (Fig. 4b). The frequency and amplitude of the electrical pulses are controlled by the LabView program, which is capable of delivering multiple protocols of mechanical stretching and various waveforms of electrical stimulation.

3.3 Bioreactor Setup and Sterilization for Acellular Tissue Construct

1. After sample washing, the naked end of the positive and negative electrodes are inserted into the two opposite edges of the acellular myocardial scaffolds (20 mm × 20 mm × ~3 mm), and the other ends of the Teflon-coated silver wires are dangled for later bioreactor connection.
2. The acellular myocardial scaffolds (with electrodes mounted) are transferred into the culture chamber of the bioreactor and mounted between a fixed clamp and a movable clamp using the surgical Ti-corn blue sutures (#0). Two to four pieces of the acellular scaffold samples can be placed in the bioreactor at one time.

3. The other ends of the Teflon-coated silver wires are then connected with the electrical-control module of the bioreactor.
4. The acellular myocardial scaffolds are then sterilized in 70% ethanol in the tissue culture chamber for 2 h (*see* **Note 9**).
5. After ethanol sterilization, all the samples are rinsed thoroughly four times with sterilized PBS while still sitting in the culture chamber (*see* **Note 9**).
6. For sterilizing the other parts of the bioreactor that could not be immersed in ethanol, the whole tissue culture chamber with the mounted acellular scaffold samples is further treated with UV light for 20 min (*see* **Note 9**).

3.4 MSC Culture, Reseeding, Differentiation, and Bioreactor Conditioning Protocol

1. Well-characterized Lewis rat mesenchymal stem cells (MSCs, fourth passage) are obtained from the Stem/Progenitor Cell Standardization Core (SPCS) at the Texas A&M Health Science Center (NIH/NCRR grant) (*see* **Note 10**).
2. After receiving the cells, MSCs are resuspended in mesenchymal stem cell medium and seeded onto 175-mm flasks at a density of 2×10^3 cells/cm^2. The medium is changed twice a week.
3. The fifth to eighth passages of MSCs are resuspended after HyQtase (Thermo Scientific) treatment and two washes with Tyrode's balanced salt solution (Sigma).
4. Next, the MSCs are used for scaffold recellularization. The density of MSCs used for scaffold recellularization was 10^6 cells/mL in mesenchymal stem cell medium.
5. After the completion of the sterilization protocols for the bioreactor and tissue constructs, each scaffold sample is injected with totally 10^6 MSCs (1 mL MSC solution used; cell density of MSC solution: 10^6 cells/mL) .
6. An 1-mL syringe with 26G permanent needle (BD) is used for cell injection, and 1-mL MSCs are injected evenly at nine injection points (~0.1 mL/point) located as a 3×3 array within the middle region of the square sample (Fig. 4a) (*see* **Note 11**).
7. The movable clamp is adjusted to make sure no stretch applied on the reseeded scaffold sample, and this clamp-to-clamp distance is set as the reference distance to calculate the applied strain (*see* **Note 12**).
8. The tissue culture chamber is filled with differentiation medium until the medium fully immersed all the reseeded scaffold samples.
9. The tissue culture chamber is covered and the top edges sealed with sterilized parafilm.

10. The bioreactor chamber is carefully moved into the incubator with incubation condition set at 37 °C in a humid atmosphere with 5% CO_2 (*see* **Note 13**).
11. After wires are connected to the bioreactor control modules, the LabView program initiates application of mechanical stretch (20% strain) and electrical pulse (5 volt).
12. Both the triangular strain waveform and square wave electrical pulse are set at a frequency of 1 Hz (Fig. 4b), which simulates the physiological frequency experienced by the heart muscles (*see* **Note 14**).
13. Note that the differentiation medium is added in the bioreactor chamber to facilitate the cardiomyocyte differentiation during the first 24 h of tissue culture.
14. After the 24-hour differentiation medium treatment, the medium is changed to the complete medium for the remaining bioreactor conditioning protocol, and the medium is changed every 3 days.

4 Notes

1. For the square myocardium patch dissection, one edge was aligned along the muscle fiber preferred direction (PD), and the other edge aligned along the cross-fiber preferred direction (XD); note that the PD direction was determined based on overall muscle fiber texture and heart anatomy (Fig. 1b).
2. Snap freezing can disrupt cellular membranes by forming intracellular ice crystals and cause cell lysis [43, 44]; therefore, we deep-froze the myocardium samples before decellularization treatment in order to increase the efficiency of decellularization.
3. 5-azacytidine is a member of the cytosine analog, which had been reported to induce uncontrolled myogenic specification by random [45–47]. It was reported that treating mesenchymal stem cells with 5-azacytidine could generate a cardiomyocyte differentiation rate at ~30% [46, 47].
4. Frame-pin supporting system was applied for preventing scaffold contraction; this design well preserved the 3D cardiomyocyte lacunae during decellularization procedures (Fig. 5b).
5. For cyclic mechanical stretching, a stepper motor was chosen because its motion could be precisely controlled and easily programmed.
6. To monitor the real-time tension level in the tissue construct, load cells can be applied in the bioreactor design to oversee the

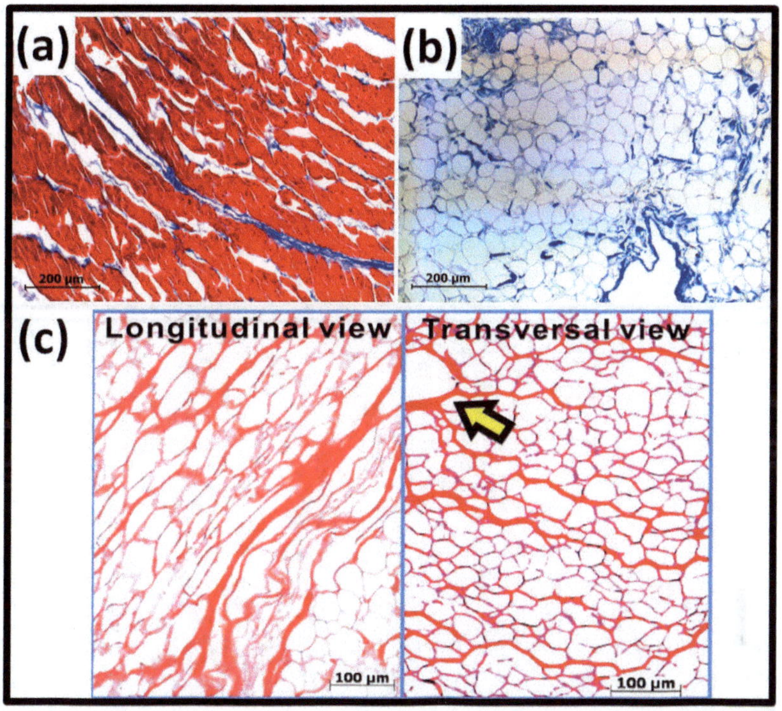

Fig. 5 Details of the aligned 3D cardiomyocyte lacunae. (**a**) Edge-to-edge view of the acellular myocardial scaffold revealed by H&E staining; thorough decellularization and preservation of cardiomyocyte lacunae (porous structures) were verified. (**b**) 3D topography of the acellular myocardial scaffold revealed by SEM; enlarged view showed more details of the aligned 3D cardiomyocyte lacunae; note that arrows highlight the interconnecting openings inside the cardiomyocyte lacunae. Figures reproduced with permission [31, 32]

mechanical forces experienced by the construct during the tissue remodeling process.

7. The efficiency of cardiomyocyte removal could be verified by histology (Figs. 5a and 6b, c) and quantitative DNA analysis by ~2.5 weeks [31, 32]. Xenogeneic antigens, porcine a-Gal, were found being completely removed from the acellular myocardial scaffolds [32]. We also showed that the vasculature templates (acellular blood vessel structure) were preserved in the acellular myocardial scaffolds [31, 32].
8. The decellularization protocol described here generated acellular myocardial scaffolds with thorough decellularization and ECM preservation; however, it took a relatively long treatment time (~2.5 weeks). To achieve effective decellularization within a shorter time period, the concentration of SDS can be increased to 0.5%.
9. Effective sterilization is an essential step for tissue culture, especially for a bioreactor that has many complicated

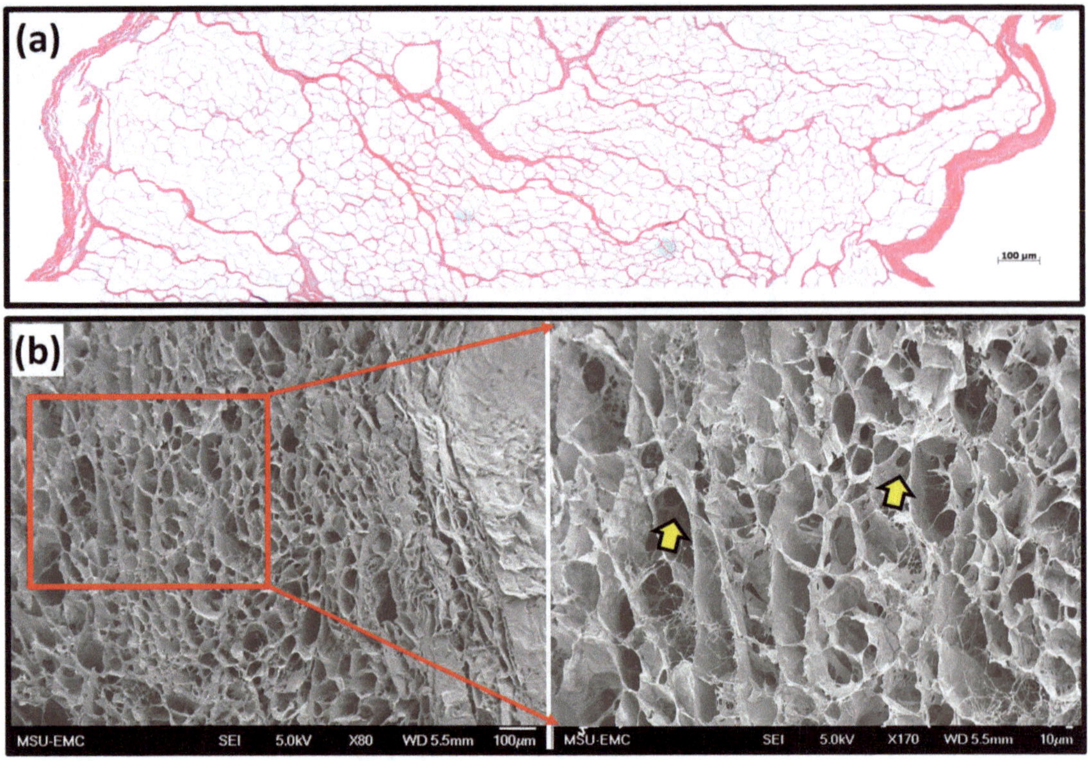

Fig. 6 Acellular myocardial scaffolds show well-preserved cardiomyocyte lacunae. Mason's trichrome staining of (**a**) the native myocardium and (**b**) the acellular myocardial scaffold (red: cardiomyocytes; blue: collagen). (**c**) H&E staining of the longitudinal and transversal views of the acellular myocardial scaffolds showed structural anisotropy (red: collagen); arrow indicates vasculature channel was preserved after decellularization. Figures reproduced with permission [31, 32]

components and was constructed for repeated use. For our application, a combined method that used both 70% ethanol sterilization and UV light sterilization was adopted. Yet, the sterilization protocol can be further optimized to reduce sterilizing treatments to a minimal level [33]. Moreover, for thinner samples, the duration for the 70% ethanol treatment should be largely reduced, and a thorough rinse must be performed to remove residual ethanol.

10. For better cardiomyocyte differentiation, other cell sources that can be adopted include cardiac stem cells, embryonic stem cells (ESCs), or induced pluripotent stem (iPS) cells [48–52].

11. To obtain the tissue engineered cardiac construct with a high cell density, we employed a needle injection method for cell implanting with a total cell amount of 10^6 cells/scaffold. However, due to the porous structure of the acellular myocardial scaffold, a small amount of leakage happened during the

process of the cell injection. For this kind of situation, we injected the leaked medium back into the same region of the acellular myocardial scaffold.

12. The applied tissue construct strain was estimated by normalizing the displacement of the movable clamp to the reference distance of two sample mounting clamps. The use of the clamp-to-clamp displacement in strain estimation was not an ideal method to accurately measure the tissue construct strain. To achieve more accurate measurement of the tissue construct strain, a camera can be used for real-time tracking of markers pasted on the tissue construct.

13. Temperature is another important parameter in bioreactor conditioning. Heat exchange has to be carefully designed to maintain the incubator/bioreactor at a constant temperature (37 °C). In our application, we placed both the culture chamber and the stepper motor inside the incubator. We noticed that the heat generated by the stepper motor after long hours of operation greatly affected the motor performance. This problem was solved by designing a water refrigeration system in which cold copper coils were wrapped around the step motor, and cooling water was circulated by a rotation pump outside of the incubator. The water refrigeration system dissipated the heat generated by the step motor effectively, and the temperature was maintained in a reasonable range without causing any motor malfunction.

14. Previous studies have shown that cyclic mechanical stimulation can assist in the cell alignment, stimulate ECM formation [53, 54], and improve the cardiomyocytes development and function [55, 56]. Electrical stimulation is believed to be able to induce transient calcium levels and facilitate cell proliferation and promote the formation and localization of electric gap junctions [57, 58]. The benefit of the combined mechanical and electrical stimulations was evidenced by good cell viability, repopulation, differentiation, and positive tissue remodeling within a short period of time (2–4 days) [33].

Acknowledgments

We greatly appreciate the support from NIH National Heart, Lung, and Blood Institute grant 1R15HL097321 to JL, 1R15HL22929 to GZ, and 1R15HL140503 to YH. The authors also would like to acknowledge the support from AHA (13GRNT17150041), NIH (1R01EB022018-01), and MAFES Strategic Research Initiative (CRESS MIS-361020).

References

1. Rosamond W, Flegal K, Friday G, Furie K, Go A, Greenlund K et al (2007) Heart disease and stroke statistics—2007 update: a report from the American Heart Association Statistics Committee and Stroke Statistics Subcommittee. Circulation 115(5):e69–e171
2. Sharma R, Raghubir R (2007) Stem cell therapy: a hope for dying hearts. Stem Cells Dev 16(4):517–536
3. Grauss RW, Winter EM, van Tuyn J, Pijnappels DA, Steijn RV, Hogers B et al (2007) Mesenchymal stem cells from ischemic heart disease patients improve left ventricular function after acute myocardial infarction. Am J Physiol 293(4):H2438–H2447
4. Strauer BE, Kornowski R (2003) Stem cell therapy in perspective. Circulation 107(7): 929–934
5. Langer R, Vacanti J (1993) Tissue Engineering. Science 260(5110):920–926
6. Zimmermann WH, Melnychenko I, Eschenhagen T (2004) Engineered heart tissue for regeneration of diseased hearts. Biomaterials 25(9):1639–1647
7. Thompson RB, Emani SM, Davis BH, van den Bos EJ, Morimoto Y, Craig D et al (2003) Comparison of intracardiac cell transplantation: autologous skeletal myoblasts versus bone marrow cells. Circulation 108(Suppl 1): II264–II271
8. Bursac N, Papadaki M, Cohen RJ, Schoen FJ, Eisenberg SR, Carrier R et al (1999) Cardiac muscle tissue engineering: toward an in vitro model for electrophysiological studies. Am J Phys 277(2 Pt 2):H433–H444
9. Fujimoto KL, Guan J, Oshima H, Sakai T, Wagner WR (2007) In vivo evaluation of a porous, elastic, biodegradable patch for reconstructive cardiac procedures. Ann Thorac Surg 83(2):648–654
10. Fujimoto KL, Tobita K, Merryman WD, Guan J, Momoi N, Stolz DB et al (2007) An elastic, biodegradable cardiac patch induces contractile smooth muscle and improves cardiac remodeling and function in subacute myocardial infarction. J Am Coll Cardiol 49(23): 2292–2300
11. Li W-J, Laurencin CT, Caterson EJ, Tuan RS, Ko FK (2002) Electrospun nanofibrous structure: A novel scaffold for tissue engineering. J Biomed Mater Res 60(4):613–621
12. Smith IO, Liu XH, Smith LA, Ma PX (2009) Nanostructured polymer scaffolds for tissue engineering and regenerative medicine. Wiley Interdiscip Rev Nanomed Nanobiotechnol 1(2):226–236
13. Ozawa T, Mickle DA, Weisel RD, Koyama N, Ozawa S, Li RK (2002) Optimal biomaterial for creation of autologous cardiac grafts. Circulation 106(12 Suppl 1):I176–I182
14. Ozawa T, Mickle DA, Weisel RD, Koyama N, Wong H, Ozawa S et al (2002) Histologic changes of nonbiodegradable and biodegradable biomaterials used to repair right ventricular heart defects in rats. J Thorac Cardiovasc Surg 124(6):1157–1164
15. Grad S, Zhou L, Gogolewski S, Alini M (2003) Chondrocytes seeded onto poly (L/DL-lactide) 80%/20% porous scaffolds: a biochemical evaluation. J Biomed Mater Res A 66(3):571–579
16. Weber B, Emmert MY, Schoenauer R, Brokopp C, Baumgartner L, Hoerstrup SP (2011) Tissue engineering on matrix: future of autologous tissue replacement. Semin Immunopathol 33(3):307–315
17. Badylak SF, Record R, Lindberg K, Hodde J, Park K (1998) Small intestinal submucosa: a substrate for in vitro cell growth. J Biomater Sci Polym Ed 9(8):863–878
18. Badylak SF, Tullius R, Kokini K, Shelbourne KD, Klootwyk T, Voytik SL et al (1995) The use of xenogeneic small intestinal submucosa as a biomaterial for Achilles tendon repair in a dog model. J Biomed Mater Res 29(8):977–985
19. Voytik-Harbin SL, Brightman AO, Kraine MR, Waisner B, Badylak SF (1997) Identification of extractable growth factors from small intestinal submucosa. J Cell Biochem 67(4):478–491
20. Baptista PM, Vyas D, Soker S (2012) Liver regeneration and bioengineering-the emergence of whole organ scaffolds. In: Baptista PM (ed) Liver regeneration. InTech, Rijeka, Croatia
21. Badylak SF, Taylor D, Uygun K (2011) Whole-organ tissue engineering: decellularization and recellularization of three-dimensional matrix scaffolds. Annu Rev Biomed Eng 13:27–53
22. Crapo PM, Gilbert TW, Badylak SF (2011) An overview of tissue and whole organ decellularization processes. Biomaterials 32(12): 3233–3243
23. Gilbert TW, Sellaro TL, Badylak SF (2006) Decellularization of tissues and organs. Biomaterials 27(19):3675–3683
24. Chan BP, Leong KW (2008) Scaffolding in tissue engineering: general approaches and tissue-specific considerations. Eur Spine J 17 (Suppl 4):467–479

25. Badylak SF (2004) Xenogeneic extracellular matrix as a scaffold for tissue reconstruction. Transpl Immunol 12(3–4):367–377
26. Badylak SF, Freytes DO, Gilbert TW (2009) Extracellular matrix as a biological scaffold material: Structure and function. Acta Biomater 5(1):1–13
27. Ott HC, Matthiesen TS, Goh SK, Black LD, Kren SM, Netoff TI et al (2008) Perfusion-decellularized matrix: using nature's platform to engineer a bioartificial heart. Nat Med 14(2):213–221
28. Wainwright JM, Czajka CA, Patel UB, Freytes DO, Tobita K, Gilbert TW et al (2010) Preparation of cardiac extracellular matrix from an intact porcine heart. Tissue Eng 16(3): 525–532
29. Vunjak-Novakovic G, Tandon N, Godier A, Maidhof R, Marsano A, Martens TP et al (2010) Challenges in cardiac tissue engineering. Tissue Eng Part B Rev 16(2):169–187
30. Steinhauser ML, Lee RT (2011) Regeneration of the heart. EMBO Mol Med 3(12):701–712
31. Wang B, Borazjani A, Tahai M, Curry AL, Simionescu DT, Guan J et al (2010) Fabrication of cardiac patch with decellularized porcine myocardial scaffold and bone marrow mononuclear cells. J Biomed Mater Res A 94(4):1100–1110
32. Wang B, Tedder ME, Perez CE, Wang G, de Jongh Curry AL, To F et al (2012) Structural and biomechanical characterizations of porcine myocardial extracellular matrix. J Mater Sci Mater Med 23(8):1835–1847
33. Wang B, Wang G, To F, Butler JR, Claude A, McLaughlin RM et al (2013) Myocardial scaffold-based cardiac tissue engineering: application of coordinated mechanical and electrical stimulations. Langmuir 29(35): 11109–11117
34. Rakusan K, Flanagan MF, Geva T, Southern J, Van Praagh R (1992) Morphometry of human coronary capillaries during normal growth and the effect of age in left ventricular pressure-overload hypertrophy. Circulation 86(1): 38–46
35. Korecky B, Hai CM, Rakusan K (1982) Functional capillary density in normal and transplanted rat hearts. Can J Physiol Pharmacol 60(1):23–32
36. Zhang S, Crow JA, Yang X, Chen J, Borazjani A, Mullins KB et al (2010) The correlation of 3D DT-MRI fiber disruption with structural and mechanical degeneration in porcine myocardium. Ann Biomed Eng 38(10): 3084–3095
37. Weber KT (1989) Cardiac interstitium in health and disease: the fibrillar collagen network. J Am Coll Cardiol 13(7):1637–1652
38. Holmes JW, Borg TK, Covell JW (2005) Structure and mechanics of healing myocardial infarcts. Annu Rev Biomed Eng 7:223–253
39. Humphery JD (2002) Cardiovascular solid mechanics. Springer Verlag, New York
40. Kc P, Shah M, Liao J, Zhang G (2017) Prevascularization of decellularized porcine myocardial slice for cardiac tissue engineering. ACS Appl Mater Interfaces 9(3):2196–2204
41. Shah M, Kc P, Copeland KM, Liao J, Zhang G (2018) A thin layer of decellularized porcine myocardium for cell delivery. Sci Rep 8(1): 16206
42. Shah M, Kc P, Zhang G (2019) In vivo assessment of decellularized porcine myocardial slice as an acellular cardiac patch. ACS Appl Mater Interfaces 11(27):23893–23900
43. Jackson DW, Grood ES, Cohn BT, Arnoczky SP, Simon TM, Cummings JF (1991) The effects of in situ freezing on the anterior cruciate ligament. An experimental study in goats. J Bone Joint Surg Am 73(2):201–213
44. Roberts TS, Drez D Jr, McCarthy W, Paine R (1991) Anterior cruciate ligament reconstruction using freeze-dried, ethylene oxide-sterilized, bone-patellar tendon-bone allografts. Two year results in thirty-six patients. Am J Sports Med 19(1):35–41
45. Chiu CP, Blau HM (1985) 5-Azacytidine permits gene activation in a previously noninducible cell type. Cell 40(2):417–424
46. Tomita Y, Makino S, Hakuno D, Hattan N, Kimura K, Miyoshi S et al (2007) Application of mesenchymal stem cell-derived cardiomyocytes as bio-pacemakers: current status and problems to be solved. Med Biol Eng Comput 45(2):209–220
47. Fukuda K (2003) Regeneration of cardiomyocytes from bone marrow: Use of mesenchymal stem cell for cardiovascular tissue engineering. Cytotechnology 41(2–3):165–175
48. Carrier RL, Papadaki M, Rupnick M, Schoen FJ, Bursac N, Langer R et al (1999) Cardiac tissue engineering: cell seeding, cultivation parameters, and tissue construct characterization. Biotechnol Bioeng 64(5):580–589
49. Birla RK, Borschel GH, Dennis RG (2005) In vivo conditioning of tissue-engineered heart muscle improves contractile performance. Artif Organs 29(11):866–875
50. Birla RK, Borschel GH, Dennis RG, Brown DL (2005) Myocardial engineering in vivo: formation and characterization of contractile,

vascularized three-dimensional cardiac tissue. Tissue Eng 11(5–6):803–813

51. Borschel GH, Dow DE, Dennis RG, Brown DL (2006) Tissue-engineered axially vascularized contractile skeletal muscle. Plast Reconstr Surg 117(7):2235–2242
52. Vouyouka AG, Powell RJ, Ricotta J, Chen H, Dudrick DJ, Sawmiller CJ et al (1998) Ambient pulsatile pressure modulates endothelial cell proliferation. J Mol Cell Cardiol 30(3): 609–615
53. Rubbens MP, Driessen-Mol A, Boerboom RA, Koppert MM, van Assen HC, TerHaar Romeny BM et al (2009) Quantification of the temporal evolution of collagen orientation in mechanically conditioned engineered cardiovascular tissues. Ann Biomed Eng 37(7): 1263–1272
54. Akhyari P, Fedak PW, Weisel RD, Lee TY, Verma S, Mickle DA et al (2002) Mechanical stretch regimen enhances the formation of bioengineered autologous cardiac muscle grafts. Circulation 106(12 Suppl 1):I137–I142
55. Zimmermann WH, Melnychenko I, Wasmeier G, Didie M, Naito H, Nixdorff U et al (2006) Engineered heart tissue grafts improve systolic and diastolic function in infarcted rat hearts. Nat Med 12(4):452–458
56. Fink C, Ergun S, Kralisch D, Remmers U, Weil J, Eschenhagen T (2000) Chronic stretch of engineered heart tissue induces hypertrophy and functional improvement. FASEB J 14(5): 669–679
57. Wikswo JP Jr, Lin SF, Abbas RA (1995) Virtual electrodes in cardiac tissue: a common mechanism for anodal and cathodal stimulation. Biophys J 69(6):2195–2210
58. McDonough PM, Glembotski CC (1992) Induction of atrial natriuretic factor and myosin light chain-2 gene expression in cultured ventricular myocytes by electrical stimulation of contraction. J Biol Chem 267(17): 11665–11668

Chapter 5

FRESH 3D Bioprinting a Ventricle-like Cardiac Construct Using Human Stem Cell-Derived Cardiomyocytes

Brian D. Coffin, Andrew R. Hudson, Andrew Lee, and Adam W. Feinberg

Abstract

Here we describe a method to engineer a contractile ventricle-like chamber composed of human stem cell-derived cardiomyocytes using freeform reversible embedding of suspended hydrogels (FRESH) 3D bioprinting. To do this, we print a support structure using a collagen type I ink and a cellular component using a high-density cell ink supplemented with fibrinogen. The gelation of the collagen and the fibrinogen into fibrin is initiated by pH change and enzymatic crosslinking, respectively. Fabrication of the ventricle-like chamber is completed in three distinct phases: (i) materials preparation, (ii) bioprinting, and (iii) tissue maturation. In this protocol, we describe the method to print the construct from a high-density cell ink composed of human stem cell-derived cardiomyocytes and primary fibroblasts (~300 × 10^6 cells/mL) using our open-source dual-extruder bioprinter. Additional details are provided on FRESH support preparation, bioink preparation, dual-extruder needle alignment, print parameter selection, and post-processing. This protocol can also be adapted by altering the 3D model design, cell concentration, or cell type to FRESH 3D bioprint other cardiac tissue constructs.

Key words FRESH, 3D Bioprinting, 3D Printing, Stem Cells, Cardiac, Ventricle, Engineered heart tissue (EHT), Engineering, Collagen, Replistruder

1 Introduction

The heart is a structurally complex organ with a hierarchical three-dimensional (3D) architecture composed of aligned cardiomyocytes. Formed into laminar sheets, the myocardial muscle helically wraps to form the ventricular chambers of the heart and is critical to the ability to achieve wall thickening and a decrease in ventricular volume during contraction [1]. Researchers have developed a range of techniques to recapitulate this anisotropic cardiac muscle organization in vitro in order to study these structure–function relationships, including engineered heart tissues (EHTs) formed into rings, strips, and wires [2–4]. These EHTs rely on cell-mediated self-assembly and compaction to generate mechanical forces that drive cellular alignment [5] and can be interrogated using electrical,

Kareen L. K. Coulombe and Lauren D. Black III (eds.), *Cardiac Tissue Engineering: Methods and Protocols*,
Methods in Molecular Biology, vol. 2485, https://doi.org/10.1007/978-1-0716-2261-2_5,

mechanical, and chemical stimuli. The anisotropic myofiber architecture improves muscle contraction [6], and EHTs have been used to measure contractile force and electrophysiology using cardiomyocytes derived from various species [7]. However, the complexity of EHTs is limited by the shape of the mold in which they are typically formed, making it difficult to assess the properties of cardiac tissues that are more representative of the ventricles and other anatomical structures of the heart. To address the need for more realistic cardiac tissue models, several research groups have developed additive manufacturing and electrospinning techniques to generate 3D ventricle-like chambers with integrated stem cell-derived human cardiomyocytes [8–10]. These constructs have been used to measure beat rate, calcium wave propagation velocity, cellular alignment, contractile pressure, and ejection volume [11]. However, these ventricle models are challenging to fabricate, requiring specialized equipment, and changing the size, shape, and other parameters necessitates new molds. It can also be difficult to seed with enough cardiomyocytes, fibroblasts, and other cells types and achieve uniform cell density and alignment throughout the scaffold, which has motivated the continued development of improved biofabrication approaches.

Here we describe a method to 3D bioprint a contractile ventricle-like chamber using human stem cell-derived cardiomyocytes that achieves myofiber alignment, anisotropic action potential propagation, and wall thickening during contraction. To do this, we utilize Freeform Reversible Embedding of Suspended Hydrogels (FRESH) 3D printing, which greatly improves fidelity by extruding hydrogels and cells into a sacrificial support bath instead of printing in air, where these materials would otherwise collapse [12, 13]. The support bath enables multiple crosslinking chemistries to be used simultaneously including pH, ionic, and enzymatically driven gelation for collagen, alginate, and fibrinogen, respectively. Once the print is complete, the part is gently released from the support bath by raising the temperature to 37 °C, which melts the gelatin-derived support [12, 13]. To form the ventricle-like chamber, we FRESH print human stem cell-derived cardiomyocytes and cardiac fibroblasts in a fibrinogen ink that is embedded between inner and outer layers of collagen that provide structural support. Once printed, the cardiomyocytes and fibroblasts organize into an anisotropic, electrically coupled, contractile ventricle-like construct. While this protocol describes the preparation of a ventricle model, it can be adapted to create rings, strips, sheets, chambers, and other 3D geometries of interest by modifying the CAD model.

2 Materials

Prepare all stock solutions and buffers using ultrapure water (18 MΩ-cm at 25 °C) unless otherwise noted. Reagents involved in cell culture should be sterile to avoid infection. Prepare and store all reagents at room temperature (unless indicated otherwise). All equipment and reagents should be sterilized prior to use. All steps should be carried out in a sterile environment using aseptic technique unless otherwise stated.

1. Dual material bioprinter (*see* **Note 1**) that uses a mechanical syringe pump design. A pneumatic syringe extruder can be used but will require modification of these instructions due to inherent hardware differences.
2. Two luer lock glass syringes—2.5 mL (Hamilton, 81420).
3. Two Becton Dickinson luer lock syringes—10 mL.
4. Two luer lock stainless steel blunt tip needles, 30 gauge (Jensen global, JG30-0.5HPX).
5. Two female-female luer lock couplings (Hamilton, 86505).
6. Two luer lock syringe caps (Hamilton, 16801).
7. Vacuum grease (Dow Corning, 1597418).
8. FRESH support material.
   (i) LifeSupport® support material for FRESH printing (Advanced Biomatrix, 5244; Allevi, LIFES; CellINK, D16110022047) (*see* **Note 2**).
   (ii) 4-(2-hydroxyethyl)-1-piperazineethanesulfonic acid (HEPES): 50 mM in ultrapure water, pH 7.4 (*see* **Note 3**).
   (iii) Thrombin, 100 U/mL in sterile ultrapure water (Sigma, T4648).
   (iv) Chemically defined medium with three component (CDM3): RPMI-1640 medium (Thermofisher, 21870076), 1% (v/v) L-glutamine (Thermofisher, 25030081), 500 μg/mL human albumin (Sigma, A9731), 213 μg/mL L-Ascorbic acid 2-phosphate sesquimagnesium salt hydrate >95% (Sigma, A8960).
   (v) Thiazovivin (Selleck Chemicals, S1459): 10 mM in dimethyl sulfoxide.
   (vi) Fetal Bovine Serum (FBS).
9. Cellular Ink.
   (i) Spontaneously contracting cardiomyocytes, 50 μL or ~15×10^6 per model (*see* **Note 4**).
   (ii) Human ventricular cardiac fibroblasts, ~400×10^3 per model (CC-2904, Lonza).

(iii) 60 mg/mL fibrinogen, bovine plasma (Millipore-Sigma, 341573) in sterile distilled water, 1 mL.

10. Acidified collagen type I bioink (Advanced Biomatrix, LifeInk 200).
11. Acetic acid, 0.24 M.
12. Syringe centrifuge adapter–10 mL (*see* **Note 5**).
13. Recommended software: use the following software packages or an equivalent software package.
    (i) Autodesk Inventor 2019: Computer-aided design (CAD) program (https://www.autodesk.com/products/inventor/overview).
    (ii) Slic3r v1.3.0: Model slicing and G-code pathing generator (https://slic3r.org/).
    (iii) Pronterface v1.6: open-source printer control software (https://www.pronterface.com/).

3 Methods

3.1 Prepare Bioprinting Reagents

1. Prepare 500 mL of CDM3 by combining components in the reservoir of a sterile filter unit and applying vacuum to the bottle to sterile filter the media.
2. Prepare 24 mg/mL collagen ink by mixing 35 mg/mL collagen type I (Advanced Biomatrix, LifeInk 200) in 2:1 ratio (v/v) with acetic acid (0.24 M) between two coupled 10 mL BD syringes. Mix back and forth approximately 40 times to ensure collagen is completely solubilized and allow to come to room temperature by placing on bench for ~1 h to allow entrapped gases to expand. Place collagen syringe into BD Syringe centrifuge adapter and centrifuge at 3000 × *g* for 5 min to remove any air bubbles. Air bubbles should never be present in any ink when transferred to the bioprinter. If bubbles are present, centrifuge the syringe to collect them and expel them from the syringe. Acidified collagen ink can be prepared ahead of time and stored at 4 °C. The prepared ink should be loaded into a 2.5 mL Hamilton gastight syringe just prior to printing.

3.2 Design Cardiac Ventricle Model

1. Two models were designed and then combined to create the ellipsoidal ventricle construct. The first model is an ellipsoidal shell with a maximum outer diameter of 6.6 mm, height of 8 mm, and 900 μm wall thickness. The second model is an ellipsoidal insert with a maximum outer diameter of 6.3 mm, height of 7.85 mm, and 450 μm wall thickness. Wall thickness is always a multiple of the needle diameter to avoid perimeters

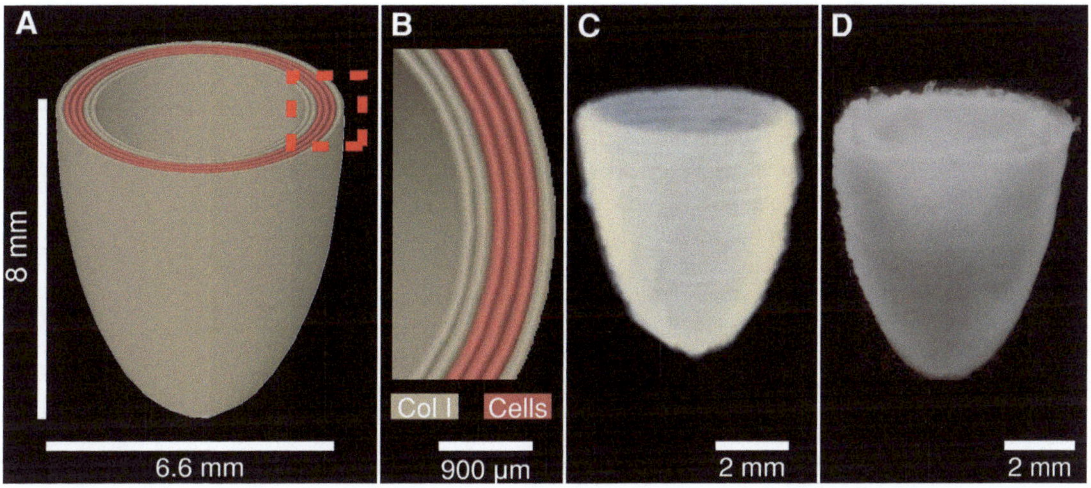

Fig. 1 (**a**) Ellipsoidal CAD model of a ventricle-like chamber shown after the model has been sliced into G-code (brown indicates collage type I ink, red indicate cellular ink). (**b**) Zoomed in view of the wall of the ventricle-like chamber showing two inner layers of collagen, three middle layers of cells, and one outer layer of collagen. (**c**) An example of a FRESH printed inner layer of collagen after release form the support bath (Scale bar 2 mm). (**d**) An example of the complete FRESH printed ventricle-like chamber after multiple weeks of culture

thinner than the needle diameter from being pathed. For best results, we recommend using a 150 μm inner diameter needle to print the model described in this protocol. Hence, a wall 450 μm thick requires exactly three perimeters.

2. Load the 8 × 6.6 mm ventricle model into Slic3r or other model slicing software of choice and assign the collagen extruder to the part.
3. Load the second model as a modifier to the first part by right clicking on the model then selecting "add modifier -> load..." and select the 7.85 × 6.3 mm ellipsoid model. Assign the second extruder to this modifier. Position the cell layer 150 μm off of the build plate so that the top of the collagen model and cell model are aligned. Standard slicing programs force placement of the part on the build platform which is why part modifiers are employed to fabricate this ventricle model.
4. Assign the cell ink extrusion tool head to the part modifier. Consult Fig. 1 for clarification.
5. Next create a block modifier by right clicking on the model and create a block with dimensions of 7 × 7 × 1.5 mm (L × W × H) and place on the bed of the print platform. Set the printer speed for all features (walls, infill, and perimeters) to 3 mm/s on the modifier. This will reduce the print speed for the ventricle apex which reduces defects in this region of the ventricle model.

Table 1
FRESH print parameters optimized for fabricating human ventricle scale model using Collagen I and hESC ink

Print Parameter	Typical Value	
First 25 layer print speed	3	mm/sec
Regular print speed	10	mm/sec
Layer height	60	μm
Extrusion width	150	μm
Infill pattern	Concentric	
Infill density	100%	
Overlap	25%	
Tool change z hop	20	mm

6. Consult Table 1 for the recommended printer parameters for fabricating the cardiac ventricle model with collagen type I and cellular ink constructs.

3.3 Preparation of Cellular Ink

1. Differentiate cardiomyocytes from human embryonic stem cells (hESC) using differentiation methods previously described by Burridge et al. [14]. Purify cardiomyocytes by metabolic selection using a lactate supplemented growth medium (CDM3L). Alternatively, cardiomyocytes can be generated by differentiation from human induced pluripotent stem cells (hiPSCs). Cardiomyocytes can also be differentiated from hESCs or hiPSCs and then purified using other established protocols in the literature.
2. Remove the spontaneously beating cardiomyocytes from the culture plastic using TrypLE Express (Gibco, 12604) enzymatic digestion and transfer into CDM3 medium. Filter the cardiomyocyte suspension through a 37 μm cell strainer (Stemcell technology, 27250) to remove larger aggregates.
3. Count the number of cardiomyocytes recovered after filtration using a hemocytometer or other standard cell quantification method.
4. The ink needs to be supplemented with human ventricular cardiac fibroblasts (CC-2904, Lonza) in order for the cardiomyocytes to compact and form an interconnected network (syncytium). We recommend adding 2% human ventricular cardiac fibroblasts, based on cell number, to the cardiomyocytes purified by metabolic selection. If no fibroblasts are added to the ink, the ventricle model will not compact and the cardiomyocytes will tend to form disconnected aggregates with poor electrical connectivity. At 5% fibroblasts or higher, we have found the construct will significantly compact and deform once printed (Fig. 2i).

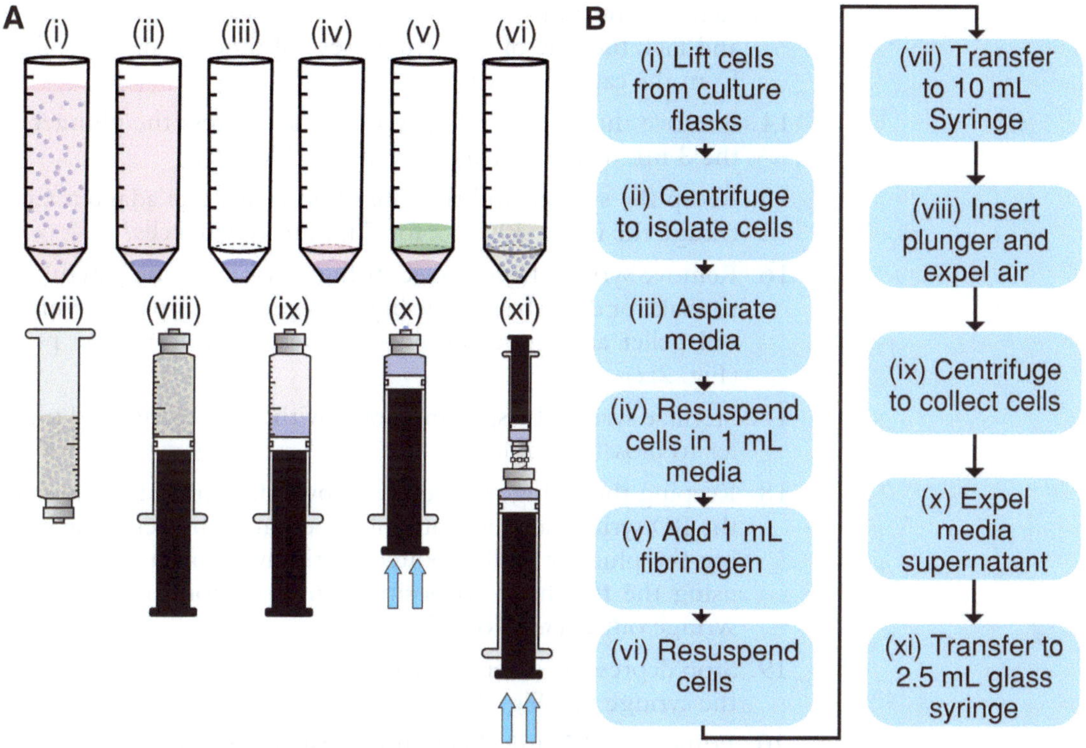

Fig. 2 (**a**) Schematic of steps required to prepare cells into a 20 mg/mL fibrinogen-based cellular ink (~300 × 10^6 cells/mL) ready for printing. Lifted cells are initially suspended in CDM3 in a 50 mL centrifuge tube then collected by centrifugation. Cells are resuspended in stock solution of 60 mg/mL fibrinogen and additional CDM3 to achieve a suspension of cells with a fibrinogen concentration of 20 mg/mL. Cells are then transferred from a centrifuge tube to a 10 mL disposable syringe for final cell collection and transfer to 2.5 mL glass syringe used for printing. (**b**) Steps to preparing a cellular fibrinogen ink for FRESH printing

5. Isolate cell pellet by centrifuging at 200 × *g* for 7 min and aspirate supernatant (Fig. 2ii, iii).
6. Add 1 mL of CDM3 to resuspend cell pellet to a working volume of ~1.5 mL (Fig. 2iv).
7. Add 1 mL of 60 mg/mL fibrinogen to the resuspended cells (Fig. 2v).
8. Mix the fibrinogen and cells by gently pipetting up and down ~5 times with a P1000 pipette (Fig. 2vi).
9. Cap a 10 mL BD luer lock syringe and remove the plunger from the syringe and set aside. Keep syringe plunger sterile, as it will be used later in this procedure.
10. Transfer the cell suspension to the 10 mL BD syringe.
11. Rinse Falcon tube used to collect the cells with ~1 mL of CDM3 media and top off the BD syringe to a working volume of 3 mL (Fig. 2vii).
12. Mix the cell ink by pipetting up and down ~5 times with a P1000 pipette.

13. Reinsert the plunger enough to create a seal. Invert the syringe and flick the side of the syringe several times to dislodge fluid from the cap (Fig. 2viii).
14. Remove the cap from the syringe and depress the plunger to the 3 mL mark, expelling the air (*see* **Note 6**).
15. Recap the syringe and then install in centrifuge adapter. Centrifuge the ink at 200 × g for 7 min to isolate cells.
16. Remove syringe from the centrifuge adapter being careful not to shake or disturb ink. The high cell density ink will resemble a cell pellet and there should be supernatant above the pellet (Fig. 2ix).
17. Sterilize syringe by spraying down with 70% ethanol and transfer to biosafety cabinet.
18. Keeping the syringe tip pointed upward, remove the cap from the BD syringe and aspirate the supernatant by gently depressing the plunger while pointed vertically with one hand while using the free hand to position an aspiration line above the syringe outlet (*see* **Note 7**).
19. Stop depressing when an opaque cell pellet phase is observed at the syringe tip (Fig. 2x).
20. Prime the 2.5 mL Hamilton glass syringe with CDM3 to prevent any air from being introduced to the cell ink.
    (i) Draw ~0.5 mL of CDM3 into 2.5 mL glass syringe.
    (ii) Invert syringe so that tip is pointed upward.
    (iii) Flick side of syringe to dislodge any air bubbles from plunger.
    (iv) Depress plunger completely so that no air is entrapped in the syringe.
21. Attach a narrow capillary female to female luer adapter to the 10 mL BD syringe. Depress the plunger until the adapter channel is filled with ink. Then attach to primed 2.5 mL Hamilton glass syringe.
22. Gently mix back and forth between syringes approximately five times to homogenize the cellular ink (Fig. 2xi).
23. Remove the 10 mL BD syringe first and then carefully withdraw the plunger on the 2.5 mL syringe until ink is evacuated from the luer adapter. This ensures as much cell ink as possible is transferred into the 2.5 mL syringe. Remove the female-female luer adapter and cap syringe.
24. The cellular ink is only stable for approximately 1 h, so printing should be performed as soon as possible after ink preparation. After approximately 1 h, cells begin to agglomerate in the syringe and will clog the needle.

3.4 Setup of Bioprinter for Dual Extrusion Fabrication

1. Install acidified collagen type I ink (24 mg/mL) syringe in the tool 0 position as defined by the printer firmware.
2. Install the high cell density ink syringe in the tool 1 position as defined by the printer firmware.
3. Connect bioprinter to computer using Pronterface application.

3.5 Alignment of Extruder Needle Tips

1. In order to reliably print with multiple materials, the offset between needle tips in the *x* and *y* dimensions should be the same prior to the start of each print and the height of the needles in the *z* dimension must be equal. To achieve this, the bioprinter needs to have mechanical correction to achieve the same *x* and *y* position correction and an independent *z* height adjustment. The most important part of this procedure regardless of the specific bioprinter used is that the offset of the needles is accounted for in the slicing so that an extrusion head does not disturb a previously printed layer, or print in the wrong location. If the needle tips are misaligned, the structure will not have sufficient structural integrity for handling and use in scientific experiments. *See* Fig. 3 for an example of a print completed with properly aligned and misaligned needle tips.
2. When slicing the ventricle model, set the offset between the needles in the *x* dimension.
3. Once syringes are loaded onto the printer, manually adjust the position of the needles so that they are within ±50 μm of the desired coordinates in the *x* and *y* directions. We accomplish this by aligning the needles to a ruler that has been adhered or taped to the printer build plate.
4. The needle tips need to be exactly at the same *z* height (±10 μm) in order to prevent dragging the needle through the previously printed layers. This should be accomplished by adjusting the height of the build plate using the electronic control interface until the first needle just contacts the platform.
5. Adjust the *z* placement of the second needle tip by manually adjusting the height of the syringe pump assembly in the *z* dimension until the needle contacts the build plate.
6. Raise extruders from the build platform and immerse the needle tips in PBS to prevent them from drying out before the start of printing.

3.6 Preparation of FRESH Support Material

1. Prepare the LifeSupport following the instructions provided. Briefly, hydrate the powder by adding 40 mL of cold 50 mM HEPES buffer.° It is important to note that LifeSupport is temperature sensitive and should not be heated above the melting temperature of the gelatin microparticles (greater than ~25 °C).

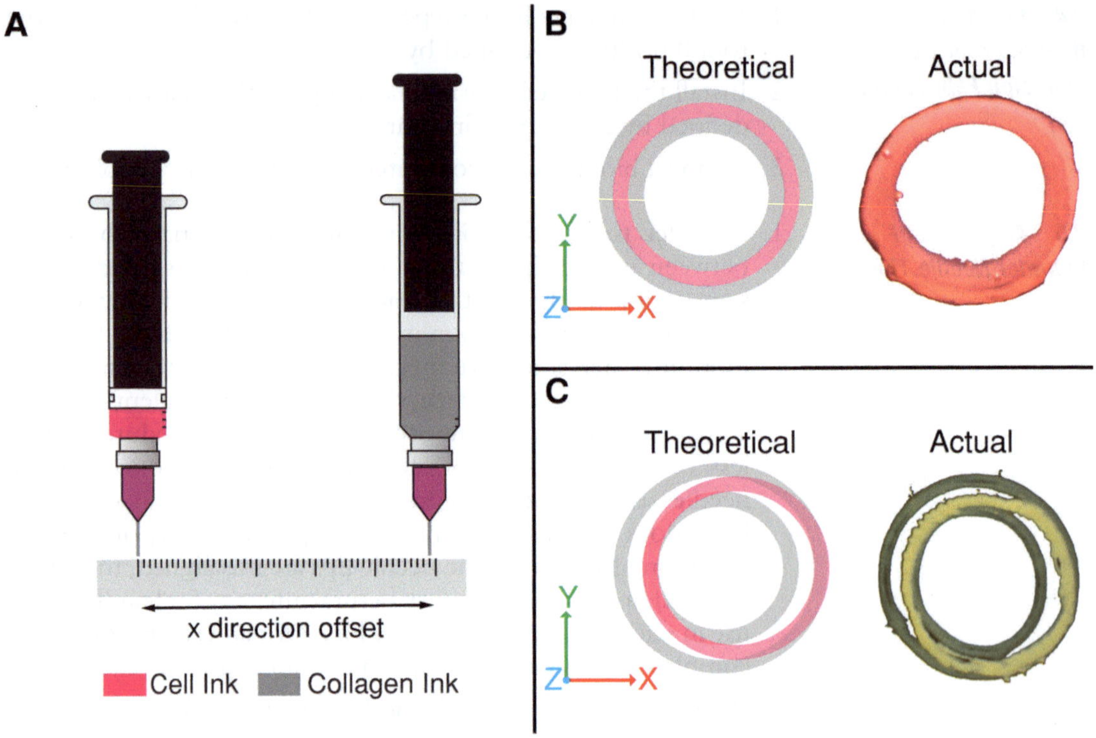

Fig. 3 Needle offset is a critical parameter to ensure dual material prints are fabricated as designed. Needle tips must be in the exact same location after swapping extruders. (**a**) Extruder needle tips should be aligned to a ruler affixed to the print platform in the *x* direction to ensure needle tips are at the predefined offset prior to printing. (**b**) When needles are properly aligned the collagen walls and cellular ink infill will form concentric rings. (**c**) When needles are misaligned in the *x* or *y* direction, the outer ventricle walls and cellular infill will be offset creating an unstable construct which will not hold together when released from the gelatin microparticle support

2. Vortex for 1 min until no clumps are visible.
3. Centrifuge at 2000 × *g* for 5 min.
4. If bubbles are visible in the support material after first centrifugation place support container in a vacuum chamber for 30 min at room temperature to allow powder to fully rehydrate while removing any dissolved gases. Degassing prior to final centrifugation removes dissolved gases that may nucleate during printing but is unnecessary if the support has been sufficiently rehydrated and homogenized at room temperature.
5. Aspirate supernatant (*see* **Note 8**).
6. Add cold CDM3 to a working volume of 40 mL and resuspend by vortexing for 1 min.
7. Centrifuge at 2000 × *g* for 5 min.
8. Aspirate supernatant.

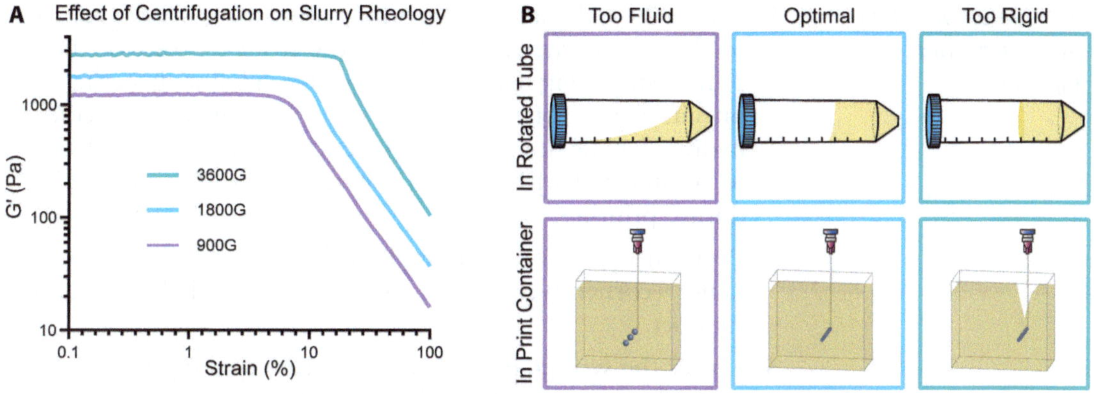

Fig. 4 The support bath should have the appropriate rheology with a yield stress when properly prepared. (**a**) Representative rheology amplitude sweeps for support material compacted at various centrifugation forces. Higher centrifugation forces correspond to a higher degree of compaction, resulting in a higher yield stress of the support. This increased yield stress is seen in the higher plateau regions before the material begins to flow, which is the region where the line deviates from the horizontal. The resulting yield stresses dictate the behavior of the bath and therefore print fidelity. (**b**) Support that is compacted too gently (purple) has too low of a yield stress and is too fluid as seen in the compacted support flowing easily when the tube is turned on its side after compaction (top left). This in turn does not provide a sufficient yield stress to stabilize FRESH printed inks, causing beading and agglomerations (bottom left). Support that is compacted too firmly (green) has too high of a yield stress and is too rigid, as seen in the compacted support not moving at all when the tube is turned on its side after compaction (top right). This results in a bath that does stabilize FRESH printed filaments, but does not sufficiently self-heal, causing chasms to be carved into the bath during printing, compromising the support of subsequent layers. Support that is compacted optimally (blue) has a minimal amount of flow when the tube is turned on its side after compaction (top middle). This results in a bath that can both self-heal and stabilize FRESH printed filaments

9. Add 4 mL FBS, 400 μL thrombin stock solution, 8 μL thiazovivin stock solution, and enough CDM3 to come to a working volume of 40 mL. Thiazovivin is a Rho-associated kinase (ROCK) inhibitor that enhances the survival of enzymatically dissociated cardiomyocytes and promotes adhesion to extracellular matrix mediated by β1 integrins. Addition of this molecule increases the number of cardiomyocytes that survive after printing is completed [15].
10. Resuspend by vortexing for 1 min.
11. Centrifuge at 2000 × *g* for 7 min.
12. Check that support is sufficiently compacted by tilting the centrifuge tube parallel to the floor and verifying the material does not flow. *See* Fig. 4 for how optimally compacted support should behave.
13. Carefully transfer the compacted FRESH support material into the corner well of a 24-well plate with a sterile metal spatula. Avoid incorporation of air bubbles into the support material (*see* **Note 9**).

3.7 Setup of Printer Stage and Print Initiation

1. Place a 24-well plate on the print platform at a 45-degree angle so that the corner protrudes to the front or back of the 3D printer stage (Fig. 5). Place petri dishes filled with sterile water on either side of well plate to keep the inactive needle tip hydrated while not in use to prevent clogging. Secure all dishes to printer bed with vacuum grease or tape to prevent sliding during printing.
2. Position needle of extruder 1 in the center of the print container (corner well of the 24-well plate) filled with gelatin support.
3. Lower the syringe needle to the bottom of the print container using the control software 1 mm at a time, then move needle in 1 mm in the *x* or *y* direction. The needle will deflect when the needle is at the bottom of the print container.
4. Back the needle off the bottom of the print container by 1 mm in the *z* direction.

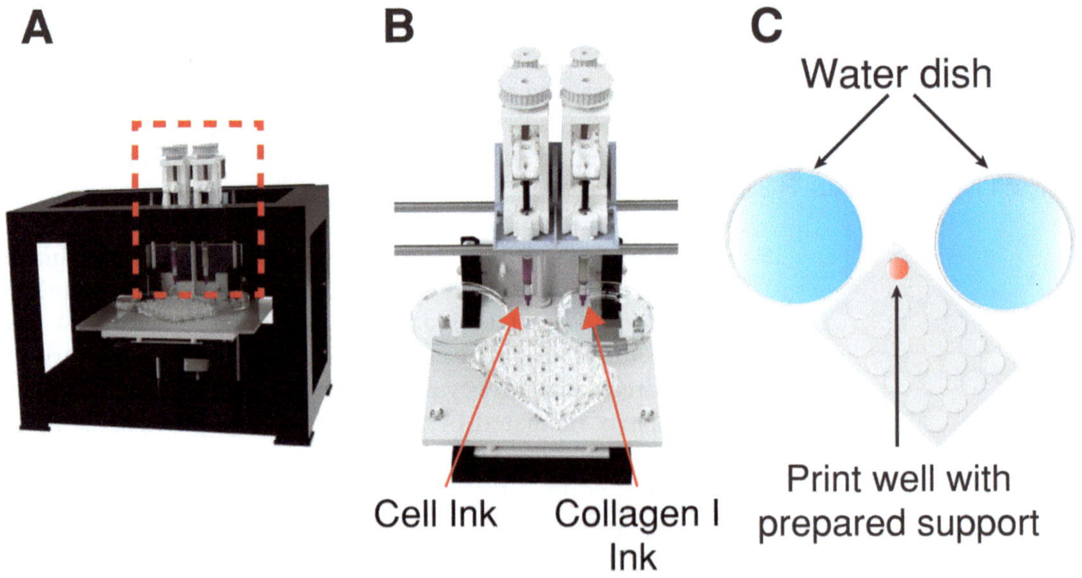

Fig. 5 Setup of the 3D bioprinter. (**a**) Open-source bioprinter setup for fabrication of the ventricle model. (**b**) Detailed view of the dual-extruder hydrogel printer with the cell ink in the left extruder and the collagen ink in the right extruder. The stage is set below the printer with a 24-well plate to hold the support material with petri dishes set to the left and right of the corner of the print well. (**c**) Top down view of the printer stage layout with a 24-well plate print container angled at 45 degrees so that the corner well will be used to print into. A 24-well plate is used to print the ventricle model because it is deep enough to contain the entire model. Water dishes are set to either side of the well plate to keep the needle tip hydrated while it is not actively being used to print the construct. If the material in the needle tip dried out, it will clog the needle tip and material will not be deposited when printing

5. Load the ventricle model G-Code on Pronterface and start the print. A single ventricle takes approximately 10 min to print using the parameters in this protocol.

3.8 Release and Culture of Ventricle Model

1. After completing the print, allow the fibrinogen to gel in the biosafety cabinet at room temperature for 10 min.
2. Transfer the well plate to a 37 °C incubator and allow gelatin support material to melt for 1 h.
3. Aspirate melted support material and replace with CDM3 media supplemented with 10% FBS, taking care not to aspirate the ventricle model.
4. Exchange media every 2 days to maintain tissues. We have cultured ventricles for up to 28 days after manufacture though they should be stable for longer periods.
5. Ventricle models should begin spontaneously beating approximately 1–2 days after fabrication.

4 Notes

1. We have prepared this manuscript based on the use of an open-source extrusion-based bioprinter. If a bioprinting system that relies on pneumatic pressure for ink deposition will be used to fabricate this model, we recommend that the user first prints a simple ring structure to determine the optimal printing parameters with the cellular ink and collagen ink. Instructions pertaining to the preparation of the cellular ink, support, and stage layout should not be affected.
2. Sterile in-house coacervation of FRESH support material is challenging as gelatin manufacturer and lot-to-lot variability affects the efficiency of the coacervation process. Due to this variation, we recommend using LifeSupport® FRESH support material composed of gelatin microparticles manufactured by FluidForm, Inc. LifeSupport® is a sterile and lyophilized powder, making it straightforward to prepare and print on demand.
3. A 10× stock of HEPES buffer can be made at 500 mM by dissolving 119.1 g HEPES powder in 1 L of water. Adjusting this strong buffer to pH 7.4 is made faster by slowly adding NaOH pellets instead of a weaker base like 1 N NaOH. Regular 50 mM HEPES buffer can then be made by diluting 10× concentrate in ultrapure sterile water at a 1:10 ratio.
4. While this manuscript was prepared based on the use of human stem cell-derived cardiomyocytes, the cardiomyocyte species and/or disease phenotype can readily be adapted. The key changes would be the passaging and base culture media used in the ink preparation. If using cardiomyocyte spheroids,

be sure to use a needle with a diameter greater than the spheroid diameter.

5. Use a fused deposition modeling (FDM) printer or equivalent to fabricate a centrifuge adapter for the BD syringe out of PLA with >50% infill. The 5 and 10 mL BD syringe centrifuge adapter design that we use is available for download on the NIH 3D print exchange website (https://3dprint.nih.gov/discover/3dpx-014806, https://3dprint.nih.gov/discover/3dpx-014805). Both models are designed to hold a working volume of 3 mL in the syringe.
6. When expelling air from partially filled syringes, be sure to depress the plunger with steady pressure as the plunger can slip if pressed with too much force and eject the syringe contents into the air. The syringe plunger should not be withdrawn once air is evacuated to prevent air bubbles from entering the ink. We use a working volume of 3 mL in the disposable plastic syringe because that is the volume the centrifuge adapter is designed to hold.
7. Glass syringes should be sterilized with the plunger pulled out of the syringe barrel. Prior to reassembling glass syringes be sure to lubricate the plunger with sterile water or PBS and help the plunger move freely.
8. When properly compacted, the FRESH support material should stay in place when the tube is slowly turned on its side. If the support material flows very easily when the tube is on its side, resuspend the support material by adding HEPES buffer back, vortexing for 30 s and centrifuge at 200 × *g* higher than the previous attempt. Due to differences in the design of centrifuges, optimization may be required. Repeat this process of incrementing the centrifugation speed after resuspension until the FRESH support material is properly compacted. FRESH support material should be degassed prior to centrifugation to remove dissolved gases that may nucleate during printing, potentially causing a void in the print.
9. Firmly tapping the well plate (or any print container) can pop large bubbles in the support material. To fully pop any small bubbles that may have formed, centrifuge the well plate in a centrifuge with plate adapters at 3000 × *g* for 30 s.

References

1. Torrent-Guasp F, Ballester M, Buckberg GD et al (2001) Spatial orientation of the ventricular muscle band: physiologic contribution and surgical implications. J Thorac Cardiovasc Surg 122:389–392. https://doi.org/10.1067/mtc.2001.113745
2. Goldfracht I, Protze S, Shiti A et al (2020) Generating ring-shaped engineered heart tissues from ventricular and atrial human pluripotent stem cell-derived cardiomyocytes. Nat Commun 11:1–15. https://doi.org/10.1038/s41467-019-13868-x

3. Hansen A, Eder A, Bönstrup M et al (2010) Development of a drug screening platform based on engineered heart tissue. Circ Res 107:35–44. https://doi.org/10.1161/CIRCRESAHA.109.211458
4. Zhao Y, Rafatian N, Feric NT et al (2019) A platform for generation of chamber-specific cardiac tissues and disease modeling. Cell 176:913–927.e18. https://doi.org/10.1016/j.cell.2018.11.042
5. Savoji H, Mohammadi MH, Rafatian N et al (2019) Cardiovascular disease models: a game changing paradigm in drug discovery and screening. Biomaterials 198:3–26. https://doi.org/10.1016/j.biomaterials.2018.09.036
6. Weinberger F, Mannhardt I, Eschenhagen T (2017) Engineering cardiac muscle tissue: a maturating field of research. Circ Res 120:1487–1500
7. Tzatzalos E, Abilez OJ, Shukla P, Wu JC (2016) Engineered heart tissues and induced pluripotent stem cells: macro- and microstructures for disease modeling, drug screening, and translational studies. Adv Drug Deliv Rev 96:234–244
8. Ding S, Feng L, Wu J et al (2018) Bioprinting of stem cells: interplay of bioprinting process, bioinks, and stem cell properties. ACS Biomater Sci Eng 4:3108–3124. https://doi.org/10.1021/acsbiomaterials.8b00399
9. Maiullari F, Costantini M, Milan M et al (2018) A multi-cellular 3D bioprinting approach for vascularized heart tissue engineering based on HUVECs and iPSC-derived cardiomyocytes. Sci Rep 8:1–5. https://doi.org/10.1038/s41598-018-31848-x
10. Jun I, Han HS, Edwards JR, Jeon H (2018) Electrospun fibrous scaffolds for tissue engineering: viewpoints on architecture and fabrication. Int J Mol Sci 19:745
11. MacQueen LA, Sheehy SP, Chantre CO et al (2018) A tissue-engineered scale model of the heart ventricle. Nat Biomed Eng 2:930–941. https://doi.org/10.1038/s41551-018-0271-5
12. Hinton TJ, Jallerat Q, Palchesko RN et al (2015) Three-dimensional printing of complex biological structures by freeform reversible embedding of suspended hydrogels. Sci Adv 1:e1500758. https://doi.org/10.1126/sciadv.1500758
13. Lee A, Hudson AR, Shiwarski DJ et al (2019) 3D bioprinting of collagen to rebuild components of the human heart. Science (80-) 365:482–487. https://doi.org/10.1126/science.aav9051
14. Burridge PW, Holmström A, Wu JC (2015) Chemically defined culture and cardiomyocyte differentiation of human pluripotent stem cells. Curr Protoc Hum Genet 87:21.3.1–21.3.15. https://doi.org/10.1002/0471142905.hg2103s87
15. Xu Y, Zhu X, Hahm HS et al (2010) Revealing a core signaling regulatory mechanism for pluripotent stem cell survival and self-renewal by small molecules. Proc Natl Acad Sci U S A 107:8129–8134. https://doi.org/10.1073/pnas.1002024107

Chapter 6

Engineered Heart Tissues for Contractile, Structural, and Transcriptional Assessment of Human Pluripotent Stem Cell-Derived Cardiomyocytes in a Three-Dimensional, Auxotonic Environment

Samantha Bremner, Alex J. Goldstein, Ty Higashi, and Nathan J. Sniadecki

Abstract

Three-dimensional, human engineered heart tissue promotes maturation of human induced pluripotent stem cell-derived cardiomyocytes (hiPSC-CMs) and provides a useful platform for in vitro cardiac development and disease modeling. This protocol describes the generation of fibrin-based engineered heart tissues (EHTs) containing hiPSC-CMs and human stromal cells. The platform makes use of racks of silicone posts that fit a standard 24-well dish. Stromal cells and hiPSC-CMs are cast in a fibrin hydrogel suspended between two silicone posts, forming an engineered tissue that generates synchronous contractions. The platform described herein is amenable to various measures of cardiac function including measurement of contractile force and calcium handling, as well as molecular biology assays and immunostaining.

Key words Engineered heart tissue, Induced pluripotent stem cells, Cardiomyocytes, Cardiac tissue engineering, Fibrin

1 Introduction

Human induced pluripotent stem cell-derived cardiomyocytes (hiPSC-CMs) have emerged as a powerful tool for modeling cardiac development and disease [1, 2]. However, their utility can be limited by relative cardiac immaturity and difficulty of assessing relevant measures of cardiac function [3]. To address these shortcomings, three-dimensional, engineered heart tissues have been developed that provide a more physiologically relevant cell environment as compared to two-dimensional in vitro platforms [4, 5]. Such platforms have been shown to improve hiPSC-CM

Samantha Bremner, Alex J. Goldstein and Ty Higashi contributed equally with all other contributors.

Kareen L. K. Coulombe and Lauren D. Black III (eds.), *Cardiac Tissue Engineering: Methods and Protocols*, Methods in Molecular Biology, vol. 2485, https://doi.org/10.1007/978-1-0716-2261-2_6,

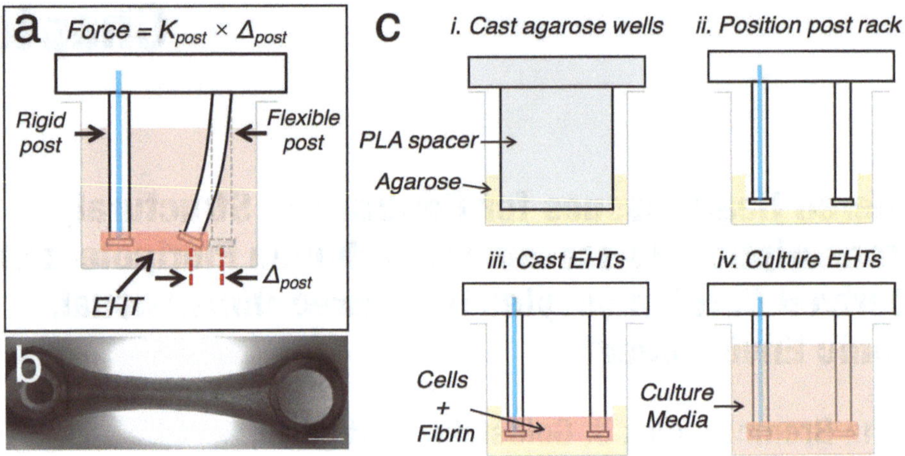

Fig. 1 Schematic of EHT platform. Cardiomyocytes and stromal cells are suspended in a fibrin gel suspended between two polydimethylsiloxane (PDMS) posts, one flexible and one rigid. The force of EHT contraction can be measured using the deflection of the flexible post (Δ_{post}) and the post stiffness (K_{post}) (**a**). Still image of an EHT shown with the flexible post on the right. Scale bar is 1 mm (**b**). Outline of EHT fabrication procedure depicting casting agarose wells using a 3D-printed polylactic acid (PLA) spacer (i), aligning the PDMS post rack in the agarose wells (ii), casting a suspension of cardiomyocytes and stromal cells in a fibrin hydrogel (iii), and culturing the EHTs (iv) (**c**)

maturation, while providing a useful method for assessment of a variety of measures of cardiac function including contractile capability and calcium handling [6–9]. Using these constructs may improve the relevance of in vitro stem cell-derived cardiac studies.

Here, we describe the generation of fully human, contractile, three-dimensional engineered heart tissue (EHT) (Fig. 1). The platform consists of racks of six pairs of polydimethylsiloxane (PDMS) posts, one flexible and one rigid, fit to the dimensions of a standard 24-well plate (Fig. 2a). Compared to other static culture methods, the flexible post provides the EHTs with a physiologically relevant auxotonic force cycle. Using agarose wells, hiPSC-CMs and human stromal cells are encapsulated in a fibrin hydrogel and cast between the PDMS posts. EHTs reorganize the fibrin matrix and compact over 1 to 2 days to form visibly contracting tissues capable of deflecting the flexible post by 1 week in culture.

After 3 weeks in culture, EHTs reach their maximum generation of contractile force. To measure contractility, the movement of the flexible post can be tracked to calculate the magnitude and kinetics of force generation using cantilever mechanics. Additionally, EHTs can be loaded with a variety of fluorescent calcium indicator dyes and imaged to assess calcium handling. Herein, we describe a protocol for generating fibrin-based, hiPSC-CM-derived EHTs as well as a brief explanation of downstream EHT analysis including assessment of contractility and calcium handling, isolation of RNA, and tissue processing for histology.

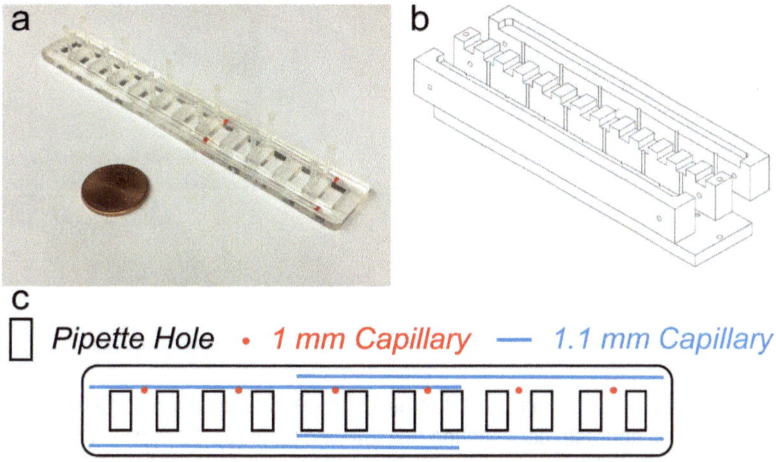

Fig. 2 PDMS post rack with six pairs of posts made to fit a standard 24-well plate (**a**). Exploded view of the four-part acrylic mold used to cast PDMS post racks (**b**). Schematic of the bottom of a post rack indicating proper placement of 1 mm outer diameter capillary tubes (into plane of image) and 1.1 mm outer diameter capillary tubes and location of holes for pipette insertion during EHT casting (**c**)

2 Materials

2.1 PDMS Post Racks and PLA Spacer Arrays

1. Four-part acrylic mold (Fig. 2b): Designed in-house. Manufactured by Limited Productions Inc. (LPI), a machine shop based in Bellevue, WA. A full, dimensioned CAD drawing of the acrylic mold is available upon request to the corresponding author.
2. PDMS (Sylgard 184): Base and curing agent.
3. Glass capillary tubes: 1 mm outer diameter (rigid posts), 1.1 mm outer diameter (supporting structure of posts).
4. Polylactic acid (PLA) spacer array (Fig. 3a): Designed and 3D printed in-house. Each spacer measures 12 mm in length, 4 mm in width, and 13 mm in height. The STL file for the spacer array is available upon request to the corresponding author.

2.2 Cell Culture

1. WTC11 (Coriell Institute for Medical Research) human-induced pluripotent stem cell-derived cardiomyocytes (hiPSC-CMs).
2. HS27a human bone marrow stromal cells (American Type Culture Collection, ATCC).
3. Dulbecco's phosphate-buffered saline (DPBS).
4. Cell lifting solution: 0.05% Trypsin + EDTA (*see* **Note 1**).

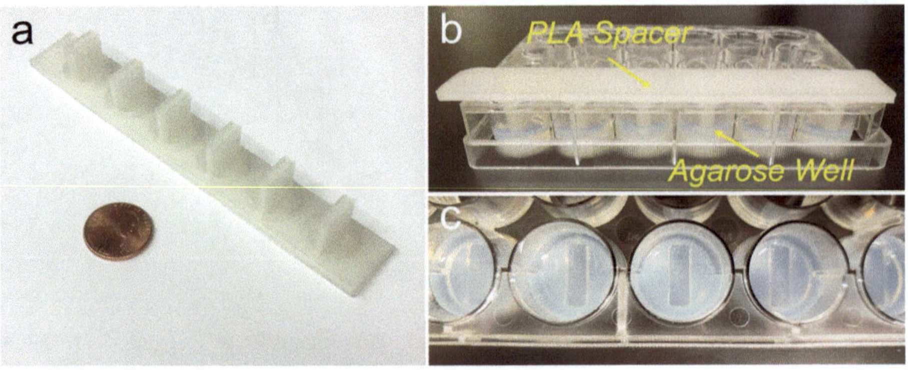

Fig. 3 3D-printed PLA spacer array used for generating agarose EHT casting molds (**a**). PLA spacer array shown during agarose mold casting in a 24-well plate (**b**). 12 mm × 4 mm × 4 mm agarose molds generated after removal of PLA spacer array (**c**)

5. Stromal culture media: DMEM + L-glutamine, 10% fetal bovine serum (FBS), 1% MEM non-essential amino acids, penicillin-streptomycin (100 U/mL).
6. Cardiomyocyte culture media: RPMI 1640 + L-glutamine, 2% B27 supplement, 100 U/mL penicillin-streptomycin.
7. Cardiomyocyte stop solution: RPMI 1640 + L-glutamine, 2% B27 supplement, 10% FBS, 100 U/mL penicillin-streptomycin.
8. 10 mM Y-27632 in deionized H_2O.
9. 200 U/mL DNAse in 5 mM $CaCl_2$ with 1% bovine serum albumin (BSA).

2.3 EHT Casting

1. 10× fibrinogen stock: 50 mg/mL fibrinogen from bovine plasma in cardiomyocyte culture media. Aliquoted and stored at −20 °C.
2. Thrombin: 100 U/mL thrombin from bovine plasma in a 3: 2 ratio of DPBS to sterile deionized H_2O. Aliquoted and stored at −20 °C.
3. EHT culture media: RPMI 1640 + L-glutamine, 2% B27 supplement, 100 U/mL penicillin-streptomycin, 5 g/L aminocaproic acid, sterile-filtered with 0.22 μm porosity filter.
4. 2% agarose solution in DPBS.
5. 24-well tissue culture dishes.
6. 0.6 mL microcentrifuge tubes.

2.4 EHT Analysis

1. 10× Tyrode's buffer stock: 1.8 mM $CaCl_2$, 1 mM $MgCl_2$, 5.4 mM KCl, 140 mM NaCl, 0.33 mM NaH_2PO_4, and 10 mM HEPES.
2. Glucose.

3. Fluo-4 AM or Fura-4 AM (Invitrogen).
4. 20% Pluronic F-127 in DMSO.
5. 140 mM KCl in deionized H_2O.
6. 4% paraformaldehyde (PFA) in DPBS.
7. 30% w/v sucrose in deionized H_2O.
8. Optimal cutting temperature (O.C.T.) compound.
9. 20 mg/mL Proteinase K in deionized H_2O.
10. RNeasy Micro or Mini Kit (QIAGEN).

3 Methods

3.1 PDMS Post Racks and Preparation Prior to Casting

1. Cut one 1 mm outer diameter capillary tube into six sections of 14 mm length. Prepare rigid posts by placing one 14 mm section into each of the six post channels on one side of the assembled mold (Fig. 2c).
2. Place two full-length 1.1 mm outer diameter capillary tubes on either side of the mold base (for a total of 4) to provide structural support to the base (Fig. 2c).
3. Prepare 11 g of Sylgard 184 PDMS by thoroughly mixing 10 g of the base component with 1 g of the curing agent for 10 min. Degas the PDMS under vacuum until all bubbles are removed. Pour the PDMS into the mold and bake for 18 h at 65 °C. Remove the mold from the oven, allow to cool to room temperature, and carefully cut the post rack from the mold using a razor blade, removing as much excess PDMS as possible.
4. The day before EHT casting, sterilize the PDMS post rack and PLA spacer with 5 min of UV/Ozone treatment, followed by 5 min submersion in 70% ethanol solution and two 5 min washes in deionized H_2O. Leave in biosafety cabinet to dry at least 1 h or preferably overnight to ensure complete evaporation of residual ethanol that may have absorbed into the PDMS.
5. Thaw fibrinogen and thrombin at 4 °C prior to EHT casting.

3.2 Cell Preparation

1. Rinse cardiomyocytes and stromal cells with DPBS.
2. Add 1 mL pre-warmed (37 °C) cell lifting solution per well of a 6-well plate to cardiomyocytes and stromal cells and incubate at 37 °C.
3. After 3 min, check if the cells have detached by lightly tapping the plate. If the cells are still attached, wait an additional 2 min.
4. Triturate cells in the cell lifting solution and transfer to separate conical tubes containing the appropriate stop solution. Ensure the amount of stop solution used is at least twice that of the cell

lifting solution (i.e., 2 mL of stop solution per well of a 6-well plate).

5. Spin down the cardiomyocytes and stromal cells at 300 × *g* for 5 min.
6. Aspirate the media and resuspend the cells in 10 mL of appropriate culture media.
7. Count the cardiomyocytes and stromal cells with a hemocytometer.
8. Determine the number of EHTs to be cast and combine the correct number of cardiomyocytes (e.g., 5×10^5 cells per EHT) and stromal cells (e.g., 5×10^4 cells per EHT) in a new "Cell-Tissue Mixture" conical tube (*see* **Notes 2** and **3**).
9. Place the cell tissue mixture vial as well as the extra cardiomyocyte and stromal cell vials on ice (*see* **Note 4**).

3.3 Agarose Wells, Thrombin Aliquots, and Fibrinogen

1. Make a 2% agarose solution by mixing 1 g of agarose powder with 50 mL of DPBS in a sterile glass bottle. Cap loosely and microwave for 1 to 2 min, or until the powder is fully dissolved.
2. Add 1.5 mL of the 2% agarose solution into each well of a row of a 24-well plate that will be used for casting EHTs (*see* **Note 5**). Position the PLA spacers in the center of each of the agarose-filled wells (Fig. 3b, c). Remove the spacers after 15–20 min or once the agarose has solidified (*see* **Notes 6** and **7**).
3. While agarose is cooling, aliquot 3 μL of thrombin into 0.6 mL microcentrifuge tubes so that there is one microcentrifuge tube for each EHT to be cast. Place tubes on crushed ice until ready to cast EHTs.

3.4 EHT Preparation

1. Spin down the cell-tissue mixture at 300 × *g* for 5 min, aspirate the media, and resuspend the cells with the appropriate volume of cardiomyocyte culture media and fibrinogen according to Table 1 (*see* **Note 8**).
2. Place the cell-tissue mixture back on ice.

3.5 EHT Casting

1. Position the PDMS post rack so the tips of the posts are centered in the agarose wells, checking the plate from below to ensure proper placement.
2. Mix the cell-tissue mixture and draw up 97 μL of the cell-tissue mixture using a 200 μL micropipette tip. Increase the pipette volume to 100 μL, then quickly triturate four times with a 3 μL thrombin aliquot. Moving carefully but quickly, transfer the mixture to an agarose well within 15 s (before gelation occurs) by inserting the pipette through the hole in the post rack on either side of the posts (Fig. 2c) (*see* **Notes 9** and **10**).

Table 1
Volumes for EHT casting, multiply each volume by the number of EHTs being prepared

Per EHT	
Cell-tissue mixture in cardiomyocyte culture media:	87 μL
Fibrinogen (50 mg/mL):	10 μL
Thrombin (100 U/mL):	**3 μL**
Total:	100 μL

3. Repeat **step 2** with a fresh 200 μL micropipette tip and thrombin aliquot until all EHTs have been cast.
4. Place the lid on the plate and incubate the EHTs at 37 °C for 80 min (*see* **Note 11**).
5. Gently add 300 μL of EHT media to the edge of the wells and incubate the EHTs at 37 °C for 10 min (*see* **Note 12**).
6. Fill a fresh 24-well plate with 2.5 mL/well of EHT media and gently transfer EHTs.

3.6 EHT Culture

1. Replace the EHT media every 2–3 days by filling an empty row with warmed (37 °C) EHT media and gently transferring post rack.
2. EHTs should compact within 48 h of casting, and depending on the cell line, will usually begin beating within 4–7 days (*see* **Note 13**).

3.7 EHT Analysis

3.7.1 Optical Force Measurement

1. Make 1× Tyrode's buffer working solution: dilute 10× stock in deionized H_2O, add 5 mM glucose, and adjust pH to 7.4 using 10 N NaOH. Sterilize final solution using a 0.22 μm porosity filter and store at 4 °C for up to 1 week.
2. Transfer EHTs to warmed (37 °C) 1× Tyrode's buffer in a 24-well plate.
3. EHTs can be electrically stimulated using an electric field stimulation of 5 V/cm and 10 ms duration at a frequency higher than the spontaneous beating rate of the EHTs (usually 1–2 Hz) (*see* **Note 14**).
4. Bring EHTs to a wide field microscope with a heated chamber (37 °C) and record videos of EHT contractions (*see* **Note 15**).
5. To measure force production, threshold the images and track the position of the centroid of the flexible post relative to the centroid of the rigid post to calculate the deflection of the flexible post (Δ_{post}) (Fig. 1). Contraction force and kinetics can be calculated from the bending stiffness of the post ($E = 2.5$ MPa for PDMS mixed at 1:10 and baked at 65 °C

for 18 h) and the dimensions of the flexible post (length L and diameter D) as follows [10] (*see* **Note 16**):

$$\text{Force} = \Delta_{\text{post}} \cdot \frac{3\pi E D^4}{64 L^3}$$

3.7.2 Calcium Handling Assessment

1. Prepare 1 mM stock solution of calcium indicator dye of choice (Fluo-4 AM or Fura-2 AM) in DMSO.
2. Incubate EHTs in 5 μM Fluo-4 or Fura-2 dye in EHT media for 1 h at 37 °C (*see* **Note 17**), followed by a washout period in EHT media for 30 min at 37 °C.
3. Transfer EHTs to Tyrode's buffer and image as described in Subheading 3.7.1 using fluorescent excitation at wavelengths specified by the manufacturer for the calcium indicator of choice.

3.7.3 Gene Expression Analysis

1. Total cellular RNA can be extracted from EHTs using available RNA Mini and Micro Kits according to manufacturer's instructions with modifications as follows.
2. Pre-digest individual EHTs in 100 μL Proteinase K (20 mg/mL) for 10 min at 56 °C with agitation.
3. Lyse cells by adding 350 μL RLT Plus Buffer (with 1% 2-mercaptoethanol) and homogenize.
4. Add 250 μL 200 proof ethanol and pipette until there is no visible bilayer (*see* **Note 18**).
5. Perform subsequent steps as recommended by the manufacturer, followed by reverse transcription and desired RT-qPCR applications (*see* **Note 19**).

3.7.4 Histology and Immunocytochemistry

1. Submerge EHTs in 140 mM KCl for 30 s to induce EHT relaxation.
2. Fix EHTs in 4% PFA in DPBS for 15 min then wash once with DPBS for 5 min.
3. Dehydrate EHTs in a 30% w/v sucrose solution overnight at 4 °C.
4. Carefully remove EHTs from posts and embed in O.C.T. compound in a plastic biopsy mold. Multiple EHTs can be embedded in a single O.C.T. preparation.
5. Obtain sections of desired thickness using a cryostat and proceed with immunocytochemistry protocol of choice (*see* **Note 20**).

4 Notes

1. We have noticed that hiPSC-CMs cultured longer than 30 days tend to clump during lifting. To mitigate this issue, use the following cell lifting solution: 2:1 ratio of Versene to 0.25% Trypsin+EDTA, 10 μM Y27 rock inhibitor, 200 mU/mL DNAse.
2. Prepare 1–2 extra EHTs to ensure sufficient cell mixture volume.
3. The optimal number of cardiomyocytes and stromal cells may vary depending on the cell line used. We suggest 5×10^5 to 1×10^6 cardiomyocytes and 5×10^4 to 1×10^5 stromal cells per EHT.
4. It is important that the mixture be cool, otherwise the reaction between the thrombin and the fibrinogen will occur too quickly, and casting will be difficult.
5. It is better to use the outer rows of the 24-well plate for improved visibility when casting.
6. Do not rotate or wiggle the spacers when removing them as it can deform the agarose wells. If necessary, gently rotate the agarose wells into the correct alignment using a pipette tip.
7. It is important that the agarose step is done just before casting the EHTs to prevent the solidified agarose from drying out. Agarose that has dried out will dehydrate the EHT during casting and ruin their morphology.
8. When bringing the cell-tissue mixture volume up to the calculated amount, aspirate the supernatant, add a small amount of cardiomyocyte culture media to resuspend the pellet, use the pipette to measure the cell volume, and add the difference between the measured volume and the desired volume using more cardiomyocyte culture media.
9. Avoid introducing air bubbles during EHT casting. Do not invert the pipette at any point and try to avoid pressing the plunger all the way to the second stop, which often produces air bubbles.
10. Hold the PDMS post rack in place during casting to prevent movement which may damage the EHTs.
11. Take great care when placing and removing the plate lid to avoid ripping the EHTs.
12. At this point, the EHTs should release easily from molds. If the EHTs resist transfer after 10 min, add another 100 μL of EHT media and incubate the EHTs at 37 °C for another 5 min.

13. EHTs that do not sufficiently compact will not generate synchronized contractions. Should EHTs fail to compact, try increasing the number of stromal cells per EHT.
14. Electrical field stimulation can be accomplished using a custom-built 24-well array of carbon electrode pairs, with an electrode on both sides of each EHT, attached to an electrical stimulator device, such as the Astro Med Grass S88X Stimulator [7].
15. A variety of low-magnification optical configurations can be used to enable capture of an entire EHT, we have had success using a 2× objective with a 0.7× coupler between the microscope and the camera. Use a sufficiently high capture frame rate to accurately resolve EHT contraction (i.e., 50–100 frames per second).
16. This equation for cantilever beam deflection applies for relatively small deflections (i.e., less that 10% of the post length). Error can be introduced into the post stiffness experimentally via the elastic modulus of the PDMS, which can vary greatly depending on mixing ratio as well as curing temperature and time. We recommend characterizing the stiffness of the cured PDMS and standardizing manufacturing protocols. Additionally, the tolerances of post diameter and length can greatly affect the calculated stiffness due to their fourth- and third-order effects, respectively.
17. An equal volume of 20% Pluronic F-127 in DMSO can be mixed with the stock Fluo-4 or Fura-2 solution before adding to EHT media to assist with dye dispersion.
18. A smaller amount of higher proof ethanol is added as compared to the manufacturer's instructions to account for the 100 μL of Proteinase K solution.
19. If there are difficulties obtaining sufficient RNA from single EHTs, multiple EHTs can be pooled together in a single Proteinase K digestion reaction.
20. We recommend blocking and permeabilizing with 1% BSA and 0.1% Triton X-100 and performing primary antibody incubation overnight at 4 °C.

References

1. Mummery CL (2018) Perspectives on the use of human induced pluripotent stem cell-derived cardiomyocytes in biomedical research. Stem Cell Reports 11:1306–1311. https://doi.org/10.1016/j.stemcr.2018.11.011
2. Eschenhagen T, Carrier L (2019) Cardiomyopathy phenotypes in human-induced pluripotent stem cell-derived cardiomyocytes-a systematic review. Pflugers Arch 471:755–768. https://doi.org/10.1007/s00424-018-2214-0
3. Karbassi E, Fenix A, Marchiano S et al (2020) Cardiomyocyte maturation: advances in knowledge and implications for regenerative medicine. Nat Rev Cardiol 17(6):341–359.

https://doi.org/10.1038/s41569-019-0331-x

4. Feric NT, Radisic M (2016) Maturing human pluripotent stem cell-derived cardiomyocytes in human engineered cardiac tissues. Adv Drug Deliv Rev 96:110–134. https://doi.org/10.1016/j.addr.2015.04.019
5. Mills RJ, Hudson JE (2019) Bioengineering adult human heart tissue: how close are we? APL Bioeng 3:010901. https://doi.org/10.1063/1.5070106
6. Bielawski KS, Leonard A, Bhandari S et al (2016) Real-time force and frequency analysis of engineered human heart tissue derived from induced pluripotent stem cells using magnetic sensing. Tissue Eng Part C 22:932–940. https://doi.org/10.1089/ten.tec.2016.0257
7. Leonard A, Bertero A, Powers JD et al (2018) Afterload promotes maturation of human induced pluripotent stem cell derived cardiomyocytes in engineered heart tissues. J Mol Cell Cardiol 118:147–158. https://doi.org/10.1016/j.yjmcc.2018.03.016
8. Bertero A, Fields PA, Smith AST et al (2019) Chromatin compartment dynamics in a haploinsufficient model of cardiac laminopathy. J Cell Biol 218:2919–2944. https://doi.org/10.1083/jcb.201902117
9. Zaunbrecher RJ, Abel AN, Beussman K et al (2019) Cronos titin is expressed in human cardiomyocytes and necessary for Normal sarcomere function. Circulation 140:1647–1660. https://doi.org/10.1161/CIRCULATIONAHA.119.039521
10. Sniadecki NJ, Chen CS (2007) Microfabricated silicone elastomeric post arrays for measuring traction forces of adherent cells. Methods Cell Biol 83:313–328. https://doi.org/10.1016/S0091-679X(07)83013-5

Chapter 7

High-Throughput Analysis of Drug Safety Responses in Induced Pluripotent Stem Cell-Derived Cardiomyocytes Using Multielectrode Array

Nadjet Belbachir, Nathan Cunningham, and Joseph C. Wu

Abstract

Microelectrode array (MEA) is an electrophysiological instrument used to track activities of ion channels in excitable cells. Neurons and cardiomyocytes are seeded to form a cell monolayer on a field of sensors able to detect electrical signals, called extracellular field potentials (EFPs). This noninvasive tool allows researchers to investigate key parameters such as EFP amplitude, duration, and arrhythmias. MEA is progressively considered the gold standard for high-throughput in vitro electrophysiological evaluation, particularly for cardiac disease modeling and cardiac toxicity assessment.

Key words Microelectrode array, Arrhythmias, Human-induced pluripotent stem cells, Cardiomyocytes

1 Introduction

Microelectrode array (MEA) technology has emerged from a need to achieve rapid and user-friendly recordings of electrical activity in large cell populations. The MEA system can simultaneously monitor cells on multiple electrodes at high temporal resolution, allowing for noninvasive, long-term observation. Since introduced 30 years ago, this technology has been improved and adapted to several applications in both academic and industrial settings [1]. Recently, increasing attention has focused on cardiac ion channels and their regulatory pathways in understanding cardiac arrhythmias and drug toxicity for both preclinical and FDA-approved compounds [2, 3]. Among the numerous studies that illustrate the implementation of MEA in drug toxicity assessment, the Comprehensive in vitro Proarrhythmia Assay (CiPA) initiative leveraged the many advantages of the MEA system [4]. In the CiPA initiative, human-induced pluripotent stem cell-derived cardiomyocytes (iPSC-CMs) were used as an in vitro model

Kareen L. K. Coulombe and Lauren D. Black III (eds.), *Cardiac Tissue Engineering: Methods and Protocols*, Methods in Molecular Biology, vol. 2485, https://doi.org/10.1007/978-1-0716-2261-2_7,

to assess the proarrhythmic risk of over 28 compounds using MEA. Human iPSC-CMs are also increasingly being used in different aspects of cardiovascular research, such as disease modeling and regenerative medicine, as well as ever more intensively in personalized medicine [5, 6]. Combining this inexhaustible and genome-specific source of human tissue with high-throughput technologies such as MEA has begun to replace animal testing as a preclinical investigation model.

Here, we demonstrate that arrhythmic events can be efficiently detected, identified, and reported through MEA recordings using an iPSC-CM model. The following experiment gives an example of how MEA can assess the arrhythmogenic risk of increasing doses of Ibutilide, a Class 3 antiarrhythmic agent classified as a hERG channel blocker. This drug is known for prolonging the action potential duration (APD) and triggering arrhythmias [7]. We used iPSC-CMs reprogrammed from a patient affected with long QT syndrome (LQT) and a healthy volunteer (Ctrl) to investigate the effect of Ibutilide on iPSC-CM electrical activity. After cardiac differentiation, iPSC-CMs were plated on MEA multi-well plates. The bottom of each well is comprised of point-electrodes arranged in a two-dimensional network that measures the fluctuation of EFPs of attached cells (*see* Fig. 1).

2 Materials

All the solutions and media were prepared in a laminar flow hood and were stored at 4 °C. The prepared media was used within 2 weeks, except for Matrigel coating which needed to be renewed prior to every experiment.

2.1 Cell Culture

1. Coating solution: Add 0.1 mg/mL of Matrigel in DMEM/F12 media for a final volume of 5 mL.
2. Maintenance media: Mix 500 mL of RPMI 1640 (Life Technologies) with 10 mL B27 + insulin supplement in a filter unit (cellulose acetate, low protein-binding) and filter (*see* **Note 1**).
3. Plating media: Prepare 50 mL of 20% KnockOut Serum (KSR) supplemented with RPMI B27 + insulin maintaining media.
4. Trypsin solution 10X without phenol red and with EDTA (*see* **Note 2**).
5. Phosphate-buffered saline (PBS) solution without calcium and magnesium (*see* **Note 3**), pH = 7.4.
6. Trypan blue 0.4% solution and hemocytometer with coverslip for cell count.

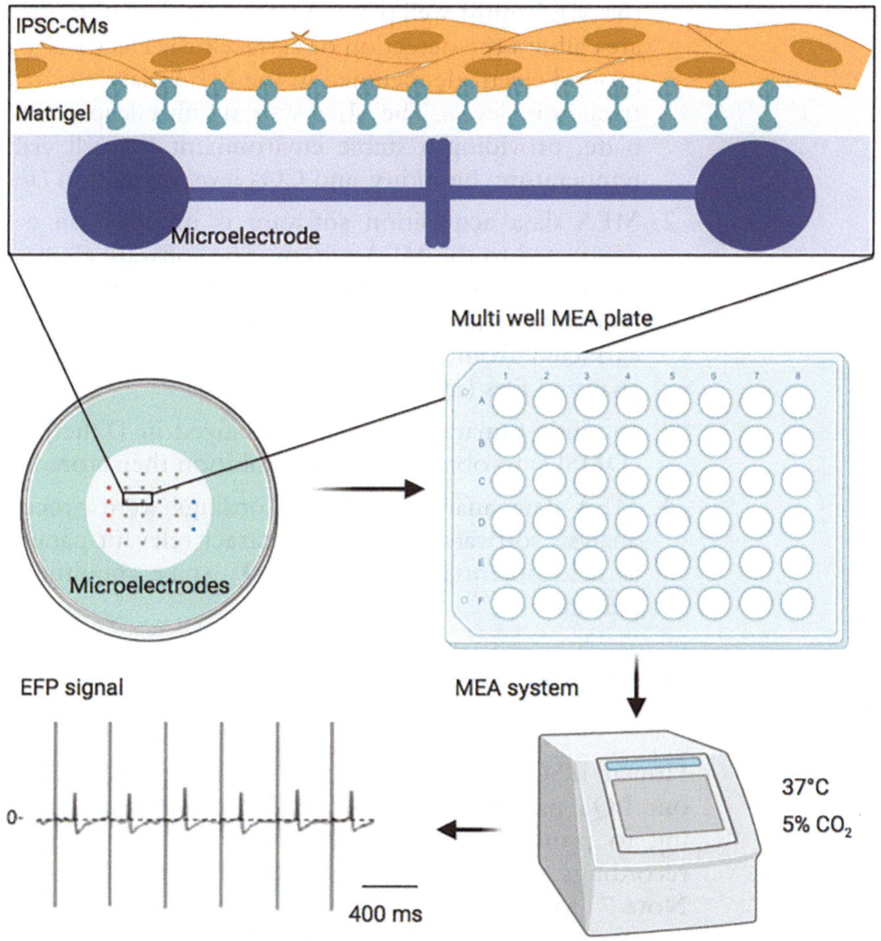

Fig. 1 Multielectrode array experiment: Top: Zoomed representation of IPSC-CMs plated across two electrodes. Middle: 48-well MEA plate with microelectrode fields at the base of each well. Bottom: MEA system with a typical EFP signal from a healthy iPSC-CM

7. Multi-well Microelectrode Array plate, 48 wells (Axion Biosystems). MEA plates contains a circuit of microelectrodes at the bottom of each well, printed with a metallic material designed to detect and conduct electrical signals from the cell monolayer to the recording area of the device. The field of 16 electrodes is located at the center of each well, with by 200 μm spacing and surrounded by a ground electrode (*see* **Note 4**). All electrodes are optimized to detect and provide pacing signals to the cells attached to the bottom of the plate.
8. Sterile water.

2.2 MEA Experiment

1. The MEA system consists of a square recording area with electrical sensors arranged asymmetrically and designed to connect with the electrodes printed at the bottom of each well of

the MEA multi-well plate. An amplifier linked to the recording area allows the detection of extracellular electrical signals. The ground electrode included in the MEA plate ensures low electrical noise levels. The MEA system fully encapsulates the MEA plate, providing a stable environment for cell viability with temperature, humidity, and CO_2 level regulation (*see* **Note 5**).

2. MEA data acquisition software is installed on a computer connected to the MEA system. The software allows control of the environmental settings, recording duration, and pacing commands. The software is also able to compile parameters extracted from the extracellular signals and generate semi-analyzed files (*see* **Note 6**).
3. Ibutilide fumarate salt was solubilized in Dimethyl Sulfoxide (DMSO) to obtain 1 M stock solution then stored at −20 °C.
4. MEA data analysis tools: Recordings were processed using analysis software in order to extract relevant parameters such as field potential duration (FPD), spike amplitude, and beat period.

3 Methods

Human iPSC lines reprogrammed from one healthy individual and one LQT patient were differentiated into cardiomyocytes according to protocol previously described by our group [8]. MEA recordings were conducted at day 30 after differentiation (*see* **Note 7**) for both lines, and cell plating on multi-well MEA plates was performed at day 20 to allow 10 days recovery from dissociation (*see* **Note 8**).

3.1 Cell Plating

1. For the coating: place a droplet of 8 μL of Matrigel covering the electrode array at the bottom of each well of the MEA plate. In order to prevent the Matrigel from evaporating, fill the edges of the plate with sterile water to ensure enough humidity, then place the plate in an incubator (37 °C, 5% CO_2) for at least 1 h (*see* **Note 9**).
2. Culture iPSC-CMs into beating-monolayers from two iPSC-CMs lines (1 healthy and 1 LQT) until day 20 of cardiac differentiation in 6-well plates. Then, add 1 mL/well of 10X Trypsin solution on the cardiomyocytes after a PBS wash, then place the plate in an incubator for 7 min (*see* **Note 10**).
3. Mechanically disperse the cells by pipetting up and down in the well then transfer them in 10 mL plating media.
4. Centrifuge the cell suspension at 500 rpm for 5 min at room temperature, discard the supernatan and resuspend the pellet in 1 mL of plating media.

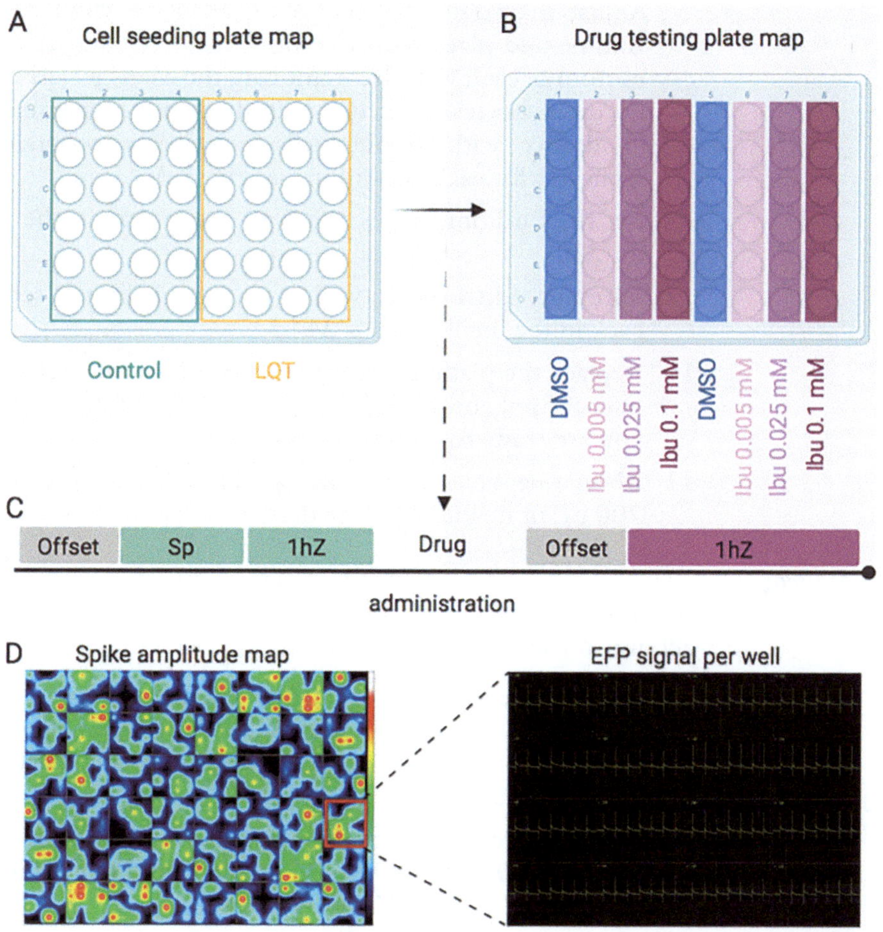

Fig. 2 Representative MEA recording: (**a**) Plate map for cell seeding for iPSC-CMs lines derived from control and LQT affected individuals. (**b**) Drug testing plate map for assessing the cardiac toxicity of increasing concentrations of Ibutilide. (**c**) Timeline of the different steps of MEA recording. (**d**) Screenshots of the spike amplitude map of 48 wells seeded with beating iPSC-CMs (left) and EFP signals recorded over 16 electrodes from 1 well of the MEA plate (right)

5. For optimal MEA recording, the cardiomyocytes need to form a syncytium on the microelectrode field. For the 48-well MEA plate, approximately 4000 cells/well are required and a total of 100,000 cells per line (*see* **Note 11**).
6. For the cell count: mix 10 μL of trypan blue solution with 10 μL of the cell suspension and load 10 μL of the mix between the hemocytometer and coverslip. Count living cells according to the following equation:

$$\text{Living cells} = \text{Number of live cells (10 squares)} \times \text{Dilution factor (2)} \times 10000$$

7. Adjust the volume of media to reach the right cell count/ volume and place a droplet of 8 μL of cell suspension at the bottom of each well after removing the Matrigel (*see* **Note 12**). Cardiomyocytes are seeded according to the plate map defined in advance, with enough wells per cell line to obtain enough replicates for each condition (*see* Fig. 2a).
8. Let the cardiomyocytes attach to the matrix in the incubator for 2 h.
9. Carefully, add 150 μL/well of plating media and set the plate back in the incubator for 24 h.
10. Replace the plating media with 500 μL of maintenance media in each well and repeat the media changing every 3 days until you are ready to record (*see* **Note 13**).
11. Make one final media change 24 h before recording with 200 μL of maintenance media.

3.2 MEA Recording and Acute Drug Testing

The MEA system used for the described experiment was the Maestro Pro system from Axion Biosystems and the corresponding data acquisition and analysis software.

1. Turn on the MEA system and set the temperature at 37 °C and the CO_2 level at 5% (the system needs to be on at least 20 min before recording). Turn on the computer and start the MEA acquisition software.
2. Prepare three different dilutions of Ibutilide in maintenance media (0.005 μM, 0.025 μM and 0.01 μM) for a final volume of 400 μL/well (*see* **Note 14**).
3. Add 5 μL of DMSO in the same volume of maintenance media as a negative control and place all the dilutions in a water bath at 37 °C (*see* **Note 15**).
4. When the environment in the MEA system is stable and the drug dilutions are at temperature, place the MEA plate on the recording area and close the lid to ensure the temperature and CO_2 level remain steady.
5. On the acquisition software, set the parameters for cardiac-specific EFP detection. This configuration includes a program to process spontaneous beating signals and electrically paced signals. Start by selecting the spontaneous configuration.
6. Allow the MEA to offset the electrical signal detected from the cardiomyocytes then record the baseline spontaneous condition for 5 min. On the acquisition screen, you can check the intensity map to gauge the baseline electrical activity of the seeded cells (*see* Fig. 2c).

7. Set the electrical pacing at a beating frequency of 1 Hz, 1 ms duration, and 20 μA current density. Record for 5 min after selecting the electrical pacing configuration.
8. Add 200 μL of drug solution following the defined plate map to obtain six replicates for each experimental condition (*see* Fig. 2b).
9. Record for 20 min under electrical pacing of 1 Hz.

3.3 MEA Analysis

The recordings generated during MEA acquisition are compiled in large files that contain EFP signals for the 16 electrodes located at the bottom of the 48 wells for the whole recording length of the MEA plate. The first step of the analysis is to cut the recording to focus on the time lapse of interest (*see* **Note 16**).

1. Launch the MEA data acquisition software and load the first file (baseline electrically paced).
2. Add the cardiac analysis configuration and select the electrically paced data setting (*see* **Note 17**).
3. The software performs basic analysis on compiled EFP signal parameters from a given time of the recording. Select the last 2 min of the recording and press re-record. The software will generate an Excel file that lists all the EFP parameters averaged per well for the most stable 30 beats. Repeat the same steps for the file generated after drug administration.
4. Launch the Cardiac Analysis Tool and load the same raw file generated during the recording and the new excel file generated in the previous step. This software gives a more detailed interpretation of the averaged EFP signal per electrode per well, defines the wavelength of the EFP and flags arrhythmic events (*see* Fig. 3a).
5. At the end of this second analysis, export the data as an Excel CSV file and extract the following values:
   - Field potential duration (*see* **Note 18**).
   - Spike amplitude.
   - Type of arrhythmic events.
   - Conduction velocity (*see* **Note 19**).
6. From the same software, extract representative signals for any presentation purpose. (*see* Fig. 3b).
7. Compile the extracted data according to the experimental groups and perform statistical analyses to assess any significant differences between each condition (*see* Fig. 3c).

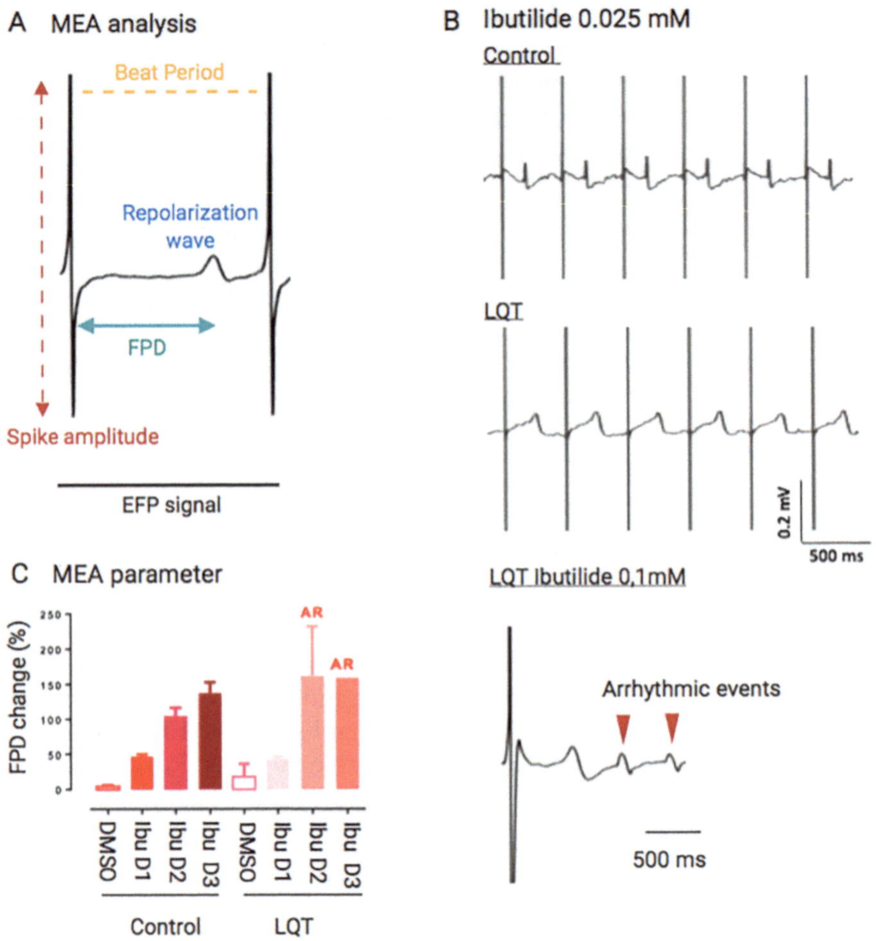

Fig. 3 Representative MEA data analysis: (**a**) Extracellular field potential (EFP) parameters typically extracted during MEA analysis. (**b**) Representative EFP signals recorded from Ctrl and LQT lines after 0.025 mM of Ibutilide administration and example of arrhythmic events observed on LQT line after application of a higher dose of Ibutilide. (**c**) Field potential duration (FPD) change after increasing doses of Ibutilide in Ctrl and the LQT lines (Mean ± SEM). *AR* arrhythmic events

4 Notes

1. Maintenance media does not include an antibiotic cocktail due the electrophysiological disturbances it can trigger when administered to iPSC-CMs, which could interfere with the MEA experiment [9].
2. Cell dissociation media should be kept at room temperature for 1 h prior to cell dissociation for fast enzyme activation when placed in the incubator.
3. Calcium and magnesium are essential for cell junction integrity, especially in cardiomyocytes, which contain many cadherins at

gap junctions and intercalated disks. Transiently depleting the media of calcium and magnesium facilitates the cell dissociation and prevents cell membrane damage [10].

4. The MEA system can read 6-, 24-, 48-, and 96-well plates. We selected the 48-well format because it is the best compromise between resolution (number of electrodes per well) and throughput. The 48-well plate can also be provided with clear bottom wells to achieve light stimulation on optogenetic cell lines and allows immunostaining after MEA experiments if needed [11].
5. Temperature and CO_2 levels can be modified and maintained for long-term recordings (mimicking hypoxia or fever). Long-term pacing can also be performed on cardiomyocytes for maturation studies [12].
6. MEA data acquisition software was optimized to record and detect EFP signals. The different configurations were designed to process signals from neurons and cardiomyocytes, spontaneous or paced. The data compiler function is efficient for signal detection and spike amplitude determination, but is not ideal for repolarization wave recognition because of biasing FPD estimation. An additional manual analysis is needed to assess the repolarization wave length and the incidence of arrhythmic events.
7. The cardiac differentiation stage at which the MEA recording is performed depends on the type of assay. We selected day 30 based on previously published studies that showed a maturation of cardiac electrical activity at day 40–45 of differentiation. This is when the SCN5A mRNA undergoes an alternate splicing which grants the final adult isoform of the main cardiac sodium channel, Nav1.5 [13]. To avoid long-term cell culture and biased results due to a mixed expression of both isoforms (fetal and adult Nav1.5), we decided to perform MEA recordings before the maturation time window.
8. Trypsin digestion was used to impair voltage gated channel function by temporarily removing segments of the channels involved in gating kinetics [14]. To ensure optimal recovery from digestion, the MEA recording was scheduled 7–10 days after enzyme-dissociation of iPSC-CMs.
9. Matrigel polymerization is reached 30 min after coating the MEA plates. Coated plates can be kept in the incubator (37 °C) for up to 7 days if water is added to the MEA plate's edges to ensure optimal humidity and avoid evaporation.
10. Depending on the iPSC-CM line, enzyme digestion can take more or less time to be achieved (5–15 min). This variability is due to differential cell junction protein expression, cell density, and the purity of the cell population.

11. The cell count is approximate and can be increased if the investigated line is fragile, which would increase the risk of cell death during the recovery phase and media changes.
12. The excess of Matrigel was removed right before adding the cell suspension droplet to prevent the Matrigel coating from drying, which would make the cell plating more difficult.
13. The electrical activity should be monitored 3 days after plating on the MEA system. This can help verify if cell attachment was complete, and if electrical activity has resumed.
14. The final dilution for each drug concentration needs to be calculated for a final volume of 400 μL/well in half the volume at the preparation. Double the concentration of the stock and avoid removing the media from the cells entirely by adding the drug preparation directly to the cells. This will help avoid mechanical stress and associated stretch-activated ion channel activation [15].
15. Drug dilutions need to be at the same temperature as the cardiomyocyte media when being recorded in the MEA system. Temperature variations are known to disturb gating properties of ion channels, especially with regard to the inactivation mechanism of Nav1.5 [16].
16. Data analysis is usually performed in time windows of 2 min of MEA recording. This interval is selected at the end of the baseline recording and within the last 5 min of the drug testing recording in order to appreciate the full effect of the drug after reaching the cells at the bottom of the wells.
17. The electrically paced configuration contains an "artifact eliminator" program that will delete the electrical signal automatically generated by the electrical pacing and detected by the electrodes.
18. FPD measurement can be considered correct if the analysis has been performed on paced cardiomyocytes, which ensures that the beating rate is steady and is at the same frequency for all experimental conditions. If the analysis is performed on spontaneously beating cells, the cardiac analysis tool offers a corrected FPD (FPDc) value that is normalized to the beat rate detected on each electrode for every well.
19. Conduction velocity is a measure of how fast an electrical impulse can be transmitted within the cardiomyocyte syncytium. Conduction velocity is calculated by pacing the syncytium with one electrode and recording on another electrode located at opposite side of the field.

Acknowledgments

This work was supported by NIH R01 HL113006, R01 HL123968, and AHA 17MERIT3610009. We also thank Axion Biosystems for documentation and allowing us to use the Maestro MEA system as an example.

References

1. Li X, Zhang R, Zhao B, Lossin C, Cao Z (2016) Cardiotoxicity screening: a review of rapid-throughput in vitro approaches. Arch Toxicol 90(8):1803–1816
2. Kussauer S, David R, Lemske H (2019) hiPSCs derived cardiac cells for drug and toxicity screening and disease modeling: what micro-electrode-array analyses can tell us. Cell 8(11): 1331
3. Takasuna K, Asakura K, Araki S et al (2017) Comprehensive in vitro cardiac safety assessment using human stem cell technology: overview of CSAHi HEART initiative. J Pharm Tox Met 83:42–54
4. Blinova K, Dang Q, Millard D et al (2018) International multisite study of human-induced pluripotent stem cell-derived cardiomyocytes for drug proarrhythmic potential assessment. Cell Rep 24(13):3582–3592
5. Wu JC, Garg P, Yoshida Y et al (2019) Towards precision medicine with human iPSCs for cardiac channelopathies. Circ Res 125(6): 653–658
6. Prondzynski M, Lemoine MD, Zech AT et al (2019) Disease modeling of a mutation in α-actinin 2 guides clinical therapy in hypertrophic cardiomyopathy. EMBO Mol Med 11(12):e11115
7. Kannankeril PJ, Norris K, Carrter S, Roden DM (2012) Factors affecting the degree of QT prolongation with drug challenge in a large cohort of normal volunteers. Heart Rhyt 8(10):1530–1534
8. Burridge PW, Matsa E, Shukla P et al (2014) Chemically defined and small molecule-based generation of human cardiomyocytes. Nat Methods 11(8):855–860
9. Giudiscessi JR, Ackerman MJ, Camilleri M (2018) Cardiovascular safety of prokinetic agents: a focus on drug-induced arrhythmias. Neur Motil 30(6):e13302
10. Citi S (2019) The mechanobiology of tight junctions. Biophys Rev 11(5):783–793
11. Clements IP, Millard DC, Nicolini AM et al (2016) Optogenetic stimulation of multiwell MEA plates for neural and cardiac applications. Clin Trans Neur:96902C
12. Kijala K, Ahola A, Pekkanen-Mattila M et al (2012) Electrical field stimulation with a novel platform: effect on cardiomyocyte gene expression but not on orientation. Int J Biomed Sci 8(2):109–120
13. Veerman CC, Mengarelli I, Lodder EM et al (2017) Switch from fetal to adult SCN5A isoform in human induced pluripotent stem cell-derived cardiomyocytes unmasks the cellular phenotype of a conduction disease-causing mutation. J Am Hear Asso 6(7):pii: e005135
14. Solaro CR, Lingle CJ (1992) Trypsin-sensitive, rapid inactivation of a calcium-activated potassium channel. Science 257(5077):1694–1698
15. Peyronnet R, Nerbonne JM, Kohl P (2016) Cardiac mechano-gated ion channels and arrhythmias. Circ Res 118(2):311–329
16. Mangold KE, Brumback BD, Angsutararux P et al (2017) Mechanisms and models of cardiac sodium channel inactivation. Channels 11(6): 517–533

Chapter 8

iPSC-Derived Micro-Heart Muscle for Medium-Throughput Pharmacology and Pharmacogenomic Studies

Daniel W. Simmons and Nathaniel Huebsch

Abstract

Micro-heart muscle arrays enable medium-throughput experiments to model the cardiac response to a variety of environmental and pharmaceutical effects. Here, we describe stem cell culture maintenance, methods for successful cardiac differentiation, and formation of micro-heart muscle arrays for electrophysiology and molecular biology assays.

Key words Induced pluripotent stem cells (iPSC), Cardiac differentiation, Micro-heart muscle (μHM), Electrophysiology, Pharmacology

1 Introduction

Induced pluripotent stem cells (iPSC) have enabled new fields of research since their creation [1]. One exciting area of investigation is the differentiation of iPSC into tissue-specific cells such as cardiomyocytes to study fundamental biology and for drug development. iPSC can be differentiated into cardiomyocytes through several different means, including small-molecule manipulation of canonical Wnt signaling [2, 3]. However, iPSC-derived cardiomyocytes (iPS-CM) have an embryonic/fetal phenotype. For example, iPSC-CM exhibit spontaneous beating, differences in ion channel currents, slowed conduction velocity, and small cell size compared to their adult counterparts [4]. Although monolayer cardiomyocytes have been used to study cardiac development and pharmacology response, 3D tissues have been shown to more closely mimic adult cardiac tissue [5]. Tissues provide necessary cues such as cell alignment and increased mechanical forces that have been shown to mature the cells and have been created using a wide variety of methods [6–13]. However, widespread use of these systems has been limited due to complicated, user-dependent manufacturing techniques, and low-throughput device fabrication and tissue

Kareen L. K. Coulombe and Lauren D. Black III (eds.), *Cardiac Tissue Engineering: Methods and Protocols*, Methods in Molecular Biology, vol. 2485, https://doi.org/10.1007/978-1-0716-2261-2_8,

formation, inhibiting their use for more extensive molecular analyses and pharmacology studies.

Here, we describe simple and scalable iPSC-derived micro-heart muscle arrays (μHM) [6]. μHM are formed by seeding iPSC-cardiomyocytes and stromal cells into a dog bone-shaped stencil atop standard tissue culture substrates. Although at present we use poly(dimethylsiloxane) (PDMS) to make these stencils, in theory it is possible to use a wide variety of materials. Stencil shape defines the boundary conditions for tissue formation and cellular alignment. Within μHM, iPSC-CM exhibit pharmacology that is more similar to that of adult heart muscle than do iPSC-CM monolayers [6]. In their current iteration, we make μHM using 3D-printed molds, which can be ordered at reasonable cost from commercial vendors, obviating the need for cleanroom access or for users to learn soft-lithography. Furthermore, to enable studies of the molecular pathways underlying μHM physiology, we describe methods for dissociating μHM for molecular assays such as Western Blot and qPCR. Because efficient generation of iPSC-cardiomyocytes is one of the most important steps in making μHM, we also include detailed protocols for pluripotent stem cell culture and differentiation.

2 Materials

2.1 Stem Cell Culture

1. Induced Pluripotent Stem Cell Media: Essential 8 (E8) Media kit (Thermo Scientific).
2. Matrigel: Corning Growth-Factor Reduced Matrigel Matrix.
3. Knock-Out (KO) DMEM.
4. Accutase.
5. Sterile Dulbecco's PBS (dPBS).
6. Peprotech Y-267632 ROCK Inhibitor: reconstitute to 10 mM in DMSO, sterile filter and aliquot.
7. Trypan Blue: Filter 0.4% trypan blue through Whatman paper to remove aggregates and store at room temperature (*see* **Note 1**).
8. Freezing Media: 90% Fetal Bovine Serum, 10% sterile filtered DMSO, 1:1000 Y27632 aliquot (final concentration: 10 μM).

2.2 Cardiac Differentiation and Maintenance

1. B12/Penicillin: Weigh 20 mg of Vitamin B12 into a 15 mL conical tube covered in aluminum foil. Separately, weigh 32.154 mg of 1555 unit/mg penicillin-G into a 15 mL conical tube. Move both into a biosafety cabinet and add 10 mL of MilliQ water to each. Mix each until dissolved; you may need to leave on a shaker overnight at 4 °C. Sterile filter both contents into a single 50 mL conical tube. Aliquot into 1 mL solutions and store at −20 °C.

2. CHIR99021: reconstitute to 6 mM in DMSO, sterile filter and aliquot.
3. IWP-2: reconstitute to 2.5 mM in DMSO, sterile filter and aliquot. While other small molecules are aliquoted at 1000× the final concentration, the solubility limit of IWP-2 in DMSO is 5 mM, which makes the 2.5 mM stock easier to prepare for routine use.
4. RPMI-C: 1:50 B-27 Supplement, 1:1000 of Y27632 aliquot (final concentration: 10 μM), 150 μg/mL L-Ascorbic Acid, and 1:500 of B12/penicillin aliquot in RPMI 1640 Media (*see* **Note 2**).
5. RPMI-I: 1:50 B-27 Supplement without insulin, 1:1000 Y-267632 aliquot (final concentration: 10 μM), 150 μg/mL L-Ascorbic Acid, and 1:500 B12/penicillin aliquot in RPMI 1640 Media (*see* **Note 2**).
6. Day 0 Differentiation Media: Dilute CHIR 99021 aliquot 1: 1000 in RPMI-I media (final concentration: 6 μM).
7. Day 2 cardio media: Dilute IWP-2 aliquot 1:500 in RPMI-I media (final concentration: 5 μM).
8. EB20 media: 20% qualified FBS, 1:100 GlutaMAX, and 1:100 MEM Non-Essential Amino Acids in Knock-Out DMEM.
9. Lactate media: Dissolve 2.2412 g of Sodium L-lactate in 18 mL of 1 M HEPES. Mix 1:250 of this aliquot, 1:100 GlutaMAX, and 1:100 MEM NEAA into RPMI 1640.
10. 0.25% Trypsin-EDTA
11. Cardiomyocyte Thawing Media: 20% Fetal Bovine Serum in RPMI-C.

2.3 Stencil Manufacturing and Seeding

1. Glass plates: available from standard hardware stores, cut to approximately 3 × 3 inch at ¼″ thickness.
2. 3D print (available from protolabs.com: we use MicroFine Green).
3. Plastic folders or laser printer paper.
4. Slygard-184 polydimethylsiloxane.
5. Trichloro(1H,1H,2H,2H-perfluorooctyl)silane.
6. Medium-sized office clips.
7. 70% ethanol—made by combining sterile water with sterile 190 proof ethanol (*see* **Note 3**).
8. Kolliphor P-188: Mix 10% w/v Kolliphor P-188 (Sigma-Aldrich) in PBS and boil to dissolve. Let cool to room temperature and sterile filter (0.2 mm mesh). Store at 4 °C.

2.4 Western Blot

1. RIPA buffer: 10 mM Tris–HCL (pH 8.0), 140 mM NaCl, 1 mM EDTA, 0.5 mM EGTA, 1% Triton X-100, 0.1% Sodium Deoxycholate, 0.1% SDS, 1:100 dilution of protease inhibitor.

2.5 qPCR

1. RNAqueous Total RNA Isolation Kit (ThermoFisher).

3 Methods

3.1 Matrigel Aliquoting

1. The day before aliquoting, move pipette tips and conical tubes to a −20 °C freezer. Move the Matrigel from the −20 °C freezer to a 4 °C fridge, storing it on an ice bucket. Keep all items in the freezer/fridge overnight.
2. The day of aliquoting move the pipette tips, conical tubes and Matrigel into the biosafety cabinet. Keep all items on ice.
3. Aliquot Matrigel using the chilled conical tubes and pipette tips. Store at −20 °C (*see* **Note 4**).
4. 2 to 3 days before desired Matrigel use, add KO DMEM to aliquoted Matrigel for a 1:100 dilution. Move to 4 °C fridge to allow Matrigel to dissolve into the media, now called Matrigel Media. Do not manually mix (*see* **Note 5**).

3.2 iPSC Thawing

1. Pre-coat 1-2 wells of a multi-well (typically 6-well) plate with Matrigel Media, for at least 2 h but preferably overnight (*see* **Note 6**).
2. Pre-warm E8 media to room temperature before use. Add 5 mL of media to a falcon tube (*see* **Note** 7).
3. Remove iPSC cryovial from liquid nitrogen storage. We typically freeze one million cells per vial, approximately 1-2 wells of nearly confluent wells of iPSC (Fig. 1). Warm in the palm of your gloved hand. When the last ice crystals melt, immediately transfer the contents of the cryovial to the falcon tube.
4. Pellet cells (3–5 min at 100 × *g*). While centrifuging cells, prepare pluripotent stem cell media with Y27632. Use

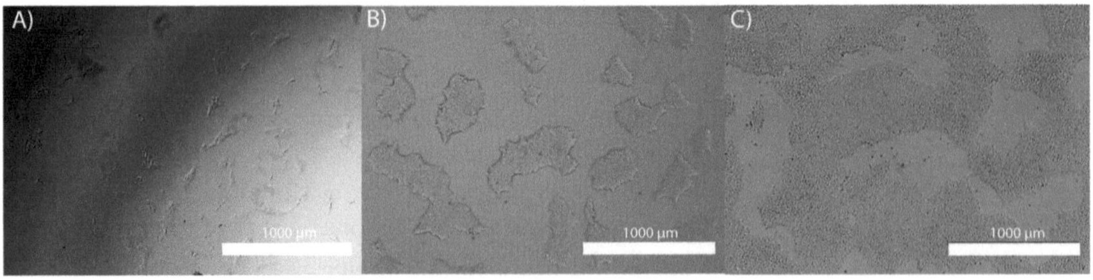

Fig. 1 Representative images of induced pluripotent stem cells. (**a**) Stem cell density 1 day after passaging (**b**) Stem cell density 2 days after passaging (**c**) High density stem cells ready for passaging. Scale bars: 1000 m

Y27632 at a final concentration of 5–10 μM, a 1:1000–1:2000 dilution of the stock. Prepare 2 mL of media per well of 6-well plate.

5. Aspirate the supernatant off the cell pellet. Add pluripotent media with Y27632 to the pellet without disturbing the pellet.
6. Wait 5+ min. This will solvate the pellet to allow you to singularize the cells without excessive shear.
7. Resuspend cells. You should have a single cell suspension. Seed these into 2 wells of a 6-well plate.
8. Wait at least 24 h before changing media to pluripotent media without Y27632.

3.3 iPSC Passaging and Maintenance

1. At least 1 h prior to passaging, and up to 1 day before, coat the desired number of wells with the matrigel media and allow to incubate at 37 °C. Use 1.25 mL per well of a 6-well plate.
2. Aspirate E8 media.
3. Rinse wells with dPBS.
4. Add accutase to the cells (*see* **Note 8**).
5. Incubate cells at 37 °C for 3 min.
6. Remove the well plate and dislodge cells by lightly hitting the side of the plate with your palm, rotating the plate so each side is hit several times.
7. Incubate the cells at 37 °C for 3 min.
8. Remove the well plate and triturate using a 1000 μL pipette several times.
9. Repeat **steps 6** and **7** one to two more times, so that the cells have been in accutase for 9–12 min. Ensure the cells are singularized.
10. Dilute the accutase with dPBS. Use 5× the amount of accutase used.
11. Centrifuge the cells at 100 × *g* for 4–5 min.
12. Aspirate the supernatant, and add E8 + 10 μM Y27632 (*see* **Note 9**).
13. Allow the cells to solvate for 8–10 min.
14. Pipette several times to singularize the cells.
15. Count the cells (*see* **Note 10**).
16. Aspirate the matrigel media from the new wells. Add new E8 media.
17. Plate the cells into the new wells. Depending on cell doubling rates and desired number of days between passages, plate between 150 and 400 × 10^3 cells per well of a 6-well plate.

18. Place the well plate back into the 37 °C incubator. Lightly move the plate North-South-East-West to ensure the cells are spread over the entirety of the wells (*see* **Note 11**).

3.4 iPSC Freezing

1. Follow **steps 2–15** from the iPSC passaging protocol above.
2. Re-spin the cells at 100 × *g* for 5–8 min.
3. Aspirate the supernatant and add freezing media, so that the cell concentration is between one and five million cells per mL.
4. Pipette several times to singularize the cells.
5. Add ½ to 1 mL of the cell suspension to each cryovial.
6. Store the cryovials in a CoolCell, Mr. Frosty, or similar device, and move to a −80 °C freezer.
7. After 48 h, move the cryovials to liquid N2 for long-term storage.

3.5 Cardiomyocyte Differentiation

1. Follow **steps 1–16** from the iPSC passaging protocol above.
2. Plate the stem cells onto a Matrigel-coated plate for differentiation. Use the following four cell densities: 50 k/cm^2, 37.5 k/cm^2, 25 k/cm^2, and 12.5 k/cm^2. Specify day of seeding as Day-3. Wait at least 24 h before the next media change (*see* **Note 12**).
3. Feed the stem cells daily with E8 for days −2 and −1.
4. Day 0, feed the cells with 6 μM CHIR in RPMI-I media. (Day 0 Differentiation Media; **Note 13**).
5. Day 1: no media change.
6. Day 2: 48 h after feeding on day 0, change the media to be 5 μM IWP2 in RPMI-I media (Day 2 cardio media).
7. Day 3: no media change.
8. Day 4: change media to RPMI-I.
9. Day 5: no media change (*see* **Note 14**).
10. Day 6: change media to RPMI-C.
11. Subsequently, replace media ever 3–4 days with new RPMI-C. We typically observe beating cells starting on days 7–9. We culture cardiomyocytes until at least day 13 before dissociation.

3.6 Cardiomyocyte Dissociation and Freezing

1. Aspirate the RPMI-C media.
2. Wash cells twice for 15 min with dPBS to stop spontaneous beating.
3. Add room temperature 0.25% trypsin to the cells.
4. Incubate the cells at 37 °C for 3 min.
5. Remove the well plate and triturate using a 1000 μL pipette several times.

6. Repeat **steps 4** and **5** twice so the cells have been in trypsin for 9 min. Ensure the cells are singularized.
7. Dilute the singularized cells with EB20. Use 2× the amount of trypsin used.
8. Centrifuge the cells at 150 × *g* for 10 min.
9. Aspirate the supernatant and add EB20 media with 10 μM Y27632 (*see* **Note 15**).
10. Allows the cells to solvate for 5–10 min.
11. Pipette several times to singularize the cells.
12. Count. Human iPS-CMs are typically 15–18 μm in diameter and are 80–95% viable after passaging. If there are clumps of cells that persist at this point, they should be removed by filtering the singularized cells through a 40–70 mm mesh cell strainer.
13. Re-spin the cells at 150 × *g* for 10 min.
14. Aspirate the supernatant and add freezing media, so that the cell concentration is between one and five million cells per mL.
15. Pipette several times to singularize the cells.
16. Add ½ mL of the cell suspension to each cryovial.
17. Store the cryovials in a CoolCell, Mr. Frosty, or similar device, and move to a −80 °C freezer.
18. After 48 h, move the cryovials to the vapor phase of liquid nitrogen dewar for long-term storage.

3.7 Cardiomyocyte Thawing

1. Pre-coat 2-4 wells of a multi-well (typically 6-well) plate with matrigel media, for at least 2 h but preferably overnight (*see* **Note 6**).
2. Pre-warm cardiomyocyte thawing media to room temperature before use. Prepare 1 mL of thawing media per well of 12-well plate that you will thaw into. Add 1uL of 10 mM Y27632 aliquot per mL of cardiomyocyte thawing media (final concentration: 10 μM).
3. Separately, prepare 5 mL of EB20 media.
4. Remove iPS-CM cryovial from liquid nitrogen storage. Typically, we freeze one–five million cells per vial, approximately 2–3 wells of iPS-CM from a 24-well plate. Warm in the palm of your gloved hand. When the last ice crystals melt, immediately transfer the contents of the cryovial to the bottom of an empty falcon tube.
5. Add the 5 mL of EB20 dropwise to the thawed cell pellet.
6. Pellet cells for 10 min at 100 × *g*.

7. Aspirate the supernatant off the cell pellet. Add Y27632-enriched cardiomyocyte thawing media (final concentration 10 μM Y27632) without disturbing the pellet.
8. Wait 5–10 min. This will solvate the pellet to allow you to singularize the cells without excessive shear.
9. Resuspend cells. You should have a single cell suspension. Seed these into 4–8 wells of a 12-well plate. Target cell density is 200,000–500,000 cells/cm^2.
10. Wait at least 24 h before changing media to RPMI-C without Y27632 (*see* **Note 16**).

3.8 Cardiomyocyte Lactate Purification

1. If using fresh cardiomyocytes, follow **steps 1–12** from the cardiomyocyte dissociation protocol above. However, use 10 μM Ri in RPMI-C media instead of cardiomyocyte thawing media for **step 9**. If using frozen cardiomyocytes, follow the cardiomyocyte thawing protocol above.
2. Plate the cells into new Matrigel-coated wells at a 1:2 split ratio (e.g., plate the cells from one well of a 24-well plate into two new wells of a 24-well plate).
3. The next day (day 1), switch the media to RPMI-C.
4. Days 2–4, no media change.
5. Day 5: aspirate media. Wash cells with dPBS. Add lactate media.
6. Day 6: no media change.
7. Repeat **steps 5** and **6** so that the cells have undergone 2–3 lactate media treatments.
8. After sufficient lactate treatment, aspirate the lactate media. Wash cells with dPBS. Add RPMI-C media.

3.9 3D-Printed Stencil Mold Design

Using Solidworks, we created our stencil mold by first extruding a rectangular base several millimeters thick to prevent warping during the polymer crosslinking process and eventual print removal from the printer. On top of this base, an individual dog bone shape is designed and extruded to a height of 500–1000 μm. Using linear patterning, we can easily create groups of dog bones, typically clustered in groups of 3 with several millimeters of spacing between groups (Fig. 2). This allows us to easily cut and place groups into wells of a 24-well plate. The larger individual dog bone shapes have a 1 mm × 1 mm knob/square on each end, with a 200 μm wide shaft of variable length. The smaller sized dog bones are half this size. Importantly, dog bones should be spaced at least 200 μm apart from one another to prevent a syncytium of tissue from "bridging" between adjacent micro-muscles.

Fig. 2 Representative image of CAD designed stencil mold for 3D printing

3.10 PDMS Stencil Manufacturing

1. Expose 3D print to Trichloro(1H,1H,2H,2H-perfluorooctyl) silane in a fume hood (silanes are very volatile and toxic), place the 3D print in a vacuum chamber, along with a large glass cover slip. Pipette 200 μL of the silane onto the cover slip, and close the chamber. Turn on the vacuum for 10–15 s, then close the vacuum ports and turn off the vacuum. Ensure the chamber remains closed and sealed. Incubate for 48 h to allow the silane to coat the 3D print via vapor deposition.
2. Open the vacuum chamber in a fume hood, and remove the 3D print.
3. Place the 3D printed part on top of a flat glass plate covered with a cutout from a plastic folder (*see* **Note 17**).
4. Prepare another glass plate with a plastic folder cutout on top. Pour the necessary amount of PDMS to form stencils along the bottom of the folder cutout.
5. Place the PDMS covered folder/glass combination above the 3D print, so that only the bottom of the glass with the PDMS is touching the print. Slowly lower the glass at an angle so the PDMS gradually flows along the print (*see* **Note 18**).
6. After the glass has been lowered and is completely flat, use office clips to clamp the glass plates together (Fig. 3). The gray squeezers on the office clips can be removed to keep the combination flat during PDMS crosslinking.
7. Place the combination in a 60 °C for at least 12 h to cure the PDMS.
8. Remove the clamps and peel the plastic folder cutout from the top of the print. The PDMS will stick to the cutout and also peel cleanly off the print as a sheet (Fig. 4).
9. Individual stencils can be cut from the stencil sheet as needed for tissue formation.

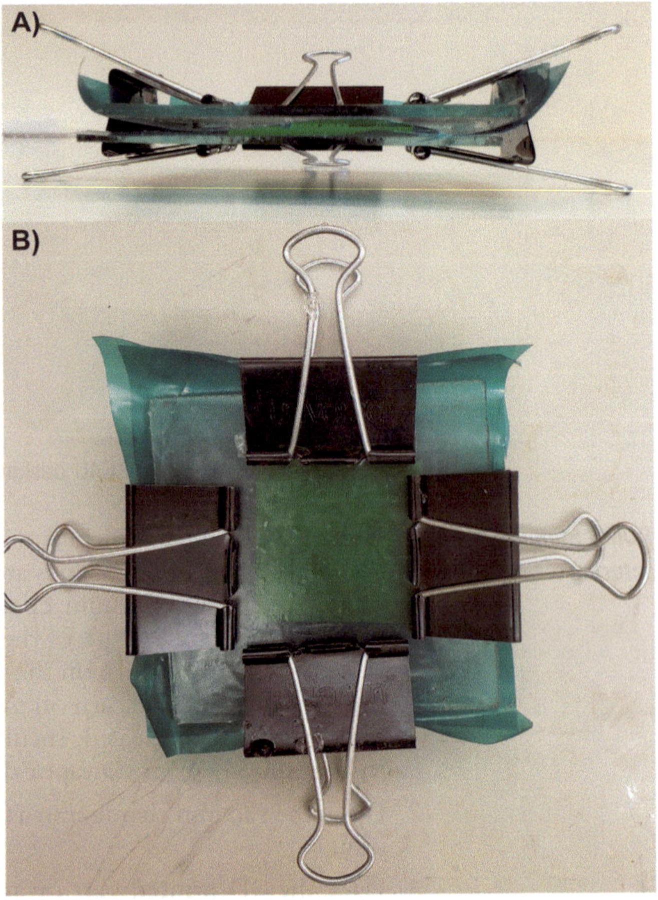

Fig. 3 PDMS stencil manufacturing shown in (**a**) side view and (**b**) top view, showing the two plastic-covered glass plates sandwiching the 3D print and PDMS, clamped together using office clips

3.11 Stencil Seeding

1. Using a razor blade, scalpel, or scissors, cut stencils from the stencil sheet so that they are an appropriate size to fit in the well plate. Stencils should be small enough to be centered in the well, without any part of the stencil touching the well's edges.
2. The stencils may have a smooth side that was touching the plastic folder during curing and a rough side that was touching the 3D print. Place all stencils rough side face up, and use a scalpel to make a small notch on a corner to signify orientation.
3. Place all stencils in methanol and allow to soak for 5 min.
4. Place the stencils in the center of each well, ensuring the notch is in the proper orientation so that the flat side of the pdms is touching the well surface.

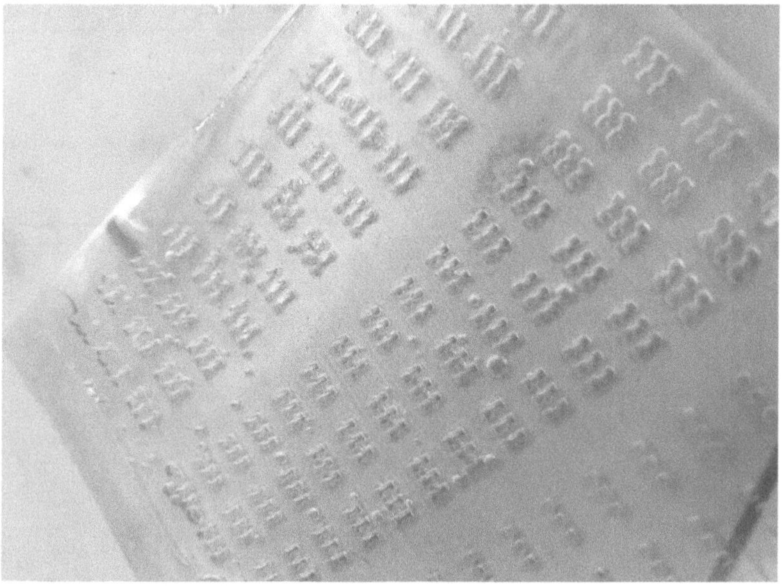

Fig. 4 Representative image of PDMS stencils after molding off 3D print

5. Move the well plate into a 60 °C oven and allow the methanol to evaporate overnight. This will create a water-tight, non-permanent bond between the stencil and the well [14].
6. Plasma treats the well plate, following the manufacturer's protocol for the system used. This will help sterilize the stencils.
7. To further ensure sterilization, overfill the well plate, including the lid and dead space between wells, with 70% ethanol (Fig. 5). After soaking for at least 3 h, aspirate any remaining ethanol and allow the residual to evaporate. From this point forward, the devices are sterile and must be handled under sterile conditions (*see* **Note 3**).

Continue here for large-sized stencils:

8. Fill the wells with dPBS, and spin at 2500 × *g* for 25 min.
9. Aspirate the dPBS. Fill each individual dog bone with 3 μL of 20 μg/mL of fibronectin in PBS.
10. Allow the fibronectin solution to incubate at room temperature overnight.
11. Place the fibronectin-coated plate into a 37 °C incubator to warm up during the dissociation.
12. Dissociate the cardiomyocytes following **steps 1–13** from the protocol above. Resuspend the cardiomyocytes into EB20 media at a concentration of 75 × 10^6 cells/mL. It is critical that you achieve a well-singularized cell suspension. If

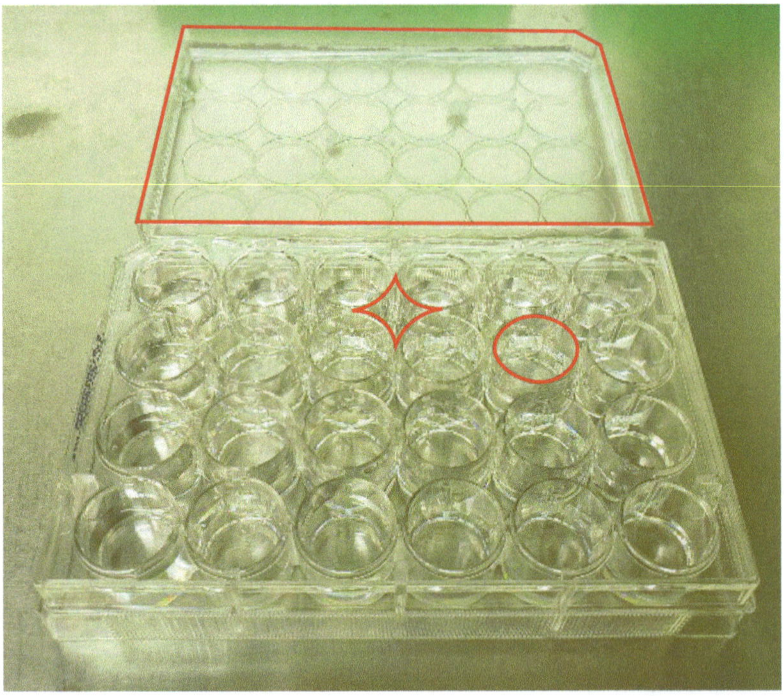

Fig. 5 Well plate used for μHM formation, with areas filled with ethanol for sterilization outlined in Red

necessary, remove clumped cells with a 45 or 70 μm mesh cell strainer (*see* **Note 19**).

13. Working one well at a time, aspirate the fibronectin solution. Then use a 2 or 10 μL pipette to inject 1.5 μL of the cell suspension into both knobs of each μHM.
14. Place the seeded well plate into the incubator and wait for 2 h before adding EB20 with 10 μM Y27632 media to the wells. Take care to avoid pipetting directly onto the cell suspension. This helps limit cell loss during the initial feeding.
15. The next day, check the μHM for spontaneous beating. If beating occurs, change the media to RPMI-C. If there is no beating, wait another day before changing media to RPMI-C.
16. Change media every 3–4 days for continued tissue maintenance (Fig. 6) (*see* **Note 20**).

Continue here for small-sized stencils:

1. Prepare fresh 3% Kolliphor P-188 solution by dilution the 10% stock with sterile dPBS. Incubate the stencils in 3% Kolliphor P-188 for 2 to 4 h at room temperature. Do not centrifuge. This will selectively deposit the Kolliphor, which blocks cell adhesion, to the sides of the PDMS but not the substrate beneath [15].

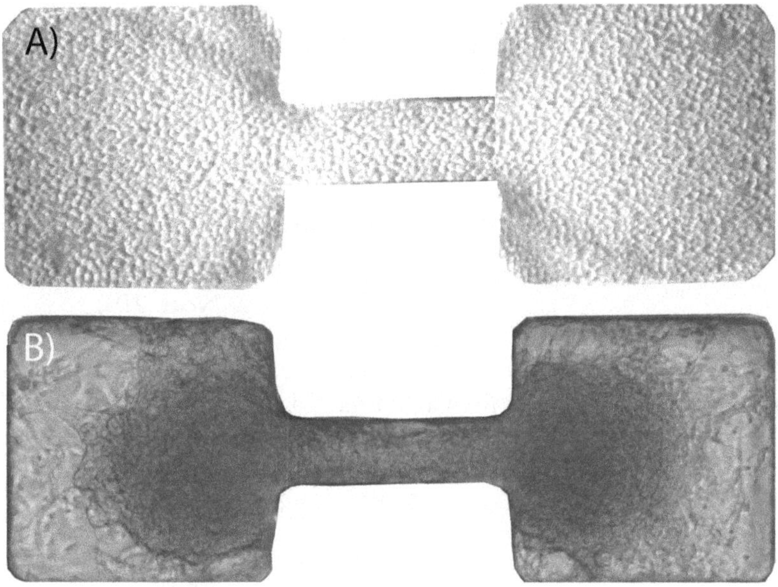

Fig. 6 Representative brightfield images of (**a**) PDMS dog bone stencil immediately after seeding with iPS-CM suspension and (**b**) several days after seeding, the cells have self-compacted into a μHM

2. Aspirate Kolliphor P-188 and replace with sterile PBS. Incubate for 5–10 min.
3. Repeat **step 9** twice, for a total of three PBS washes.
4. Fill the wells with fresh sterile PBS, and spin at 2500 × *g* for 20 min.
5. Aspirate the PBS, and fill with 10–20 μg/mL of fibronectin in PBS. Spin at 2500 × *g* for 25 min.
6. Allow the fibronectin solution to incubate at room temperature overnight.
7. Place the fibronectin-coated plate into a 37 °C incubator to warm up during the dissociation.
8. Dissociate the cardiomyocytes following **steps 1–13** from the protocol above. Resuspend the cardiomyocytes into EB20 media with 10 μM Y27632 at a concentration of 75×10^6 Cells/mL. It is critical that you achieve a well-singularized cell suspension. If necessary, remove clumped cells with a 45 or 70 mm mesh cell strainer (*see* **Note 19**).
9. Working one well at a time: aspirate the fibronectin solution. Then use a 10 μL pipette to place a 10 μL cell suspension bubble above the three connected dog bone wells.
10. Using the tip of the pipette, gently "paint" the cell suspension along the stencil surface so the bubble covers all three dog bone wells. Avoid touching the stencil itself.

11. Spin the plate at 200 × *g* for 4 min.
12. Place the seeded well plate into the incubator and wait for 1 to 2 h before adding media to the wells. This helps limit cell loss during the initial feeding.
13. The next day, check the μHM for spontaneous beating. If beating occurs, change the media to RPMI-C. If there is no beating, wait another day before changing media to RPMI-C.
14. Change media every 3–4 days for continued tissue maintenance (Fig. 6) (*see* **Note 20**).

3.12 Pharmacology Treatments

Cardiomyocytes are very sensitive to shear, and full washout of media can have a more dramatic effect on cardiomyocyte physiology than the drugs under study. Thus, we make concentrated drug stocks and then add these into cardiomyocyte culture media without doing full media changes.

1. Dissolve the desired pharmaceutical at 100–1000 times the desired final concentration in the appropriate solvent.
2. Dilute the concentrated drug stock into RPMI-C media down to a concentration of 10× the desired final concentration. Pre-incubate this drug-containing media in the same incubator where cells are cultured, to equilibrate to CO_2 and temperature (*see* **Note 21**).
3. Remove 10% of the media from the μHM-containing well, and replace with an equal volume of the 10× drug stock in RPMI-C media. For example, in a standard 24-well plate we feed with 500 μL of media per well. For this well plate, you would remove 50 μL of media and replace with 50 μL of the 10× drug stock (*see* **Note 21**).
4. Incubate the plates at 37 °C for 10–15 min to allow for the pharmaceutical effect to take place.
5. Perform the physiology measurements.
6. Repeat **steps 1–5** if you are using increasing concentrations for your study.
7. Fix tissues if a terminal experiment. Otherwise drug can be washed out with serial partial media changes.

3.13 Imaging-Based Physiology Measurements

We use a Nikon® Eclipse Tsr2-inverted microscope equipped with a Hamamatsu® ORCA Flash 4.0 V3 digital CMOS camera and a Lumencor® AURA light engine®. Together these are controlled using Nikon's NIS-Elements software. The μHM are imaged for action potentials by incubating overnight with 500 nM BeRST-1 dye [15] and calcium dynamics using GCaMP-6f [16]. To properly capture the dynamics of these two waveforms, imaging at a frame rate of at least 250 Hz is preferred. If conduction velocity values are to be determined, imaging at 500–1000 Hz is recommended.

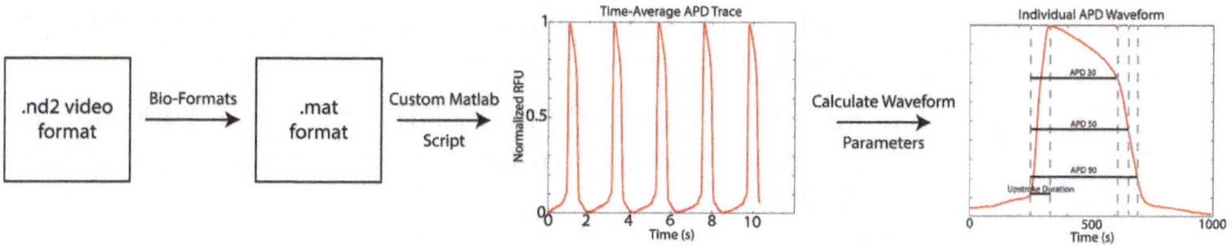

Fig. 7 Schematic of waveform parameter analysis

Using the Hamamatsu® ORCA Flash 4.0 and 4× objective, the entirety of an individual μHM can be seen when imaging at 500 Hz using line-scanning mode on the camera. If imaging only the shaft region of the μHM, imaging frame rates can reach 1000 Hz using this setup.

3.14 Image Analysis

We use the Matlab Bio-Formats [17] package to convert our videos from Nikon's .nd2 file format to MATLAB's .mat format for analysis. With custom MATLAB software, we obtain average fluorescence intensity tracing for both action potentials and calcium dynamics. Similar time-intensity tracings can be obtained with the "Plot Z Axis Profile" function of ImageJ. From these tracings we can calculate a variety of waveform parameters including amplitude, upstroke duration, spontaneous beat rate, APD90, and calcium decay τ75 (Fig. 7). The original Matlab version of this APD and calcium analysis software is available at https://github.com/sboggess04/ips_cardio_analysis. To calculate conduction velocity, we use a modified version of the open-source MATLAB software Rhythm 1.0 and is available at https://github.com/optocardiography/Rhythm-1.0 [18]. To track contractile motion, we use an open-source motion tracking software [16]. This is available for download off the Huebsch Lab website https://huebschlab.wustl.edu/resources/.

3.15 Lysing Tissues for qPCR/Western Blot

1. Pre-chill lysis buffers on ice for at least 20 min. Prepare an ice bucket. Do not process more than three different conditions at once. This will give sufficient time for rapid lysis and flash freezing.
2. Aspirate media and rinse twice with dPBS.
3. Using a 200 or 1000 μL pipette, transfer tissues to the same empty well, separating based on experimental design and whether the tissues are to be lysed for qPCR or western blot.
4. For RNA: after all tissues have been collected, add 200 mL of RNA lysis buffer. We use buffer from the RNAqueous Micro Kit. Triturate at least ten times and then transfer to a labeled microcentrifuge tube, on ice. After processing remainder of RNA samples, flash freeze by carefully moving the tube in liquid nitrogen for approximately 30 s. Transfer tubes to

−80 °C or vapor phase of liquid nitrogen for storage until further RNA processing.

5. For protein: add 200 μL of RIPA buffer +1:1000 dilution of a protease inhibitor cocktail. Process samples as for RNA and flash freeze.
6. Transfer contents to a 1.5 mL Eppendorf tube and flash freeze in liquid nitrogen.
7. Store in −80 °C.

3.16 Tissue Fixation and OCT Embedding

1. Aspirate media.
2. Wash cells in dPBS 2 × 15 min. Tissue contractions should be completely absent.
3. Fix cells by incubating in progressively increasing concentrations of PFA in PBS [12]: first, incubate in 1% PFA in dPBS at 4 °C overnight, then 2% PFA in dPBS for 1 h, 3% PFA for 1 h, and finally at 4% PFA in PBS for 1 h. All incubations should be done at 4 °C.
4. Wash cell in dPBS 3 × 30 min. The contents of the first two PBS washes are likely to include excess PFA washed out of the tissue and thus should be disposed of in the same hazardous waste stream as the PFA. The fixed tissues can be stored up to 1 week in cold PBS at this point.
5. Aspirate dPBS.
6. Use tweezers to carefully remove PDMS stencil, ensuring that μHM remains intact and attached to substrate.
7. Prepare 1% agar solution using MilliQ water. When agar beaker is cool to touch, pour enough into each well to cover all μHM.
8. Let agar cool on bench for 10 min.
9. Move agar to 4 °C fridge and let cool for 15+ min.
10. Use a spatula to remove agar-embedded μHM from well.
11. Move the tissue blocks into individual new wells of a well plate.
12. Add several mL of 15% saccharose (sucrose) into each well. The blocks should initially float.
13. Leave the plate on a shaker at 4 °C for at least 1 h or until the blocks begin to sink.
14. Aspirate the 15% sucrose and replace with 30% sucrose. Leave on shaker at 4 °C overnight.
15. Lay out plastic sectioning molds, one for each agar block. Pour a thin layer of OCT into each mold, enough to cover the bottom. Avoid bubbles.
16. Use a spatula to transfer one agar block into the center of the plastic mold (*see* **Note 22**).

17. Add OCT to the mold until the agar block is completely covered and the mold is full.
18. Flash freeze the mold in liquid nitrogen until completely white, but do not fully submerge. Keep the mold level so the OCT does not spill or bubble.
19. Store the frozen embedded tissues in −80 °C indefinitely.

3.17 Tissue Sectioning

We use standard cryosectioning techniques to section our tissue for staining. μHM do not have the same coloration as native tissues extracted from a mammal, which makes it difficult to see them in the OCT block once sectioning is started. To prevent slicing through the tissue, it is best to orient the block so that the tissue is closer to the chuck. With this orientation, you will first begin to slice through the agar where there is no tissue. To determine when you have begun slicing into the tissue, it is best to section onto a slide and use a microscope to visually determine when you have reached the tissue and want to begin saving sections. While you may be able to see the larger μHM sections on the slide, the smaller sized tissues are difficult to see without magnification.

3.18 Slide Staining

We use standard tissue staining protocol, typically staining the primary antibody overnight at 4 °C. When staining for highly abundant proteins such as sarcomeric α-actinin (Fig. 8), we use blocking buffers containing 5% normal goat serum and 5% bovine

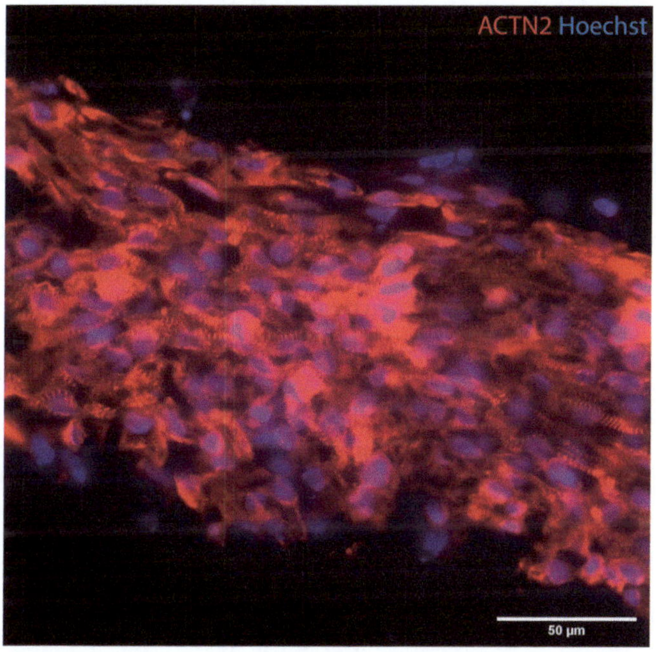

Fig. 8 Representative image of sectioned μHM stained for sarcomeric α-actinin (ACTN2; red) and nuclei (Hoechst; blue). Scale bar: 50 m

serum albumin, or higher, to help limit secondary antibody binding to enhance image quality. For these highly abundant proteins, we also use the blocking buffer as a replacement for the washing buffer for all steps before nuclear staining. Additionally, μHM may not stick to the glass slide as well as native tissues. To prevent tissue detachment, it is best to handle with care and limit any fluid shear when adding new buffers during staining.

4 Notes

1. Filtering the trypan blue removes aggregates that will appear as dead cells and negatively alter automated cell counting.
2. It is best to add the ROCK inhibitor, L-Ascorbic acid, and B12/Penicillin fresh the day of usage to prevent their degradation.
3. In our experience, using 70% ethanol comprised of non-sterile water led to a non-trivial rate of contamination. To prevent this, we use ethanol that has only been opened in a biosafety cabinet, and autoclaved water. Additionally, we overflow the well plate with 70% ethanol to fill in the areas between wells, and also soak the inside surface of the well plate lid.
4. If using 15 mL tubes, aliquot 130 μL. If using 50 mL tubes, aliquot 450 μL.
5. When adding the KO DMEM, pouring the media onto the frozen Matrigel and letting it dissolve at 4 °C over several days provides the most consistent results. If using 15 mL tubes add to 13 mL total. If using 50 mL tubes, add to 45 mL total.
6. Matrigel-coated plates can be kept at 37 °C for up to 48 h. If Matrigel-coated plates need to be stored longer term, aspirate off the Matrigel media, replace with sterile PBS and parafilm wrap the plate. This can be stored at 4 °C for up to 1 week.
7. We warm media to room temperature in the biosafety cabinet instead of a water bath. This helps limit contamination sources. Additionally, if media is left in a water bath for an extended period of time it may degrade growth factors in the media.
8. Allow accutase to warm in the biosafety cabinet for several minutes before use. This will help minimize time necessary to detach cells from the well. Use 1 mL per well of a 6-well plate, or scale appropriately.
9. We add 1 mL of E8 + 10 μM Ri for every well of 6-well plate that is being passaged.
10. We use a Countess II Automated Cell Counter. Important output parameters to record from counting include average cell size, cell viability percentage, and live cell concentration.

iPSCs are typically 17–20 μm in diameter and are 70–95% viable after passaging.

11. As described by Kim et al. Biomaterials Research (2015) 19:6, it is important to routinely test stem cell media and cultures for mycoplasma. Mycoplasma can significantly hinder/prevent successful differentiations and should be considered if differentiation quality begins to suffer. Media should also be routinely tested for sterility. This can be done by incubating media in sterile wells of a tissue culture plate and checking for growth at 37 °C.

12. To optimize differentiation for new cell lines or individuals new to cardiac differentiation, it is typically best to start with these four densities. For routine differentiation, you will likely choose the 1–2 densities that provide the best and most consistent results.

13. CHIR concentrations and protocols can vary between labs. Some may use a much higher concentration for just 24 h. With our experience and cell lines we found 6 μM for 48 h to give the most consistent differentiations.

14. Typically, between days 2 and 6 you will see large levels of cell detachment, with the cell monolayer appearing white and wispy. To prevent large cell loss during this stage, it is best to minimize shear due to media changes by leaving a small amount of residual media behind and to add new media along the wall of the well so it flows onto the cells instead of being dropped directly on top. After the day 6 media change to RPMI-C, the cell monolayer should fully reattach to the well bottom and beating begins between days 8 and 12.

15. We typically add 1 mL of EB20 with 10 μM Ri for every 6 wells dissociated from a 24-well plate.

16. If the cardiomyocytes do not look healthy, have poor attachment or minimal beating, wait until 48 h after plating to change media.

17. Un-crosslinked PDMS is very sticky and takes several days to crosslink at room temperature, making it very difficult to remove of unwanted objects. To protect the balance when measuring PDMS, covering the balance with cling wrap will protect the balance from damage. For making PDMS stencils, using a piece of plastic folder or overhead transparency prevents the PDMS from sticking to glass and makes it easier to remove the PDMS stencil from the 3D print due to preferential adhesion.

18. At this point, you need to keep steady pressure on the stencil. If you absolutely must reposition the glass plates, you will need to pour over fresh PDMS. Failure to maintain pressure at this point will result in large holes throughout the stencil.

19. Aliquot the cell suspension into 1.5 mL Eppendorf tubes before spinning down the final time. This will make the resuspension easier. Use a 1000 or 200 μL pipette to manually aspirate the supernatant after centrifuging instead of using a vacuum to prevent cell loss. When resuspending, estimate the volume of the cell pellet, then subtract this from your resuspension volume to determine how much media to add. Your calculated resuspension volume based on the desired concentration should include the volume of the cells + media, not just the media itself.
20. To prevent potential cross-contamination between wells, use a 200 μL pipette tip over the glass aspirating pipette to remove media. Use a different 200 μL pipette tip for each well.
21. Many pharmaceuticals are not directly soluble in water, but must be dissolved in solvents such as DMSO. This method limits the amount of DMSO exposed to the tissues, which is cytotoxic. Additionally, this method minimized the amount of media changed during a pharmacology study. Large media changes that require briefly exposing the μHM to air can have transient but strong effects on physiology that make it difficult to discern the effects of drugs. Pre-warming the drug-containing media is also essential, to avoid artifacts caused by transient temperature changes in media used to bathe the μHM.
22. When placing the agar-embedded tissues into the mold for OCT embedding, be sure to note the orientation. It is best to place all tissue in at the same orientation, so that all have the tissues closer to the same side of the OCT block.

References

1. Takahashi K, Yamanaka S (2006) Induction of pluripotent stem cells from mouse embryonic and adult fibroblast cultures by defined factors. Cell 126(4):663–676. https://doi.org/10.1016/j.cell.2006.07.024
2. Lian X et al (2013) Directed cardiomyocyte differentiation from human pluripotent stem cells by modulating Wnt/β-catenin signaling under fully defined conditions. Nat Protoc 8(1):162–175. https://doi.org/10.1038/nprot.2012.150
3. Kattman SJ et al (2011) Stage-specific optimization of activin/nodal and BMP signaling promotes cardiac differentiation of mouse and human pluripotent stem cell lines. Cell Stem Cell 8(2):228–240. https://doi.org/10.1016/j.stem.2010.12.008
4. Musunuru K et al (2018) Induced pluripotent stem cells for cardiovascular disease modeling and precision medicine: a scientific statement from the american heart association. Circ: Gen Prec Med 11(1):e000043. https://doi.org/10.1161/HCG.0000000000000043
5. Feric N, Radisic M (2016) Towards adult-like human engineered cardiac tissue: maturing human pluripotent stem cell-derived cardiomyocytes in human engineered cardiac tissues. Adv Drug Deliv Rev 96:110–134. https://doi.org/10.1016/j.addr.2015.04.019
6. Huebsch N et al (2016) Miniaturized iPS-cell-derived cardiac muscles for physiologically relevant drug response analyses. Sci Rep 6:24726. https://doi.org/10.1038/srep24726
7. Zhao Y et al (2019) A platform for generation of chamber specific cardiac tissues and disease modeling. Cell 176(4):913–927.e18. https://doi.org/10.1016/j.cell.2018.11.042

8. Boudou T et al (2012) A microfabricated platform to measure and manipulate the mechanics of engineered cardiac microtissues. Tissue Eng Part A 18(9-10):910–919. https://doi.org/10.1089/ten.tea.2011.0341
9. Thavandiran N et al (2019) Functional arrays of human pluripotent stem cell-derived cardiac microtissues. bioRxiv. https://doi.org/10.1101/566059
10. Mathur A et al (2015) Human iPSC-based cardiac microphysiological system for drug screening applications. Sci Rep 5:8883. https://doi.org/10.1038/srep08883
11. Goldfracht I et al (2020) Generating ring-shaped engineered heart tissues from ventricular and atrial human pluripotent stem cell-derived cardiomyocytes. Nat Commun 11(1):1–5. https://doi.org/10.1038/s41467-019-13868-x
12. Ronaldson-Bouchard K et al (2018) Advanced maturation of human cardiac tissue grown from pluripotent stem cells. Nature 556(7700):239–243. https://doi.org/10.1038/s41586-018-0016-3
13. Shadrin IY et al (2017) Cardiopatch platform enables maturation and scale-up of human pluripotent stem cell-derived engineered heart tissues. Nat Commun 8(1):1–5. https://doi.org/10.1038/s41467-017-01946-x
14. Osutni E et al (2000) Patterning mammalian cells using elastomeric membranes. Langmuir 16(20):7811–7819. https://doi.org/10.1021/la000382m
15. Huang YL, Walker AS, Miller EW (2015) A photostable silicon rhodamine platform for optical voltage sensing. J Am Chem Soc 137(33):10767–10776. https://doi.org/10.1021/jacs.5b06644
16. Huebsch N, Loskill P et al (2015) Automated video-based analysis of contractility and calcium flux in human-induced pluripotent stem cell-derived cardiomyocytes cultured over different spatial scales. Tissue Eng Part C 21(5):467–479. https://doi.org/10.1089/ten.TEC.2014.0283
17. Linkert M et al (2010) Metadata matters: access to image data in the real world. J Cell Biol 189(5):777–782. https://doi.org/10.1083/jcb.2010041
18. Laughner JI et al (2012) Processing and analysis of cardiac optical mapping data obtained with potentiometric dyes. Am J Phys-Heart Circ Phys 303(7):H753–H765. https://doi.org/10.1152/ajpheart.00404.2012

Chapter 9

Quantifying Propagation Velocity from Engineered Cardiac Tissues with High-Speed Fluorescence Microscopy and Automated Analysis Software

Andrew P. Petersen and Megan L. McCain

Abstract

Many acquired or inherited forms of heart disease as well as drugs are known to increase the susceptibility of patients to arrhythmias. To predict arrhythmogenic events and discover new therapeutic strategies to mitigate them, approaches to efficiently quantify the velocity of propagation in engineered cardiac tissues are important research tools. In this chapter, we describe how to collect videos of propagating calcium waves in engineered cardiac tissues with a high-speed camera mounted on an inverted fluorescence microscope. We also provide instructions for downloading and using our software package to analyze these videos and calculate propagation velocity. These techniques should be compatible with a variety of voltage- or calcium-sensitive fluorescent dyes or genetically encoded sensors. Although these approaches were originally developed for engineered neonatal rat cardiac tissues, the same procedures can likely be used with human-induced pluripotent stem cell-derived cardiac myocytes, paving the way for patient-specific analysis of propagation due to features such as tissue architecture, substrate rigidity, genetic mutations, or drug treatments.

Key words Calcium, Voltage, Arrhythmias, Cardiomyocytes, Micropatterning, Electrode

1 Introduction

Native myocardium is characterized by elongated cardiac myocytes that are aligned into fiber-like structures and tightly coupled to each other mechanically and electrically. Because electrical resistance in the cytoplasm is lower than across cell–cell junctions, the architecture of the myocardium biases action potentials to propagate most rapidly in the direction parallel to the long axis of the myocytes [1]. This rapid propagation of action potentials is essential for synchronizing the contractions of cardiac myocytes and ensuring that the ventricle contracts as a single unit. Importantly, many different types of perturbations, including fibrosis [2], genetic mutations [3, 4], and drugs [5], are known to alter aspects

Kareen L. K. Coulombe and Lauren D. Black III (eds.), *Cardiac Tissue Engineering: Methods and Protocols*, Methods in Molecular Biology, vol. 2485, https://doi.org/10.1007/978-1-0716-2261-2_9,

of propagation, which increases the risk for arrhythmias or other rhythm abnormalities that can progress to heart failure or death.

To predict arrhythmogenic events and discover new therapeutic strategies to counteract them, in vitro models that mimic native myocardial architecture coupled with approaches to efficiently quantify velocity of propagation are important research tools. However, in vitro models of the myocardium have conventionally been generated by plating cardiac myocytes on featureless culture surfaces, such as Petri dishes or multi-well plates coated uniformly with extracellular matrix (ECM) protein(s). On these surfaces, cardiac myocytes adopt stellate shapes and form a tissue with randomly organized myofibrils and cell–cell junctions. As a result, propagation spreads uniformly in a circular pattern [6–8], unlike the elliptical shape characteristic of native myocardium [9]. To overcome this, surface micropatterning techniques, such as photolithography [10], microcontact printing [7], and microscale [6, 11] or nanoscale [12] topography, have been implemented to align cardiac myocyte tissues in vitro. As predicted, propagation is faster in the longitudinal direction and slower in the transverse direction in aligned cardiac tissues in vitro [6–8], generating an elliptical pattern similar to native myocardium [9]. The shape of this ellipse is characterized by dividing the longitudinal velocity by the transverse velocity, a value referred to as the anisotropy ratio.

To measure propagation in engineered cardiac tissues, researchers rely primarily on microelectrode arrays (MEAs) or optical imaging techniques. One disadvantage of MEAs is that the cells must be electrically continuous with the electrodes on the surface of the MEA. Thus, MEAs can only be coated with a thin layer of ECM protein [13] or a hydrogel [14, 15] prior to cell seeding and are not compatible with the use of non-conductive biomaterials as culture substrates. MEAs are also relatively expensive and cannot be fabricated or customized by a non-specialized user. Optical imaging techniques are a more flexible option because they are compatible with any culture substrate that is optically transparent. For optical imaging of propagation, tissues are first incubated with a calcium- or voltage-sensitive dye or the cells are engineered to express genetically encoded calcium or voltage sensors [16, 17]. Videos of tissue fluorescence are then captured, usually as the tissue is electrically stimulated in a localized region to trigger a propagating wave. Original optical imaging systems required arrays of photodiodes [7, 9, 18] or optical fibers [6, 19] to achieve the sensitivity and time resolution needed to capture a propagating action potential. However, due to advances in camera technologies and microscopy systems, it is now possible to collect datasets with sufficient spatial and temporal resolution on relatively standard microscopy systems designed for live imaging, such as an inverted fluorescence microscope equipped with a high-speed camera [8, 17, 20] or a line-scanning confocal microscope [21]. However, an important step in this process is efficiently converting datasets to a quantitative value of propagation velocity, which requires software tools tailored to datasets collected on microscopes with a relatively high number of pixels.

In our previous publication, we described how to collect videos of propagating calcium waves in engineered neonatal rat cardiac tissues with a high-speed camera mounted on an inverted fluorescence microscope [8]. We also developed new software tools in MATLAB to automate the calculation of propagation velocity by compressing the three-dimensional dataset into two dimensions and applying linear regression to efficiently account for the activation times of all pixels in the field of view or a selected subset. In this chapter, we provide instructions for both recording propagating calcium waves in engineered cardiac tissues and implementing our software to calculate propagation velocity from these recordings. In our experiments, we used the intracellular calcium dye Fluo-4 because its signal-to-noise ratio is relatively high and calcium waves are a suitable proxy for action potentials [22, 23]. However, these techniques should be compatible with other types of fluorescent dyes or genetically encoded sensors as well as other cardiac cell types, such as human-induced pluripotent stem cell-derived cardiac myocytes. Thus, the procedures described here are valuable tools for efficiently quantifying propagation velocity in models of healthy and diseased engineered myocardium.

2 Materials

2.1 Live Imaging of Calcium Wave Propagation in Engineered Cardiac Tissues with Fluo-4

1. Engineered cardiac tissue construct: Cardiac myocytes adhered to any surface compatible with imaging on a fluorescence microscope (*see* **Note 1**).

1. Fluo-4 solution: Dissolve a lyophilized vial of 50 μg Fluo-4 AM (Thermo Fisher Scientific) in 100 μL of 20% Pluronic F-127 in DMSO (Thermo Fisher Scientific) and lightly vortex and centrifuge using benchtop equipment. Distribute the solution into 5 μL aliquots in 1.7–2 mL centrifuge tubes so that 1.5 mL culture media can be added directly to the aliquot on the day of experiments. Store at −20 °C, protected from light. Immediately before use, add 1.5 mL culture media to a Fluo-4 aliquot and lightly vortex for a final concentration of 1.7 μg/mL Fluo-4.

2. Tyrode's solution: 135 mM Sodium Chloride, 5.4 mM Potassium Chloride, 5.0 mM Glucose, 5.0 mM HEPES, 1.8 mM Calcium Chloride, 1.0 mM Magnesium Chloride, and 0.33 mM Sodium Phosphate in ultrapure water (*see* **Note 2**). Adjust pH to 7.4 at 37 °C.

3. Inverted fluorescence microscope with a stage-mounted micromanipulator and compatible objectives and camera: Standard cameras on fluorescence microscopes are usually too slow to capture propagating waves. The frame rate of the camera and the size of the field of view (i.e., power of the objective) must both be taken into consideration (*see* **Note 3**). It is recommended to perform these experiments in a live imaging chamber set to 37 °C. These experiments are also compatible with other high-speed fluorescence imaging systems, such as line-scanning confocal microscopes.
4. Stimulator: Any stimulation system used for electrophysiology is adequate. The IonOptix MyoPacer Cell Stimulator is recommended (*see* **Note 4**).
5. Stimulation electrode: Electrodes can be purchased or fabricated in-house (*see* **Note 5**). Mount the electrode securely into the micromanipulator on the stage of the microscope. Insert the two free ends of the electrode wires into the stimulator.

2.2 Calculating Propagation Velocity

1. MATLAB 2016b or newer: The Image Processing and the Statistics and Machine Learning toolboxes are needed. If the raw data files are not already readable by MATLAB, the open-source bioformats toolbox is needed to input a variety of microscope and video formats into MATLAB (*see* **Note 6**).
2. Analysis software: Download from: https://github.com/LLSExUSC/FluoProp/

   Extract the files and move them to your working MATLAB directory (*see* **Notes 7 and 8**). This chapter applies to the code published on GitHub in May 2020, version 1.0. Check the GitHub for instructions on any newer version of the code.
3. Computer: If running many large (>500 MB) videos, it is recommended to use a machine with a solid-state hard drive and adequate RAM (>8 GB).

3 Methods

3.1 Live Imaging of Calcium Wave Propagation in Engineered Cardiac Tissues with Fluo-4

1. Retrieve one engineered cardiac tissue construct from the incubator, replace the cell culture media with the Fluo-4 solution, and incubate for 30–60 min at 37 °C with 5% CO_2 (*see* **Note 9**).
2. Rinse the tissue construct three times with Tyrode's solution warmed to 37 °C and transfer it to the stage of the fluorescence microscope warmed to 37 °C, if it has a temperature-controlled chamber. If imaging a patterned cardiac tissue, rotate the construct such that the tissue is aligned parallel or perpendicular to the field of view.

3. Use the micromanipulator to slowly lower the point of the electrode into the Tyrode's solution until the tip of the electrode is a few millimeters above the tissue (*see* **Note 10**).
4. Turn on the stimulator using appropriate settings (neonatal rat ventricular myocytes can typically be stimulated with a 10 V waveform at 0.5–2 Hz). Look at the fluorescence intensity of the tissue to determine if the electrode is pacing the tissue. If not, slowly lower the electrode and/or increase the stimulation voltage until the electrode is pacing the tissue.
5. Turn off the stimulator. Increase the voltage by approximately 10% to ensure successful capture for the duration of the experiment.
6. Move the electrode laterally such that it is directly above, below, or to the side of the light path. For aligned tissues, the position of the electrode relative to the alignment of the tissue and the field of view determines if you will capture longitudinal or transverse propagation.
7. Turn on the stimulator and record fluorescence intensity above the minimum necessary frame rate. The duration of the video should be long enough to capture at least two full activation cycles (calculated as 2/pacing frequency). Changes in fluorescence intensity should be visible, similar to the images in Fig. 1a (*see* **Note 11**).
8. Turn off the stimulator after acquisition is complete.
9. Repeat **steps 6–8** while re-positioning the electrode and field of view as needed. To measure anisotropy ratio, capture videos of longitudinal propagation and transverse propagation. It is recommended to capture ≥3 field of view per tissue to ensure proper sampling.

3.2 Calculating Propagation Velocity

1. In MATLAB, open the main.m script from the downloaded Fluo_Prop folder. The script is organized into multiple sections indicated by "%%" and the title of that section. When instructed, run each section by pressing ctrl-enter while the cursor is active in that section.
2. In the "Input Parameters" section, update the following information:

   inputPath—path for the video file to be analyzed,

   parentSavePath—in this path, a subfolder matching the name of the raw data file will be generated and used to store all generated figures (*.png and *.fig) and data (*.mat),

   micronToPixel—size of each pixel in microns,

   framesPerSecond—frame rate of the video in frames per second,

   pacingFreq—pacing frequency in Hz,

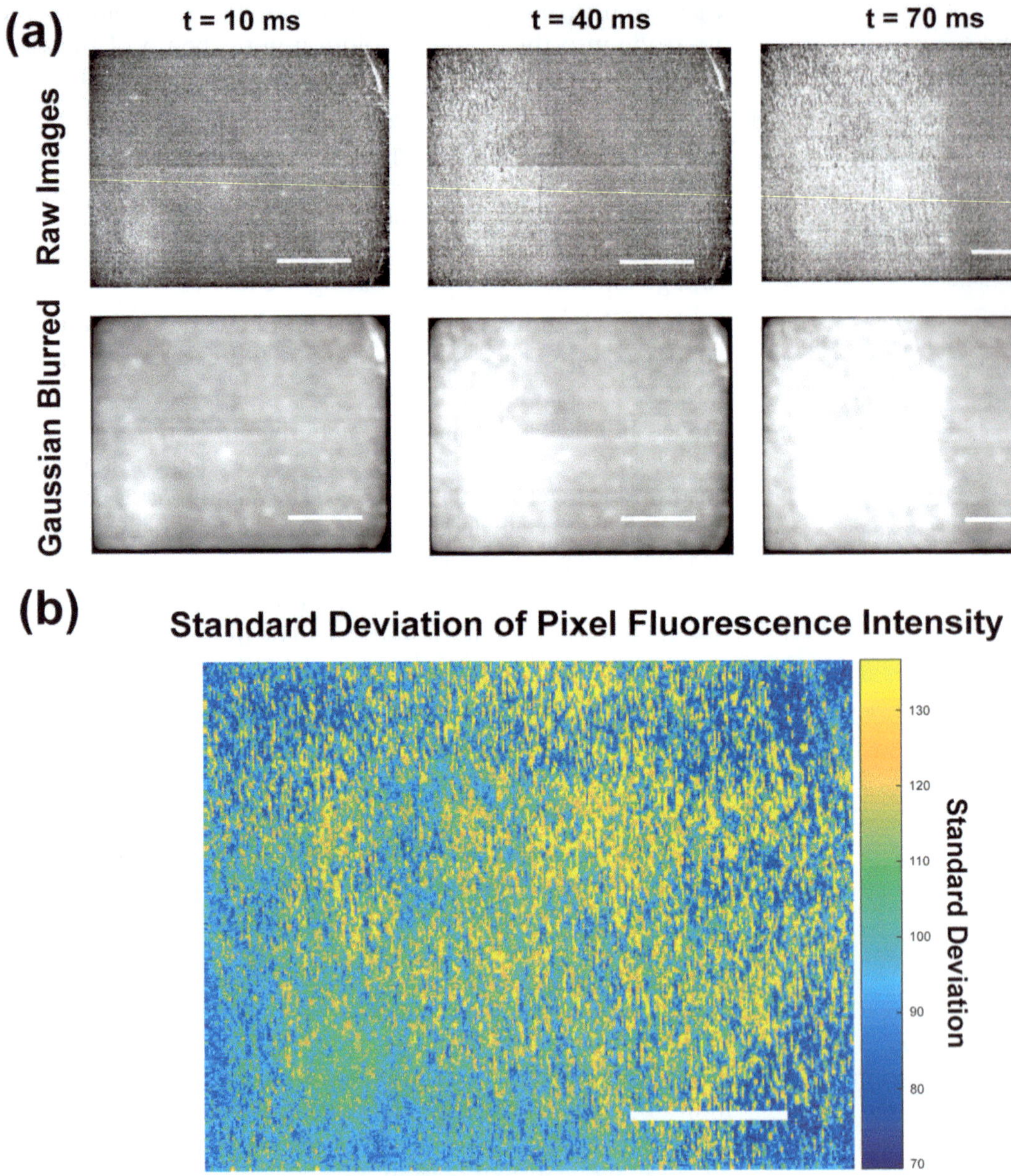

Fig. 1 (**a**) Unprocessed and Gaussian-blurred video frames of a calcium wave propagating across an engineered neonatal rat cardiac tissue, detected using Fluo-4. The tissue is aligned vertically and the electrode is to the left of the field of view. Thus, this video captured transverse propagation. Scale bars, 2 mm. (**b**) Plot of the standard deviation of fluorescence intensity for each pixel. Certain features of the tissue, such as alignment, are visible. This tissue has no major holes, which would be apparent by a very low standard deviation in a relatively large region. Scale bar, 2 mm

rejectThreshFirstBeat—fraction of first activation cycle that must be present to be included in data analysis (recommended: 0.9),

rejectThreshLastBeat—fraction of last activation cycle that must be present to be included in data analysis (recommended: 0.8).

3. Run “Input Parameters.”
4. Run “Load Video File” to input your video file into MATLAB as a three-dimensional matrix of fluorescence intensity, where the indexes indicate [pixel *x*-coordinate, pixel *y*-coordinate, frame] (*see* **Note 12**).
5. Run “Assess Video” for quick quality control of your data. The standard deviation of the signal from each pixel over the course of the video is calculated, displayed, and saved. A video with good signal will generate an image similar to Fig. 1b, where basic features of the tissue (such as alignment) can be detected. If the tissue has defects (large holes, etc.), it should be evident from this image and you may consider excluding it from your analysis.
6. Optional: Run “Rotate Video to Align” if you need to rotate the video. The image generated in the previous step will be displayed. Draw a rectangle and rotate it until it aligns with the tissue.
7. Optional: Run “Choose ROIs for Analysis” to define the regions of interest (ROIs) of the video to be analyzed for propagation velocity. Draw the ROIs manually on the image generated in the previous step by dragging and rotating rectangle(s) into place. All ROIs are saved. If you skip this section, the entire field of view will be analyzed.
8. Run “Separate Activation Cycles” to separate your dataset into individual activation cycles, as shown in Fig. 2a (*see* **Note 13**).
9. Run “Process Each Activation Cycle” to display an activation map, which should clearly show the pattern of propagation (*see* **Note 14**).
10. On the activation map, use the cursor to select a region (as large as possible) that demonstrates relatively unidirectional propagation with minimal curvature, as shown in Fig. 2b (*see* **Note 15**). The same region will be applied to every activation cycle.
11. For each activation cycle, an activation map of the entire field of view and the selected region are displayed and saved. A two-dimensional histogram showing the *x*-coordinate of each pixel plotted as a function of its activation time is also displayed (Fig. 2c). Robust linear regression is performed on these datapoints to calculate a line of best fit, the slope of which is the propagation velocity. A *.mat file is saved that reports the ROI

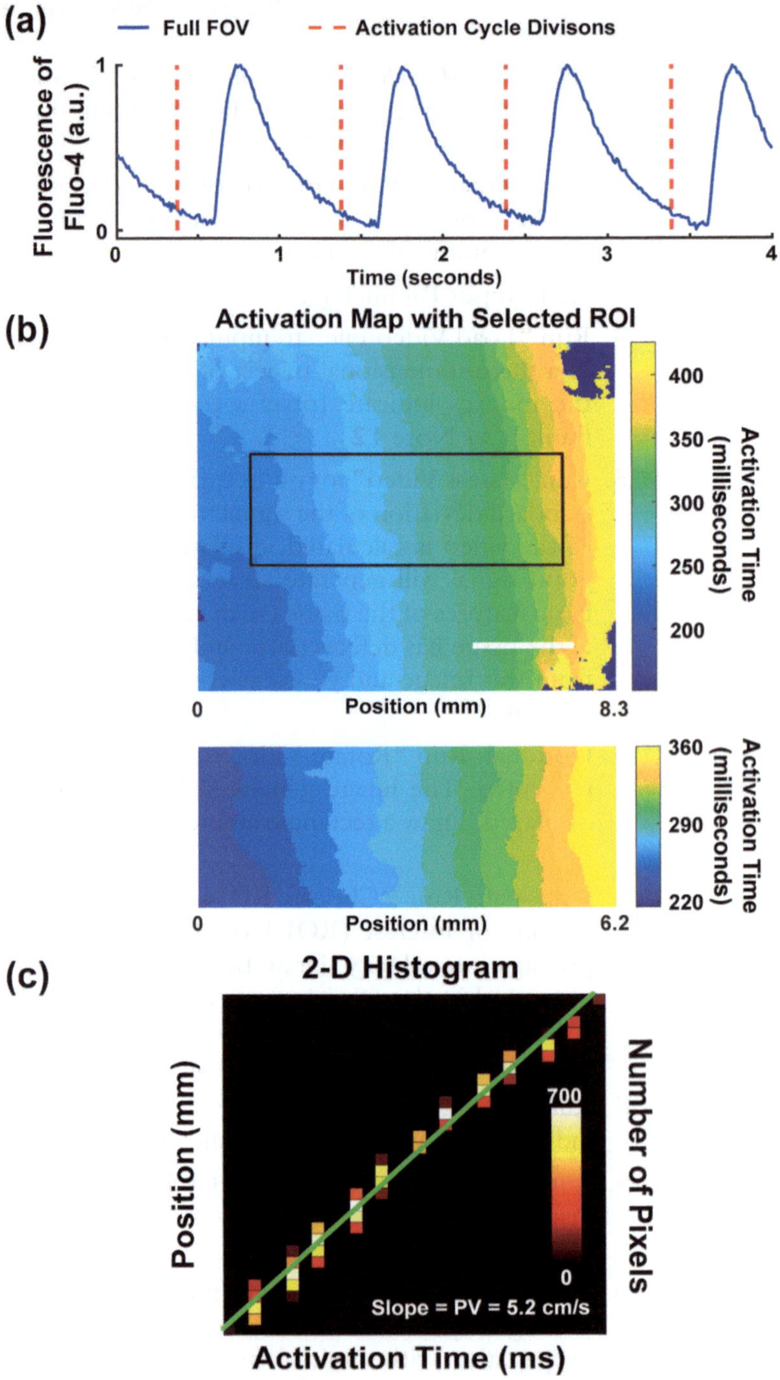

Fig. 2 (**a**) The mean fluorescence intensity for each frame of the video plotted over time (blue) overlaid with the automatically detected divisions between activation cycles (red). (**b**) Activation map of a full field of view. Note that the propagating waveform has the shape of a vertically oriented ellipse because the tissue was aligned vertically. The selected ROI is a region of the ellipse that is relatively linear and thus is appropriate to use for calculating propagation velocity. The ROI is enlarged in the lower panel with an adjusted timescale. Scale bar, 2 mm. (**c**) Two-dimensional histogram indicating the position and activation time for each pixel in the ROI. The green line indicates the robust linear regression on all datapoints, the slope of which is the propagation velocity (PV)

number, activation cycle number, propagation velocity, and robust sigma. Robust sigma is analogous to the root mean squared error of the regression line.

12. Any datasets with a weak fitting line should be excluded from further analysis because these tissues likely have major defects or other anomalies, as shown by the examples in Fig. 3. This can be done easily by accepting a dataset only if the robust sigma is above a certain threshold (such as 9.5). The cut-off value for robust sigma should be selected after evaluating a few datasets, but usually can be kept constant for multiple videos acquired from similar types of tissues with similar experimental parameters.
13. To calculate anisotropy ratio, divide the average of all longitudinal propagation velocity measurements from a single tissue by the average of all transverse propagation velocity measurements.

4 Notes

1. This technique was developed to measure calcium wave propagation velocity in neonatal rat ventricular myocytes cultured on 25 mm round glass coverslips spin-coated with PDMS [20] and microcontact printed with fibronectin or covered with a slab of micromolded gelatin hydrogel [8]. These coverslips are imaged by transferring them to an imaging chamber that holds liquid and fits on a standard microscope stage (Bioscience Tools). However, these techniques should be compatible with other cell types, culture surfaces, and imaging chambers.
2. Tyrode's solution can be stored at 4 °C for 1 week. Alternatively, a 10× concentrated Tyrode's solution without glucose and calcium chloride can be prepared and stored at 4 °C for approximately 6 months. On the day of experiments, dilute the 10× Tyrode's solution with ultrapure water (50 mL 10× Tyrode's solution plus 450 mL water, for example), add glucose and calcium chloride, and adjust pH to 7.4 at 37 °C.
3. The propagation velocity in engineered cardiac tissues ranges from 10 to 40 cm/s. To determine the minimum camera frame rate needed, divide the maximum length of your field of view by the maximum propagation velocity (40 cm/s). This number indicates the time for a propagating wave to travel the entire field of view. Because you need to capture at least two frames with the wavefront at distinct locations to measure propagation velocity, divide this number by two. The inverse of this number is the minimum frame rate needed (in frames per second). Higher frame rates will result in higher data quality. For the

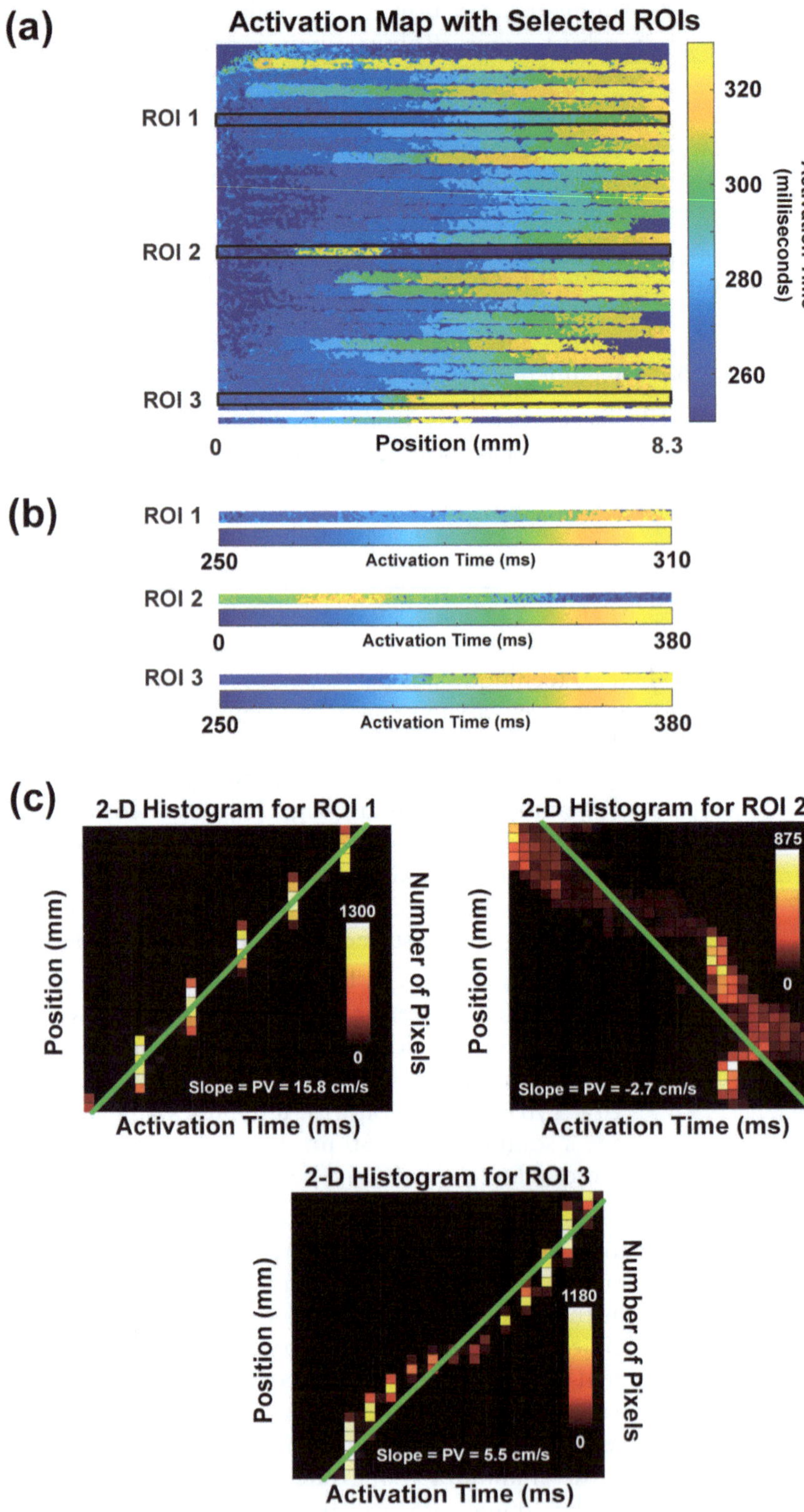

Fig. 3 (**a**) Sample activation map of a tissue with multiple isolated strips of neonatal rat cardiac tissue, similar to [20]. The electrode is to the left of the field of view. Scale bar, 2 mm. Three ROIs with distinct patterns of propagation were selected and used for further analysis. (**b**) Re-scaled activation maps demonstrate that ROI

data generated in this protocol, we used a Nikon Eclipse Ti-inverted fluorescence microscope with a 2× objective (maximum length of field of view: 8.3 mm) and an Andor Zyla sCMOS camera (frame rate: 100 frames per second).

4. The waveform of the electrical signal should be chosen to minimize build-up of charge, corrosion, and electrolysis. This is usually achieved by using a bipolar waveform, a built-in feature of the IonOptix Myopacer stimulator.
5. Purchase two wire cables with plugs compatible with your stimulator. Cut one end from each wire (the end without the plug, if there is only one plug per wire) and fray the wires. Solder a short piece (~0.5 in) of platinum wire (~25 gauge) to each frayed wire. Just above the soldering location, tightly tape the wires together to form a rigid segment that is several inches long. If needed, trim the tips of the platinum wire so that they are even and separate them by 1 mm. Clamp the taped segment of the electrode into the micromanipulator and adjust the position until the tips of the electrode are within range of the imaging area.
6. The bioformats MATLAB toolbox can be downloaded here: https://www.openmicroscopy.org/bio-formats/downloads/
7. After downloading the .zip file, extract the files into your default MATLAB working directory. This is usually "C:\Users\"Your Name"\Documents\MATLAB\" if using a PC. In MATLAB, right click on the bioformats folder, then select "Add Folder and Subfolders to Path" to give MATLAB access to the files within the toolbox. This will need to be repeated each time MATLAB is opened, unless you add the folder and subfolders permanently to the MATLAB path.
8. If you are familiar with git and version control, you can download the code directly with the following command line in MATLAB:

   git clone https://github.com/LLSExUSC/FluoProp
9. The concentration of Fluo-4 and duration of incubation will likely need to be adjusted for different cell types or tissue constructs. This protocol should also be compatible with

Fig. 3 (continued) 1 is relatively uniform and propagates from left to right, as expected. ROI 2 has unusually high noise in one location. ROI 3 has an irregular delay. (**c**) Histograms for each ROI with regression lines. The datapoints for ROI 1 are highly linear and the regression line fits the data closely. This dataset should be included. The datapoints for ROI 2 are irregularly organized and the regression line poorly fits the data. This dataset should be excluded. The datapoints for ROI 3 are mostly linear, but the delay in signal is apparent. The regression line does not closely fit the data and thus the value of propagation velocity is not accurate. This dataset should likely also be excluded

other dyes, such as Fura Red, or genetically encoded calcium or voltage sensors, which do not require dye.

10. Although the electrode should not block the light path of the microscope during data acquisition, it can be helpful to place the electrode in the light path at this step so that you can visualize its position.
11. If the fluorescence intensity is very weak, the power of the light source can be increased, but this can cause bleaching of the dye and/or phototoxicity. Consider using binning ($2 \times 2, 4 \times 4$, or 8×8) to enhance fluorescence intensity. The concentration of Fluo-4 may also need to be increased.
12. This software is designed to input file formats that are compatible with the "bfopen" function in the bioformats toolbox, such as *.nd2 files collected using Nikon software. If your video format is not compatible with the "bfopen" function, convert your video to a tiff stack (using a program such as ImageJ) and save the file. Then, comment out the line that calls the function "readFluoMovie" and write your own line of code to import your tiff stack as the variable "rawData" by using a function such as "imread."
13. First, all pixels in the video are averaged into a single intensity value for every frame. A Fourier transform is then used to identify the major frequency of the mean intensity value over time, which is used to separate the dataset into individual activation cycles, as indicated by the red dotted lines in Fig. 2a. Activation cycles with too few frames based on the pacing frequency, frame rate, rejectThreshFirstBeat, and rejectThreshLastBeat will be excluded from analysis.
14. A Gaussian filter is first applied to the dataset (Fig. 1a). The mean and standard deviation of the fluorescence intensity over time is calculated for each pixel and activation cycle. The activation threshold for each pixel is then calculated as 0.5*standard deviation/mean. The activation time is identified as the first timepoint when the intensity exceeds the activation threshold. This per-pixel thresholding approach minimizes the effects of any aberrations in the fluorescent signal over the field of view, which are common with the types of low-power objectives needed for these measurements. The activation times for each pixel are displayed as the activation map.
15. This step is needed because propagation usually has a circular or elliptical pattern, but the software is designed to measure propagation velocity in one direction. Thus, the region of the activation map corresponding to longitudinal or transverse propagation must be defined and the regions with propagation at other angles must be excluded.

References

1. Kleber AG, Rudy Y (2004) Basic mechanisms of cardiac impulse propagation and associated arrhythmias. Physiol Rev 84(2):431–488
2. Stevenson WG (2009) Ventricular scars and ventricular tachycardia. Trans Am Clin Climatol Assoc 120:403–412
3. Delmar M, McKenna WJ (2010) The cardiac desmosome and arrhythmogenic cardiomyopathies: from gene to disease. Circ Res 107(6): 700–714
4. Wehrens XH, Vos MA, Doevendans PA, Wellens HJ (2002) Novel insights in the congenital long QT syndrome. Ann Intern Med 137(12): 981–992
5. Fenichel RR, Malik M, Antzelevitch C et al (2004) Drug-induced torsades de pointes and implications for drug development. J Cardiovasc Electrophysiol 15(4):475–495
6. Bursac N, Parker KK, Iravanian S, Tung L (2002) Cardiomyocyte cultures with controlled macroscopic anisotropy: a model for functional electrophysiological studies of cardiac muscle. Circ Res 91(12):e45–e54
7. Feinberg AW, Alford PW, Jin H et al (2012) Controlling the contractile strength of engineered cardiac muscle by hierarchal tissue architecture. Biomaterials 33(23):5732–5741
8. Petersen AP, Lyra-Leite DM, Ariyasinghe NR et al (2018) Microenvironmental modulation of calcium wave propagation velocity in engineered cardiac tissues. Cell Mol Bioeng 11(5): 337–352
9. Efimov IR, Ermentrout B, Huang DT, Salama G (1996) Activation and repolarization patterns are governed by different structural characteristics of ventricular myocardium: experimental study with voltage-sensitive dyes and numerical simulations. J Cardiovasc Electrophysiol 7(6):512–530
10. Rohr S, Scholly DM, Kleber AG (1991) Patterned growth of neonatal rat heart cells in culture. Morphological and electrophysiological characterization. Circ Res 68(1):114–130
11. McCain ML, Agarwal A, Nesmith HW, Nesmith AP, Parker KK (2014) Micromolded gelatin hydrogels for extended culture of engineered cardiac tissues. Biomaterials 35(21): 5462–5471
12. Kim DH, Lipke EA, Kim P et al (2010) Nanoscale cues regulate the structure and function of macroscopic cardiac tissue constructs. Proc Natl Acad Sci U S A 107(2):565–570
13. Natarajan A, Stancescu M, Dhir V et al (2011) Patterned cardiomyocytes on microelectrode arrays as a functional, high information content drug screening platform. Biomaterials 32(18): 4267–4274
14. Kujala VJ, Pasqualini FS, Goss JA, Nawroth JC, Parker KK (2016) Laminar ventricular myocardium on a microelectrode array-based chip. J Mater Chem B 4(20):3534–3543
15. Alassaf A, Tansik G, Mayo V, Wubker L, Carbonero D, Agarwal A (2019) Engineering anisotropic cardiac monolayers on microelectrode arrays for non-invasive analyses of electrophysiological properties. Analyst 145(1):139–149
16. Mathur A, Loskill P, Shao K et al (2015) Human iPSC-based cardiac microphysiological system for drug screening applications. Sci Rep 5:8883
17. Huebsch N, Loskill P, Mandegar MA et al (2015) Automated video-based analysis of contractility and calcium flux in human-induced pluripotent stem cell-derived cardiomyocytes cultured over different spatial scales. Tissue Eng Part C Methods 21(5):467–479
18. Fast VG, Kleber AG (1993) Microscopic conduction in cultured strands of neonatal rat heart cells measured with voltage-sensitive dyes. Circ Res 73(5):914–925
19. Entcheva E, Lu SN, Troppman RH, Sharma V, Tung L (2000) Contact fluorescence imaging of reentry in monolayers of cultured neonatal rat ventricular myocytes. J Cardiovasc Electrophysiol 11(6):665–676
20. Petersen AP, Cho N, Lyra-Leite DM et al (2020) Regulation of calcium dynamics and propagation velocity by tissue microstructure in engineered strands of cardiac tissue. Integr Biol (Camb) 12(2):34–46
21. McCain ML, Sheehy SP, Grosberg A, Goss JA, Parker KK (2013) Recapitulating maladaptive, multiscale remodeling of failing myocardium on a chip. Proc Natl Acad Sci U S A 110(24): 9770–9775
22. Spencer CI, Baba S, Nakamura K et al (2014) Calcium transients closely reflect prolonged action potentials in iPSC models of inherited cardiac arrhythmia. Stem Cell Reports 3(2): 269–281
23. Kong W, Ideker RE, Fast VG (2012) Intramural optical mapping of V(m) and ca(i)2+ during long-duration ventricular fibrillation in canine hearts. Am J Physiol Heart Circ Physiol 302(6):H1294–H1305

Chapter 10

Arrhythmia Assessment in Heterotypic Human Cardiac Myocyte–Fibroblast Microtissues

Celinda M. Kofron, Bum-Rak Choi, and Kareen L. K. Coulombe

Abstract

Risk assessment assays for chemically induced arrhythmia are critical, but significant limitations exist with current cardiotoxicity testing, including a focus on single select ion channels, the use of non-human species in vitro and in vivo, and limited direct physiological translation. To be predictive of actual adverse clinical arrhythmic risk, arrhythmia assessment models for chemicals and drugs should be fit-for-purpose and suited for evaluating compounds in which the mechanism of action may not be entirely known. Here, we describe methods for efficient and reliable screening for arrhythmogenic cardiotoxicity with a 3D human cardiac microtissue model using purified human-induced pluripotent stem cell (hiPSC)-derived cardiomyocytes and human cardiac fibroblasts. Applying optical mapping of voltage and calcium-sensitive dyes—an established approach to evaluate cardiac action potentials and calcium transients—to 3D heterotypic cardiac myocyte–fibroblast tissues allows for the generation and functional analysis of a large number of individual microtissues to provide greater throughput and high statistical power in analyses. Hundreds of microtissues in standard cell culture plates can be produced with low variability beat-to-beat, microtissue-to-microtissue, and across hiPSC-cardiomyocyte differentiation batches, reducing the number of microtissues required per condition for predictive outputs. The platform described here can be used as a sensitive, efficient, and predictive preclinical model validated for the purpose of assessing human pro-arrhythmic risk.

Key words Microtissues, Arrhythmia, Cardiac fibroblast, Early afterdepolarization, Action potential, Calcium transient, Optical mapping, Spheroids

1 Introduction

Cardiotoxicity is the occurrence of electrophysiological disturbance or muscle damage in the heart, and arrhythmia, which increases the risk for stroke, heart attack, heart failure, and sudden cardiac death, is the leading manifestation of chemical toxicity [1]. However, the causal relationship between specific chemicals and cardiotoxicity is not well understood. Cardiotoxic effects can be induced by industrial chemicals, environmental toxicants, and pharmaceutical drugs. Predictive assays for chemically induced arrhythmia are critical to protecting human cardiac safety, but significant limitations exist

Kareen L. K. Coulombe and Lauren D. Black III (eds.), *Cardiac Tissue Engineering: Methods and Protocols*, Methods in Molecular Biology, vol. 2485, https://doi.org/10.1007/978-1-0716-2261-2_10,

with current cardiotoxicity testing using existing in vitro models and animal models. Human ether-go-go (HERG) channel blockade and QTc prolongation have long been used as effective and selective "biomarkers" in compound screening to identify pro-arrhythmic risk that cause Torsades de Pointes (TdP) [2, 3]. It is now widely recognized this approach does not capture all drug-induced cardiac arrhythmia mechanisms [4–8]. In vitro cell-based models are often limited to two-dimensional monolayers and lack non-cardiomyocytes, which affect the arrhythmogenic phenotype [9]. Few 3D microtissue systems are used for thorough arrhythmia assessment, preferring to focus on cellular toxicity such as through live/dead, mitochondrial, and ER imaging [10] or contractile amplitude and kinetics from microtissue edge detection or force generation [11, 12]. 3D models aimed at detecting arrhythmia risk often have limited throughput [13, 14] and focus on metrics such as beating frequency [15] or conduction velocity [16]. Animal models are complex and have limited predictability of human biological responses due to species-specific differences in ion channel content and expression that effect depolarization and repolarization kinetics of cardiac action potentials (APs) and differential sensitivity to chemical agents [17, 18]. These methods describe an efficient and reliable screening for arrhythmogenic cardiotoxicity with a 3D human cardiac microtissue model using purified hiPSC-derived cardiomyocytes and human cardiac fibroblasts.

Our 3D heterotypic cardiac myocyte–fibroblast tissue platform for arrhythmia assessment described in this chapter allows for the generation and functional analysis of a large number of individual but consistent microtissues to provide greater throughput and high statistical power in analyses. Self-assembly in commercially available molds produces organotypic heterocellular interspersion and eliminates unnatural substrate or matrix stiffness. The 3D microtissue environment enables formation of many contact sites between cells due to nonplanar distribution of gap junctional proteins in hiPSC-CMs, facilitating electrical coupling between neighboring cells to increase fidelity of arrhythmia assessment compared to 2D monolayer culture. We incorporate human cardiac fibroblasts (hCFs) to enable heterotypic cell–cell interactions characteristic of the intact myocardium [19–21] and employ culturing techniques like metabolic selection to purify and mature the hiPSC-CMs [22]. In our experiments, we use 5% primary adult normal hCF content for stable, healthy cardiac electromechanical function based on our previous work [23], yet the platform allows for alterations in cell composition and ratios to mimic physiological and pathophysiological conditions. This platform can produce hundreds of microtissues in standard cell culture plates with low variability beat-to-beat, microtissue-to-microtissue, and across hiPSC-CM differentiation batches, reducing the number of microtissues required per condition for predictive outputs [24].

To be predictive of actual adverse clinical arrhythmic risk, arrhythmia models for chemicals and drugs should evaluate both trigger and substrate mechanisms for reentry. It is generally thought that cardiac arrhythmias start from premature ventricular contractions (PVCs) due to automaticity or triggered activity such as delayed (DADs) or early afterdepolarizations (EADs), that must occur in a localized tissue mass in order to propagate and form reentry to cause ventricular tachycardia and fibrillation [25]. Therefore, successful proarrhythmic drug screening should be able to detect changes in action potential shape associated with increased triggered activity of a small mass of tissue (not just single or a few cells), which may lead to arrhythmia initiation and propagation. In this 3D microtissue platform, visual detection is used with high temporal and spatial resolution for signal extraction and automated, unbiased quantification to define electrophysiological responses to chemicals. 3D microtissues are mounted on a temperature regulated chamber and loaded with reactive fluorescence dyes (e.g., voltage- or calcium-sensitive dyes), electrically stimulated with platinum electrodes to evoke action potentials and recorded with a high-speed camera [19]. Our analysis of cardiotoxicity as presented herein generates measurements of cardiac waveforms, including excitation and calcium, with the potential for expanding to include contraction. Measurements can include action potential excitability, stimulation delay time (ms), rise time (ms), action potential duration (APD) to 30%, 50%, 80%, 90%, and maximal relaxation (APD_{30}, APD_{50}, APD_{80}, APD_{90}, APD_{MxR}, respectively), presence of EADs or DADs [24], or similar metrics from the calcium transient. Multiple doses of test compounds can be perfused to investigate changes in action potential or calcium transient metrics induced by the test compounds. Tissues can be exposed to compounds acutely (short-term exposure, typically minutes to hours) or chronically (long-term exposure, typically days). The platform described here can thus be used as a robust, predictive, and efficient preclinical model of human proarrhythmic risk.

2 Materials

2.1 Cardiomyocyte Differentiation

1. 6-, 12-, or 24-well culture plates.
2. Matrigel.
3. Human-induced pluripotent stem cell line (hiPSC).
4. CDM3 basal media.
5. 6 μM Chiron 99021.
6. 5 μM IWP2.
7. RPMI 1640 medium with B27 supplement (RPMI/B27).

8. Lactate medium: DMEM without glucose, L-glutamine, phenol red, sodium pyruvate and sodium bicarbonate; 4 mM L-glutamine; 1× Non-Essential Amino Acids; 1× Glutamax; 4 mM lactate; pH 7.4.
9. Penicillin/streptomycin (p/s).
10. 100 mm culture dish.
11. Human cardiac fibroblasts (hCFs).
12. hCF media: DMEM/F12; 10% fetal bovine serum (FBS); 1% p/s; 4 ng/mL basic fibroblast growth factor.

2.2 Fabrication of Hydrogels and 3D Culture

1. 2% (weight/volume) ultrapure agarose in 1× phosphate-buffered saline.
2. Microwave or hot plate.
3. 3D Petri Dish®.
4. Microtissue media: RPMI/B27; 10% FBS; 1% p/s.
5. Field culture stimulator, i.e., C-Pace EP (IonOptix).

2.3 Optical Mapping and Action Potential Analysis

1. Olympus Macroview system with filter set for di-4 ANEPPS.
2. Photometrics Evolve128+ EM-CCD camera (*see* **Note 1**).
3. Dual Automatic Temperature Controller TC-344B, Warner Instrument to maintain 35 ± 1 °C.
4. Motorized syringe pump.
5. Thorlabs motorized stage for automated repetitive recordings (*see* **Note 2**).
6. 5 μM di-4-ANEPPS.
7. Thorlabs Solis LED light source.
8. Platinum electrode.
9. Ionoptix BioPacer field stimulator.
10. Tyrode solution of (in mM) 140 NaCl, 5.1 KCl, 1 $MgCl_2$, 1 $CaCl_2$, 0.33 NaH_2PO_4, 5 HEPES, and 7.5 glucose.
11. No. 1.5 glass coverslip.

3 Methods

3.1 Cardiomyocyte Differentiation and Cardiac Fibroblast Maintenance

1. Culture human-induced pluripotent stem cells (hiPSCs) in a high-density monolayer using CDM3 basal medium on plates coated with Matrigel.
2. Treat hiPSCs with 6 μM Chiron 99,021 (Tocris), a glycogen synthase kinase 3 (GSK3) inhibitor at day 1, followed by 5 μM IWP2 (Tocris), a chemical Wnt inhibitor at day 3. Cardiac phenotype, expressed by beating cells, is usually visible between days 8 and 12.

3. Harvest and replate hiPSC-CM to new culture plates coated with Matrigel in RPMI 1640 medium with B27 supplement.
4. Deprive these cells of media change for 4 days and then feed with lactate medium for 4 days, refreshing medium every 2 days. After lactate purification, feed cells with RPMI +B27 + 1% penicillin/streptomycin until harvested for microtissues assembly.
5. Maintain commercially available human cardiac fibroblasts in hCF media (*see* **Note 3**).
6. Pass cells at a 1:4 ratio and incorporate into cardiac microtissues at cell passages P2-P4.

3.2 Fabrication of Hydrogels and 3D Culture

1. Heat 2% agarose with microwave or hot plate to boiling.
2. Pipet molten 2% agarose into 3D Petri dish mold (i.e., molds for 24-well plates with 800-μm-diameter rounded pegs).
3. Cool to room temperature (~5 min), separate agarose gel from the mold, and transfer to 24-well plate.
4. Equilibrate in 1 mL microtissue media for 24 h.
5. Harvest hiPSC-CM and hCF. Prepare cell seeding solution of 400,000–800,000 hiPSCs with 5% hCFs in 100 μL of microtissue media (per well) (*see* **Note 4**).
6. Remove equilibration media from molds including media in the center cell seeding chamber. Pipet 100 μL of cell seeding solution into the cell seeding chamber of each mold (Fig. 1a).
7. Allow cells to settle into recesses for 15–30 min. Add media to the well, outside of the 3D Petri dish to cover (1 mL for 24-well plate, 2–3 mL for 12-well plate).
8. Culture for 6–8 days with 1 Hz, 10.0 V, 4.0 ms duration bipolar pulse train field stimulation. (Fig. 1b, c, *see* **Note 5**).

3.3 Optical Mapping and Automated Action Potential Analysis

1. Setup for optical mapping is shown in Fig. 2a.
2. Set temperature control to 35 °C, and set constant perfusion flow (0.1–1 mL/min).
3. Transfer 3D Petri dish to temperature regulated chamber on motorized stage (Fig. 2b).
4. Place stimulation electrode, perfusion, and suction tubing in solution with 3D Petri dish. Two leads of platinum electrode should be 1 or 2 cm apart for easy calculation of stimulation strength in V/cm.
5. To reduce water vibration noise, float a glass coverslip on top of the solution in the well.
6. Load microtissues with voltage-sensitive di-4-ANEPPS in Tyrode solution for 10 min at 35 °C for measurements of membrane potential (V_m) and wash out with Tyrode solution for 10 min.

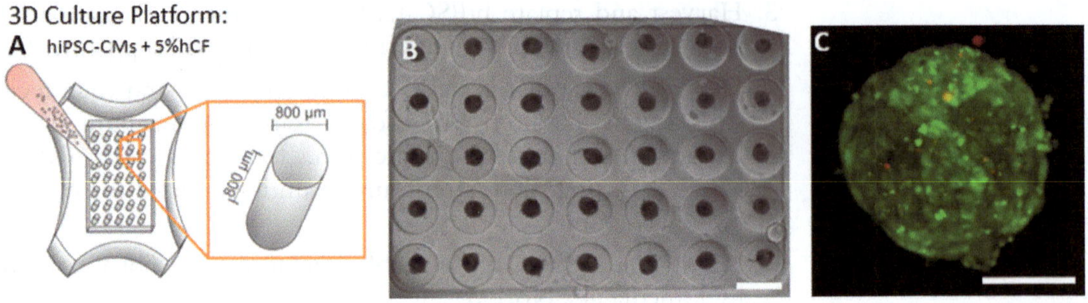

Fig. 1 Formation of cardiac microtissues. (**a**) Schematic of three-dimensional (3D) cardiac microtissue generation shows non-adhesive agarose gels with cylindrical recesses with hemispherical bottoms that guide self-assembly. (**b**) Phase contrast image shows consistent spherical microtissue formation after 5 days of 3D culture in all 35 microwells. Cardiac microtissues were cultured for 6–8 days with 1 Hz pacing. Scale bar, 800 μm. (**c**) Compressed confocal z-stack image shows a representative cardiac tissue with hiPSC-CM (green) and hCF (red) stained with CellTracker dyes with a low number of interspersed hCFs. Scale bar, 100 μm

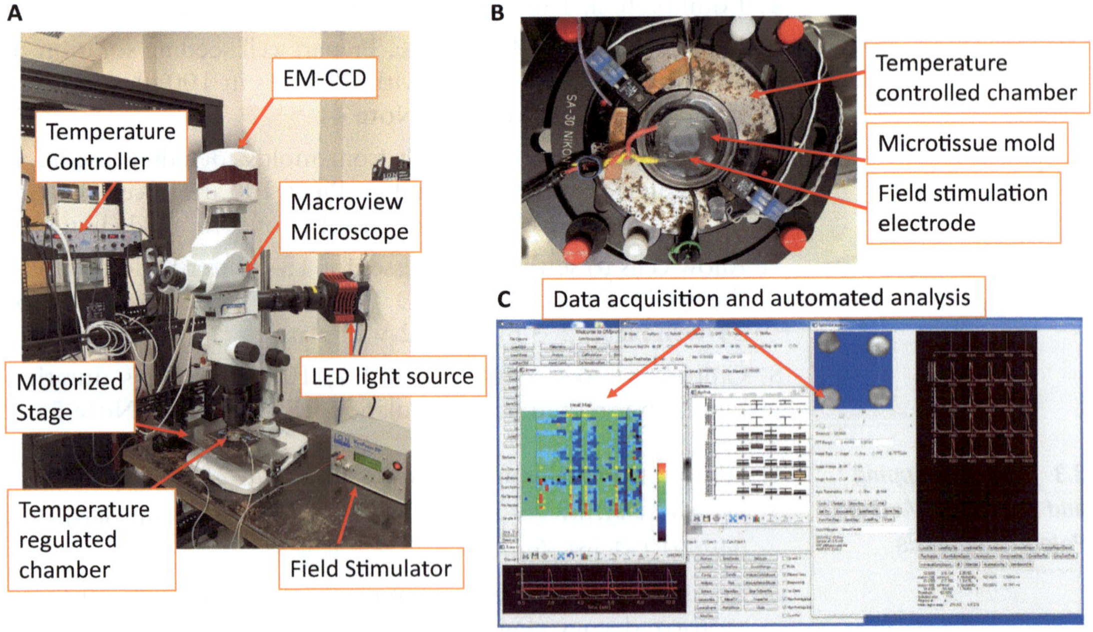

Fig. 2 Optical mapping acquisition and analysis. (**a**) Microscope and stimulation setup to acquire fluorescence images at 979 frames/s in $1.2 \times 1.2\text{-mm}^2$ field of view. (**b**) Stimulation electrode setup during optical mapping image acquisition. Two linear platinum electrodes (2 cm apart) were used to field-stimulate (20 V, 4 ms biphasic stimulation pulse with 10 V/cm amplitude) all of the microtissues simultaneously (adapted from Kofron et al. 2021, Supplemental Figures). (**c**) Example of data acquisition and custom-automated analysis user interface

7. Set up the stimulation protocol using 2 ms biphasic pulse with 10–15 V/cm strength, 2 s interval (0.5 Hz).
8. Adjust field of view and focus of the image. The locations of the four corners of the mold are registered from the data acquisition software to automate sequential recording of action potentials from all 35 microtissues.
9. Acquire fluorescence images at 979 frames/s using Photometrics Evolve +128 EM-CCD camera (2×2 binning to 64×64 pixels, 18.7×18.7-μm^2 resolution, 1.2×1.2-mm^2 field of view). Record four microtissues simultaneously per scan at this magnification (*see* **Note 6**). Record action potentials from all of the microtissues in a single mold (35 microtissues per mold). Filter fluorescence images using nonlinear bilateral filter (spatial filter: 5×5 window, temporal filter: 21-point window) to preserve AP upstrokes from blurring (Fig. 2c).
10. Check how many microtissues show action potentials evoked by electrical stimulation. For quality control, ensure more than 70% of microtissues show action potentials under 2 s pacing (> 25/35 microtissues, typically excitability of 95% is an indication of a good preparation).
11. After baseline recording of all microtissues in this mold, change perfusion solution to include drug or chemical of interest. Expose for desired time frame (i.e., 20-min exposure) to E4031, a high-risk hERG channel blocker, or other compounds of interest.
12. Repeat data acquisition (**step 8**) and perfusion (**step 10**) for each concentration of compound to be tested. Total imaging time should be no more than about 1 h to minimize effects of dye toxicity.
13. The baseline drifting caused by water level fluctuation or photobleaching of dye can be subtracted using asymmetric least square method [26].
14. Process the fluorescence signal using fast Fourier transform (FFT) to construct an FFT image of the amplitude. Apply Otsu's thresholding and then segment the image to average the fluorescence signals from the regions containing each microtissue to generate an AP signal for each microtissue. Use the first and second derivatives of the AP to quantify parameters (Fig. 3a–g).
15. Quantify proarrhythmic risk metrics including excitability, stimulation time delay between stimulation pulse and peak AP upstroke (stimulation delay time, stimΔ), rise time of AP upstroke, AP duration (APD) to 30%, 50%, and 80% repolarization (APD_{30}, APD_{50}, APD_{80}), AP duration to end of maximum repolarization rate (APD_{MxR}, determined with $d^2F/$

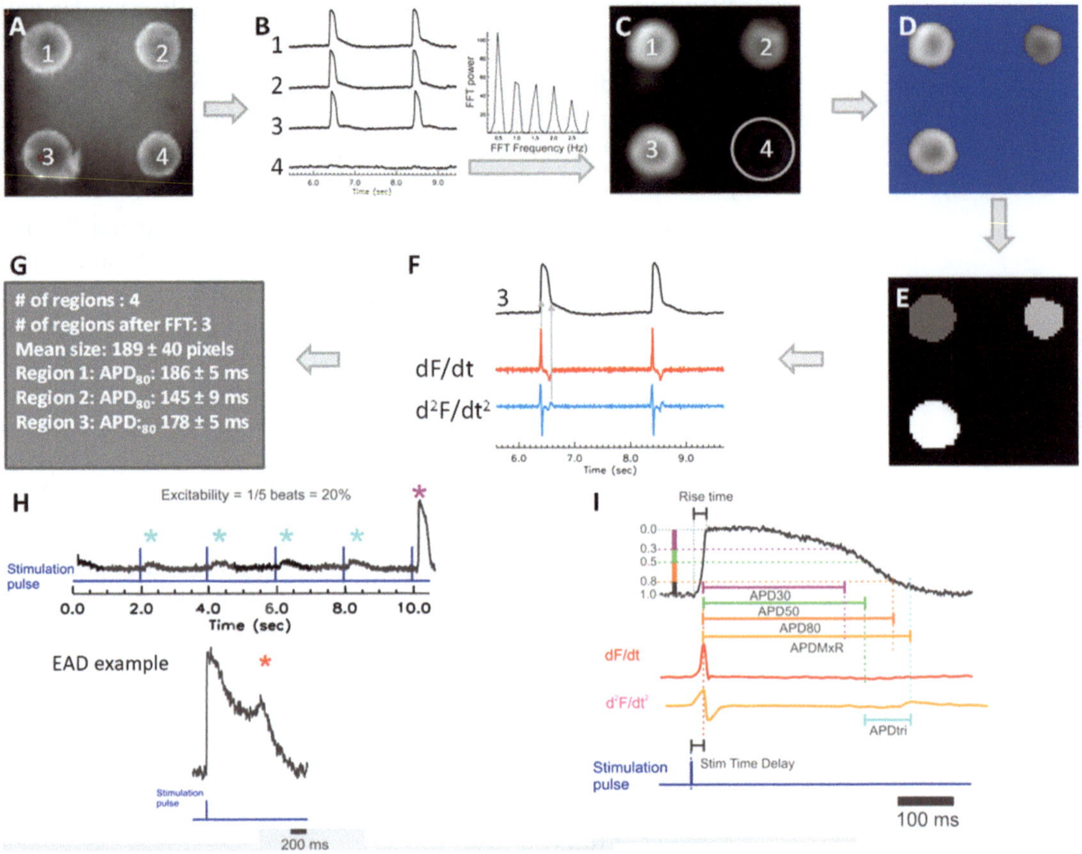

Fig. 3 Automated analysis pipeline and definition of metrics. (**a**) A grayscale snapshot of fluorescence from microtissues at 3.2× magnification during optical mapping shows individual microtissues are numbered 1–4. (**b**) Sample membrane voltage (V_m) traces recorded from the corresponding microtissues in panel **a**. Note that electrical stimulation did not evoke APs in microtissue #4 (where the fluorescence trace shows a flat line). Fast Fourier transform (FFT) was used to distinguish responsive microtissues from non-responsive microtissues lacking APs. (**c**) FFT_{max} image shows that the microtissue (#4) without APs is automatically removed in FFT_{max} (circle). (**d**) Automated thresholding using Otsu's thresholding to remove background and speed analyses. (**e**) Blob coloring algorithm to detect individual microtissues. Signals within the same microtissue are averaged to acquire high signal-to-noise ratio and fidelity of data analysis. (**f**) Automated analysis (of sample trace #3, black) uses the first derivative (dF/dt, red trace) for detecting AP upstroke to calculate AP duration to a defined recovery amplitude (*see* Fig. 1g) or uses the maximum peak of the second derivative (d^2F/dt^2, blue trace) for assessing the end of maximum repolarization rate to calculate AP duration APD_{MxR}. APD_{MxR} is useful when motion artifacts or other interference on fluorescence recordings such as fluctuation in water level is suspected to elevate fluorescence level during repolarization or reproducible APD_{80} measurement with low variation is not as reliable. (**g**) Sample output shows APD_{80} statistics from the three excitable individual microtissues. Schematics of the AP metrics that were defined (with units) as: (**h**) "excitability" (%) measured from the percentage of captured APs during 10 s duration of recording with 2 s pacing cycle length (top) and occurrence of "early afterdepolarization" (EAD) reported as (%) of microtissues showing EADs (bottom, *) and (**i**) "stimulation time delay" (ms; stim delay) between stimulation pulse and evoked AP upstroke (dF/dt_{max}), "rise time" (ms) of AP, "AP duration" (ms) to 30%, 50%, and 80% repolarization (APD_{30}, APD_{50}, APD_{80}), "APD to maximum repolarization" (ms; APD_{MxR}) defined as time between AP upstroke and the end of rapid repolarization marked by d^2F/dt^2_{max}, "APD triangulation" (ms; APD_{tri}) defined as APD_{MxR}–APD_{50}. Adapted from Kofron et al., 2021, Fig. 1 and Supplemental Figures

Table 1
Quantitative metrics and physiological targets of the AP

Metric (units)	Ion channels involved	Ion currents
Excitability (%) and stimulation time delay	Na channels ($Na_v1.5$, 1.1, others) Inward rectifier K^+ channel (Kir2.1)	I_{Na} I_{K1}
Rise time (ms)	Na channel ($Na_v1.5$)	I_{Na}
APD_{30} (ms)	Late Na channels ($Na_v1.1$, others) Ca channels ($Ca_v1.2$) K channels (Kv4.2/4.3, Kv1.4)	$I_{Na,late}$ I_{CaL} I_{to}
APD_{50} (ms)	Late Na channels ($Na_v1.1$, others) Ca channels ($Ca_v1.2$) K channels (hERG, KvLQT, Kv4.2/4.3, Kv1.4)	$I_{Na,late}$ I_{CaL} I_{to}, I_{Kr}, I_{Ks}
APD_{80}, APD_{MxR} (ms)	K channels (hERG, KvLQT, Kir2.1)	I_{Kr}, I_{Ks}, I_{K1}
APD_{tri} (ms) = APD_{MxR}-APD_{50}	Late Na channels ($Na_v1.1$, others) Ca channels ($Ca_v1.2$) K channels (hERG, KvLQT, Kv4.2/4.3, Kv1.4, Kir2.1)	$I_{Na,late}$ I_{CaL} I_{to}, I_{Kr}, I_{Ks}, I_{K1}
EADs (count/AP)	K channels (hERG, KvLQT) Late Na channel (Nav1.5)	I_{Kr}, I_{ks} $I_{Na,late}$

dt^2_{max}), APD triangulation ($APD_{tri} = APD_{MxR} - APD_{50}$), and the presence of EADs (% of microtissues showing EAD) (Fig. 3h, i). Changes in these metrics can provide insights into alteration in ion channel kinetics as demonstrated in Table 1.

16. Statistical analysis can compare data for AP recordings from the same mold before and after drug treatment with paired t-tests or for AP recordings from the different molds with unpaired t-tests. Variability between microtissues, molds, or batches of hiPSC-CMs can be quantified.

4 Notes

1. Other systems and cameras could be suitable. The Photometrics Evolve128+ EM-CCD camera records from 0.3 × 0.3 mm2 to 3 × 3 mm2 field of view with a sampling rate of 979 frames per sec.
2. In our setup, the temperature-controlled chamber is mounted on a custom-built motorized stage for automated repetitive recordings. The microscope and the perfusion system are mounted on the isolation table to eliminate possible vibration noise.

3. hCFs were maintained in 10 cm dishes and split 1:4 when 90% confluent.
4. Seeding density may need to be optimized with different phenotypes of cardiac myocytes and fibroblasts. This seeding density yielded reproducible, synchronously beating tissues that could be easily imaged.
5. Stimulation protocol and/or culture time can be adjusted for different cell types and investigations, but this protocol and timeline yields reproducible microtissues with AP similar to in vivo.
6. The data acquisition, stimulation protocol, and motorized stage are incorporated in a custom-built software (e.g., written in Interactive Data Language (IDL, Harris Geospatial Solutions) and C routines). The spatial and temporal filtering using bilateral filter, baseline adjustment using asymmetric least square algorithm, and FFT of fluorescence signals are calculated in a multicore graphics card (Nvidia Quadro P4000) to shorten calculation time.

References

1. Virani SS et al (2020) Heart disease and stroke statistics-2020 update: a report from the american heart association. Circulation 141: e139–e596. https://doi.org/10.1161/CIR.0000000000000757
2. Haraguchi Y, Ohtsuki A, Oka T, Shimizu T (2015) Electrophysiological analysis of mammalian cells expressing hERG using automated 384-well-patch-clamp. BMC Pharmacol Toxicol 16:39. https://doi.org/10.1186/s40360-015-0042-9
3. Wacker S, Noskov SY (2018) Performance of machine learning algorithms for qualitative and quantitative prediction drug blockade of hERG1 channel. Comput Toxicol 6:55–63. https://doi.org/10.1016/j.comtox.2017.05.001
4. Ferdinandy P et al (2019) Definition of hidden drug cardiotoxicity: paradigm change in cardiac safety testing and its clinical implications. Eur Heart J 40:1771–1777. https://doi.org/10.1093/eurheartj/ehy365
5. Alinejad S, Kazemi T, Zamani N, Hoffman RS, Mehrpour O (2015) A systematic review of the cardiotoxicity of methadone. EXCLI J 14: 577–600. https://doi.org/10.17179/excli2015-553
6. Heranval A et al (2016) Drugs with potential cardiac adverse effects: retrospective study in a large cohort of parkinsonian patients. Rev Neurol (Paris) 172:318–323. https://doi.org/10.1016/j.neurol.2015.11.007
7. Sun C, Brice JA, Clark RF (2018) Brugada-type pattern on electrocardiogram associated with high-dose loperamide abuse. J Emerg Med 54:484–486. https://doi.org/10.1016/j.jemermed.2017.12.040
8. Ramalho D, Freitas J (2018) Drug-induced life-threatening arrhythmias and sudden cardiac death: a clinical perspective of long QT, short QT and Brugada syndromes. Rev Port Cardiol 37:435–446. https://doi.org/10.1016/j.repc.2017.07.010
9. Kurokawa YK, George SC (2016) Tissue engineering the cardiac microenvironment: Multicellular microphysiological systems for drug screening. Adv Drug Deliv Rev 96:225–233. https://doi.org/10.1016/j.addr.2015.07.004
10. Archer CR et al (2018) Characterization and validation of a human 3D cardiac microtissue for the assessment of changes in cardiac pathology. Sci Rep 8:10160. https://doi.org/10.1038/s41598-018-28393-y
11. Beauchamp P et al (2020) 3D co-culture of hiPSC-derived cardiomyocytes with cardiac fibroblasts improves tissue-like features of cardiac spheroids. Front Mol Biosci 7:14. https://doi.org/10.3389/fmolb.2020.00014
12. Feric NT et al (2019) Engineered cardiac tissues generated in the biowire II: a platform for

human-based drug discovery. Toxicol Sci 172(1):89–97. https://doi.org/10.1093/toxsci/kfz168
13. Beauchamp P et al (2015) Development and characterization of a scaffold-free 3D spheroid model of induced pluripotent stem cell-derived human cardiomyocytes. Tissue Eng Part C Methods 21:852–861. https://doi.org/10.1089/ten.TEC.2014.0376
14. Zhao Y et al (2019) A platform for generation of chamber-specific cardiac tissues and disease modeling. Cell 176:913–927. e918. https://doi.org/10.1016/j.cell.2018.11.042
15. Bergstrom G, Christoffersson J, Schwanke K, Zweigerdt R, Mandenius CF (2015) Stem cell derived in vivo-like human cardiac bodies in a microfluidic device for toxicity testing by beating frequency imaging. Lab Chip 15: 3242–3249. https://doi.org/10.1039/c5lc00449g
16. Petersen AP et al (2018) Microenvironmental modulation of calcium wave propagation velocity in engineered cardiac tissues. Cell Mol Bioeng 11:337–352. https://doi.org/10.1007/s12195-018-0522-2
17. Tanner MR, Beeton C (2018) Differences in ion channel phenotype and function between humans and animal models. Front Biosci (Landmark Ed) 23:43–64. https://doi.org/10.2741/4581
18. Bracken MB (2009) Why animal studies are often poor predictors of human reactions to exposure. J R Soc Med 102:120–122. https://doi.org/10.1258/jrsm.2008.08k033
19. Kofron CM et al (2017) Gq-activated fibroblasts induce cardiomyocyte action potential prolongation and automaticity in a three-dimensional microtissue environment. Am J Physiol Heart Circ Physiol 313:H810–H827. https://doi.org/10.1152/ajpheart.00181.2017
20. Kofron CM, Mende U (2017) In vitro models of the cardiac microenvironment to study myocyte and non-myocyte crosstalk: bioinspired approaches beyond the polystyrene dish. J Physiol 595:3891–3905. https://doi.org/10.1113/JP273100
21. Zuppinger C (2019) 3D cardiac cell culture: a critical review of current technologies and applications. Front Cardiovasc Med 6:87. https://doi.org/10.3389/fcvm.2019.00087
22. Tohyama S et al (2013) Distinct metabolic flow enables large-scale purification of mouse and human pluripotent stem cell-derived cardiomyocytes. Cell Stem Cell 12:127–137. https://doi.org/10.1016/j.stem.2012.09.013
23. Rupert CE, Kim TY, Choi BR, Coulombe KLK (2020) Human cardiac fibroblast number and activation state modulate electromechanical function of hiPSC-cardiomyocytes in engineered myocardium. Stem Cells Int 2020: 9363809. https://doi.org/10.1155/2020/9363809
24. Kofron CM et al (2021) A predictive in vitro risk assessment platform for pro-arrhythmic toxicity using human 3D cardiac microtissues. Sci Rep 11(1):10228. https://doi.org/10.1038/s41598-021-89478-9
25. Weiss JN, Garfinkel A, Karagueuzian HS, Chen PS, Qu Z (2010) Early afterdepolarizations and cardiac arrhythmias. Heart Rhythm 7: 1891–1899. https://doi.org/10.1016/j.hrthm.2010.09.017
26. Peng J et al (2010) Asymmetric least squares for multiple spectra baseline correction. Anal Chim Acta 683:63–68. https://doi.org/10.1016/j.aca.2010.08.033

Chapter 11

Human-Engineered Atrial Tissue for Studying Atrial Fibrillation

Julia Krause, Marta Lemme, Ingra Mannhardt, Alexandra Eder, Bärbel Ulmer, Thomas Eschenhagen, and Justus Stenzig

Abstract

This chapter details the generation of atrial fibrin-based engineered heart tissue (EHT) in standard 24-well format as a 3D model for the human atrium. Compared to 2D cultivation, human-induced pluripotent stem cells (hiPSCs)-derived atrial cardiomyocytes demonstrated a higher degree of maturation in 3D format. Furthermore, we have demonstrated in previous work that the model displayed atrial characteristics in terms of contraction and gene expression patterns, electrophysiology, and pharmacological response. Here, we describe how to embed atrial cardiomyocytes differentiated from hiPSCs in a fibrin hydrogel to form atrial EHT attached to elastic silicone posts, allowing auxotonic contraction. In addition, we describe how force and other contractility parameters can be derived from these beating atrial EHTs by video-optical monitoring. The presented atrial EHT model is suitable to study chamber-specific mechanisms, drug effects and to serve for disease modeling.

Key words Atrial fibrillation, Tissue engineering, Human-induced pluripotent stem cells, Engineered heart tissue, Contraction analysis, Arrhythmia

1 Introduction

Atrial fibrillation (AF) represents a significant public health burden as it affects 1–2% of the general population. The asynchronous contraction of the atria increases the risk of secondary diseases such as stroke or heart failure, which results in an increased mortality [1]. The pathology of AF is complex and involves multiple mechanisms including electrophysiological and structural alterations, such as shortening of the action potential duration and fibrotic remodeling in atrial tissue [2]. Despite our knowledge of these fundamental mechanisms contributing to AF development and maintenance, the pathophysiology remains incompletely understood. Therapeutic interventions such as antiarrhythmic drug therapy are only modestly effective and induce serious side

Kareen L. K. Coulombe and Lauren D. Black III (eds.), *Cardiac Tissue Engineering: Methods and Protocols*, Methods in Molecular Biology, vol. 2485, https://doi.org/10.1007/978-1-0716-2261-2_11,

effects [1]. The lack of reliable disease models hampers research on the pathophysiology of AF and the evaluation of atrial-selective antiarrhythmic drugs [3]. As primary human tissue is difficult to obtain and starts to disintegrate once ex vivo, cell culture systems and animal models are commonly used. However, their predictive value is limited, due to species differences on many levels, ranging from contractile parameters to ion channel composition and Ca^{2+} handling mechanisms [4].

Recently established protocols for differentiation of patient-specific-induced pluripotent stem cells (hiPSCs) into cardiomyocytes have opened up the possibility to overcome species- as well as intra-individual differences. To investigate chamber-specific disease mechanisms and drug effects, the discrimination between the main subtypes of cardiomyocytes (such as ventricular, atrial, and nodal) is essential to consider their respective functional, molecular, and electrophysiological differences. So far, several studies suggested that hiPSC-derived cardiomyocytes consist of heterogeneous cell populations containing predominantly ventricular-like cells but also atrial- and nodal-like cells [5–8]. These findings initiated efforts to differentiate pure populations of chamber-specific cardiomyocytes. All-trans retinoic acid (RA), a derivative of vitamin A, was shown to induce atrial cell fate decision during cardiac differentiation [3, 9–11]. Based on RA usage, several protocols for atrial cardiomyocyte differentiation have been developed and yield cardiomyocytes with electrophysiological properties and gene expression patterns of atrial-like cells [3, 9, 12, 13]. Even though the advanced differentiation protocols have improved the *purity* of hiPSC-derived cardiac subtypes, they still exhibit a lack of *maturity* compared to adult cardiomyocytes.

Several concepts for further maturation of hiPSC-derived cardiomyocytes have been evaluated, including pharmacological activation of signaling pathways, metabolic substrate optimization, electrical pacing, prolonged culture time, and cultivation in a three-dimensional (3D) format [14–16]. In this protocol chapter, we are focusing on 3D cultivation. There are multiple cardiac tissue engineering strategies available, which differ regarding cell composition, geometry, and size as well as the material used to induce 3D organization. Here, we describe the generation of fibrin-based engineered heart tissue (EHT). This well-established method uses (i) fibrin as hydrogel to embed the cardiomyocytes and (ii) two elastic silicone posts between which the developing tissue is attached, providing mechanical strain [17]. Other than in static scaffold systems, the use of elastic silicone posts allows the cardiomyocytes to perform auxotonic contraction resulting in contractile work similar to that of native tissue [17, 18]. Importantly, the exposure to mechanical load has been recognized as a crucial factor for cardiomyocyte organization, maturation, and development [15]. For ventricular hiPSC-derived cardiomyocytes, it has previously been shown that compared to 2D cultivation, fibrin-based EHT has several advantages: (i) stability over several weeks [17], (ii) longitudinal cardiomyocyte alignment along the force lines, resulting in improved sarcomere organization [15, 19], (iii) maturation of energy metabolism, indicated by decreased anaerobic glycolysis and increased oxidative metabolism of fatty acids [20], and (iv) the possibility to monitor cardiac electrophysiology, for example, by sharp microelectrodes [21, 22].

In line with generally improved maturation, the EHT format has also been shown to be more suitable to reflect chamber specificity of hiPSC-cardiomyocytes. Consequently, hiPSC-derived atrial cardiomyocytes in EHT format demonstrated a higher expression of atrial-specific genes, compared to 2D cultivation [12]. Moreover, atrial EHT can be clearly distinguished from ventricular EHT with regard to further characteristics, such as contractility, electrophysiology, and pharmacological response [12]. While differences to the native atrial tissue remain, for example, in terms of automaticity and specific force, the data nevertheless suggest that the chamber-specific EHT represents a suitable tool for atrial drug testing applications and mechanistic studies [12, 13, 23]. Characteristic properties of human atrial cardiomyocytes were also confirmed by others, using different tissue engineering models and cardiomyocytes differentiated from either hiPSCs or human embryonic stem cells (hESCs) [12, 13, 23]. Because the study of arrhythmia mechanisms is limited in single cells and 2D cell layers, the generation of atrial EHT represents an essential step forward to investigate AF in vitro. This was recently underlined by Goldfracht et al. who investigated re-entry mechanisms in a ring-shaped engineered atrial tissue [23]. In addition, in our own human *ventricular* fibrin-based EHT model, arrhythmia was successfully induced by chronic optical tachypacing [24]. The induced tachycardia could be terminated by classical antiarrhythmic drugs, demonstrating that fibrin-based EHT is a valuable model to study arrhythmia. Hence, this protocol has the potential to be adapted to the human *atrial* fibrin-based EHT model to study AF. In addition to optogenetic tachypacing, there are several possibilities to further modify the model to induce and modify arrhythmia, one prime example being the addition of other cell types like fibroblasts. This would not only represent the native tissue composition more closely, but might also enable induction of structural remodeling processes such as fibrosis, which is characteristic for AF [2, 25]. Altogether, the presented atrial fibrin-based EHT model provides the opportunity to consider chamber-specific mechanisms and drug effects as well as to model diseases like arrhythmia. Furthermore, it can potentially be harnessed for a multitude of AF-related experiments and interventions.

In this chapter, we provide a detailed protocol for the generation and video-optical monitoring of fibrin-based EHT made of hiPSC-derived atrial cardiomyocytes. The differentiation of atrial cardiomyocytes from hiPSCs is based on a modified three-step protocol [26] including additional RA treatment to obtain atrial cells [12]. Differentiation of atrial cardiomyocytes can be achieved by following the protocol by Breckwoldt et al. with the modifications described below.

2 Materials

All reagents and tools used for the following procedures are standard cell culture equipment, except for spacers and silicone racks. These are custom designed and can be either manufactured individually or ordered from EHT Technologies GmbH (Hamburg, Germany).

1. Agarose: Dissolve 2% (w/v) agarose (highly pure) in PBS. Autoclave and store the agarose at 60 °C. Never use soaps/detergents to clean the bottles used for agarose storage.
2. Aprotinin solution: Dissolve 33 mg/mL aprotinin (from bovine lung) in sterile water. Sterile filter with a 0.22 μm filter, aliquot in stocks of, for example, 500 μL and store at −20 °C.
3. Fibrinogen: Dissolve 200 mg/mL fibrinogen (from bovine plasma) and add 100 μg/mL of aprotinin (33 mg/mL stock from 2.2) in sterile and pre-warmed 0.9% NaCl solution (37 °C). Mix by shaking continuously for 30 min at room temperature until all clots are dissolved and aliquot in stocks of, for example, 400 μL and store at −20 °C, or at −80 °C for long-term storage.
4. Thrombin: Dissolve 100 U/mL of thrombin (from bovine plasma) in 60% (v/v) PBS and 40% (v/v) sterile water. Mix the solution, aliquot either, for example, 450 μL as stocks or 3 μL for EHT generation in sterile 200 μL tubes (1 aliquot per EHT). Store all aliquots at −20 °C.
5. Y-27632 (10 mM): Dissolve Y-27632 in water at 10 mM, sterile filter with a 0.22 μm filter, aliquot stock solution, and store at 4 °C for several days or at −20 °C for long-term storage.
6. 2× Dulbecco's modified eagle medium (DMEM): Dissolve 26.8 mg/mL 10× DMEM powder (4.5 g/L (high) glucose) in sterile water and add heat-inactivated horse serum to a final fraction of 20% (v/v) and penicillin/streptomycin to a final fraction of 2% (v/v). Sterile filter with 0.22 μm filter and store aliquots of, for example, 450 μL at −20 °C.
7. EHT casting medium: Prepare 1% (v/v) penicillin/streptomycin, 10% (v/v) heat-inactivated horse serum or heat-inactivated fetal calf serum, and 1% (v/v) L-glutamine in DMEM (1.0 g/L (low) glucose, phenol-red, no L-glutamine).

8. EHT culture medium: Prepare 1% (v/v) penicillin/streptomycin, 10% (v/v) heat-inactivated horse serum, 10 μg/mL insulin, and 33 μg/mL aprotinin in DMEM (*see* above and **Note 1**).
9. Silicone racks: Silicone racks are used for the generation and cultivation of EHTs. Prior to experiments, racks are prepared by boiling two times in deionized water and subsequent autoclaving. Handle with care to avoid bending of the silicone posts. Do not use soaps/detergents for cleaning. After EHT culture, rinse under running water and manually remove EHT leftovers. The racks can be reused for up to ten autoclaving cycles, but should not be reused after having been exposed to drugs, which often exhibit strong binding to silicone [27]. For further information, contact the authors or EHT Technologies.
10. Teflon spacers: Teflon spacers are used to prepare the cavities in agarose casting molds, which provide the 3D form to generate EHT. Prior to experiments, spacers are prepared by boiling two times in deionized water and subsequent autoclaving. Do not use soaps/detergents for cleaning. The spacers can be reused infinite times. For further information, contact the authors or EHT Technologies.

3 Methods

3.1 Preparations

Before starting the experiment, take care of the following aspects:

1. Prepare all solutions as described in the Subheading 2.
2. Ensure that the spacers and silicone racks are sterile and ready to use (*see* Subheading 2).
3. We suggest using freshly dissociated hiPSC-derived cardiomyocytes for the generation of human atrial EHT (*see* **Note 2**). After dissociation, place the cardiomyocytes on ice (*see* **Notes 3** and **4**) and directly continue with Subheading 3.4. Alternatively, follow Subheading 3.3 for thawing of hiPSC-derived cardiomyocytes.

3.2 Differentiation of Atrial-Like hiPSC-Derived Cardiomyocytes (Optional)

The differentiation of ventricular/mixed cardiomyocytes has been described in detail by Breckwoldt et al. [26]. To obtain atrial-like cardiomyocytes, this protocol can be used with the following modifications:

1. Retinoic acid (RA): Prepare all-trans RA by dissolving 15 mg/mL (50 mM) in DMSO [3]. Further dilute the solution in sterile water to a final concentration of 100 μM, and store aliquots of, for example, 500 μL at −20 °C. Minimize ambient light exposure since RA is extremely sensitive to UV-light.

2. Based on the protocol by Breckwoldt et al. for ventricular-like cardiomyocyte differentiation, two modifications are necessary for atrial differentiation. (i) Mesodermal induction is initiated by using lower concentrations of BMP-4 (3 ng/mL) and activin A (2 ng/mL). (ii) RA treatment (1 μM) from day 4 to day 7 of cardiac lineage specification without media change during this time [3]

3.3 Thawing of hiPSC-Derived Cardiomyocytes (Optional)

The following procedure is suitable for hiPSC-derived cardiomyocytes, which were frozen in heat-inactivated fetal calf serum with 10% DMSO and stored at −150 °C. The protocol is also based on the work of Breckwoldt et al. [26].

1. Pre-warm RPMI medium (L-glutamine and phenol red, additionally supplemented with 1% (v/v) penicillin/streptomycin) to room temperature (additional material for this section).
2. Remove a cryo vial with cells from the cryo storage (−150 °C) and directly place it in a water bath at 37 °C for 2–3 min until the cell suspension is thawed. Only thaw one cryo vial at a time to avoid time-dependent cell loss.
3. In a biosafety cabinet, immediately transfer the cells into a sterile 50 mL centrifuge tube using a serological pipette.
4. Rinse the vial with 1 mL of pre-warmed RPMI and add it in a dropwise manner to the cell suspension in the 50 mL centrifuge tube over 90 s while gently swirling the 50 mL tube (*see* **Note 5**).
5. Slowly add another 8 mL of pre-warmed RPMI to the 50 mL tube. Add the first mL under gentle swirling of the tube in a dropwise manner over 30–60 s. The remaining 7 mL can be added over 30 s.
6. Invert the tube five times to ensure a homogenous cell suspension before counting the cells, using your preferred counting method (e.g., Trypan blue, Neubauer chamber, automatic counter). The cell recovery rate (living cells after thawing/ living cells before thawing) should be above 50%.
7. Centrifuge the cells for 10 min at 100 × *g* and 4 °C, resuspend them in the appropriate volume of EHT casting medium (*see* Subheading 3.4), and place them on ice.

3.4 Casting Molds and Preparation of EHT Master Mix (All Steps Carried Out in a Biosafety Cabinet, Fig. 1)

1. Prepare casting molds for EHTs by quickly pipetting 1.6 mL of liquid 2%-agarose (60 °C) into each well of a 24-well plate.
2. Place one spacer per 4 wells (6 spacers per 24-well plate). Each spacer has a small circular embossment to ensure correct positioning in the wells of a 24-well plate.
3. Let the agarose solidify during 10–15 min at room temperature (*see* **Note 6**).

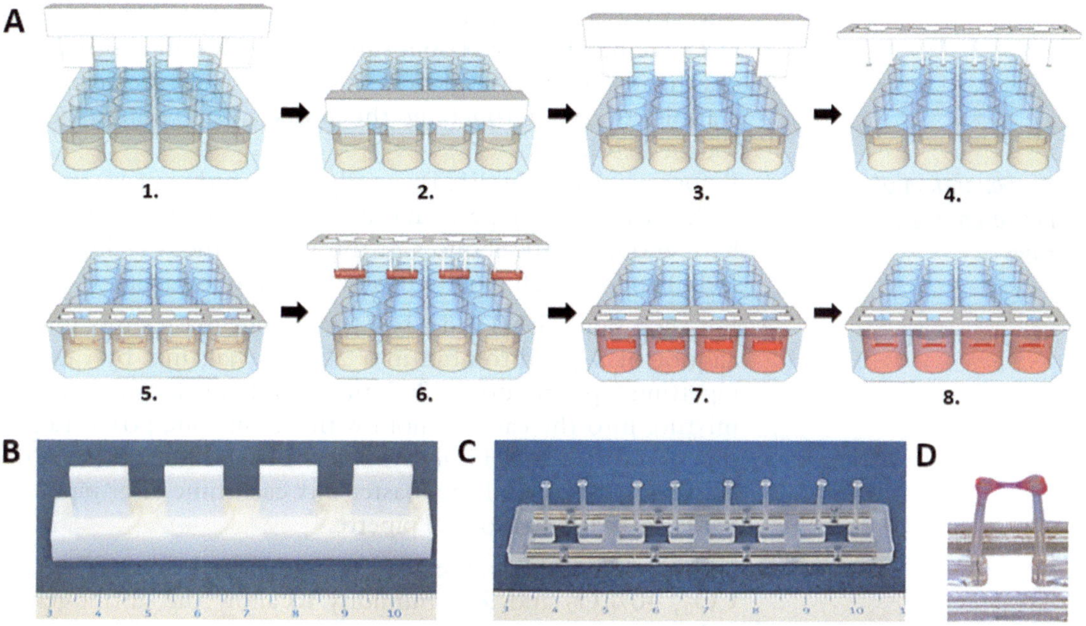

Fig. 1 EHT casting procedure. (**a**) Schematic illustration of the EHT casting process. Pipette liquid agarose into the wells of a 24-well cell culture dish (1) and place one spacer per 4 wells into the agarose to prepare EHT casting molds (2). After the agarose has solidified, remove the spacer (3) and place silicone racks into the casting molds (4). Pipette the EHT master mix into the casting mold around two silicone posts (5). After incubation for 2 h at 37 °C, 7% CO_2, 40% O_2, and 98% humidity, carefully remove the silicone racks from the agarose casting molds (6) and transfer to a new culture dish filled with warm EHT culture medium (7). After 7–10 days in culture, EHTs have remodeled and start spontaneous coherent contraction (8). (**b**) Teflon spacer, (**c**) silicone rack, (**d**) human EHT between two silicone posts

4. While the agarose solidifies, prepare the master mix on ice. Calculate the appropriate master mix for the number of EHTs you want to generate (*see* **Note 7**). Follow the table below for guidance and add fibrinogen last (volumes include 10% excess to account for pipetting error):

Component	Per EHT	Per 24 EHTs
hiPSC-derived cardiomyocytes	1.1×10^6 cells	26.4×10^6 cells
EHT casting medium (including cells)	97.6 μL	2344.4 μL
2× DMEM	6.1 μL	147 μL
Y-27632	0.1 μL	2.6 μL
Fibrinogen	2.8 μL	66.8 μL

5. When the agarose is solid, remove the spacers by carefully moving them straight upwards to avoid damaging the casting molds.

6. Place the silicone racks into the plate. Make sure that each pair of silicone posts is correctly placed within one casting mold

since they are elastic and can bend. In general, sterile EHT racks are handled by clasping them with thumb and index of one hand (always use gloves) at both ends of the reinforced scaffold to avoid touching the posts and to minimize the touched surface.

3.5 Generation of Engineered Heart Tissue

1. Before pipetting the EHTs, gently resuspend your master mix with a serological pipette about 10 times and ensure that you have enough 3 μL aliquots of thrombin on ice (1 aliquot per EHT) and enough pipette tips in the safety cabinet.
2. For each single EHT, add 100 μL of your master mix into an aliquot of 3 μL thrombin in a 200 μL vial, mix by carefully pipetting up and down 2–3 times and quickly transfer the mixture into the casting mold with the silicone posts. Repeat this step for each EHT always using a fresh filter tip (*see* **Note 8**). Gently resuspend the master mix each time after pipetting 8 EHTs with a serological pipette.
3. Incubate the EHTs in their casting molds for 1.5 h at 37 °C, 7% CO_2, 40% O_2 and 98% humidity to allow the fibrinogen to polymerize.
4. To extract the EHTs from the agarose molds, cover each well with EHT casting medium or EHT culture medium (approximately 200–500 μL) and incubate them at 37 °C for an additional 30 min (*see* **Note 9**).
5. In the meantime, prepare a fresh 24-well plate with 1.5 mL warm EHT culture medium per well.
6. Slowly transfer the silicone posts with the attached EHTs into the prepared 24-well plate filled with EHT culture medium. As described previously, hold the silicone racks by clasping them with thumb and index of one hand at both ends of the reinforced scaffold to avoid touching the posts.

3.6 Maintenance

1. Cultivate the EHT at 37 °C, 7% CO_2, 40% O_2, and 98% humidity.
2. Change the EHT culture medium three times per week. Freshly prepare EHT culture medium on the same day (*see* **Note 1**). When changing the media for the first time, prepare a second 24-well plate. Pipette the pre-warmed fresh media into the empty plate and subsequently transfer EHTs from the plate containing the old medium to the plate containing the fresh medium. Keep the old plate and follow this procedure upon every following media change, always transferring the EHT racks manually from old to fresh media.
3. Monitor the remodeling of the EHT, which starts with microscopically observable single-cell contractions followed by coherent contraction (day 5–7). A standardized method to

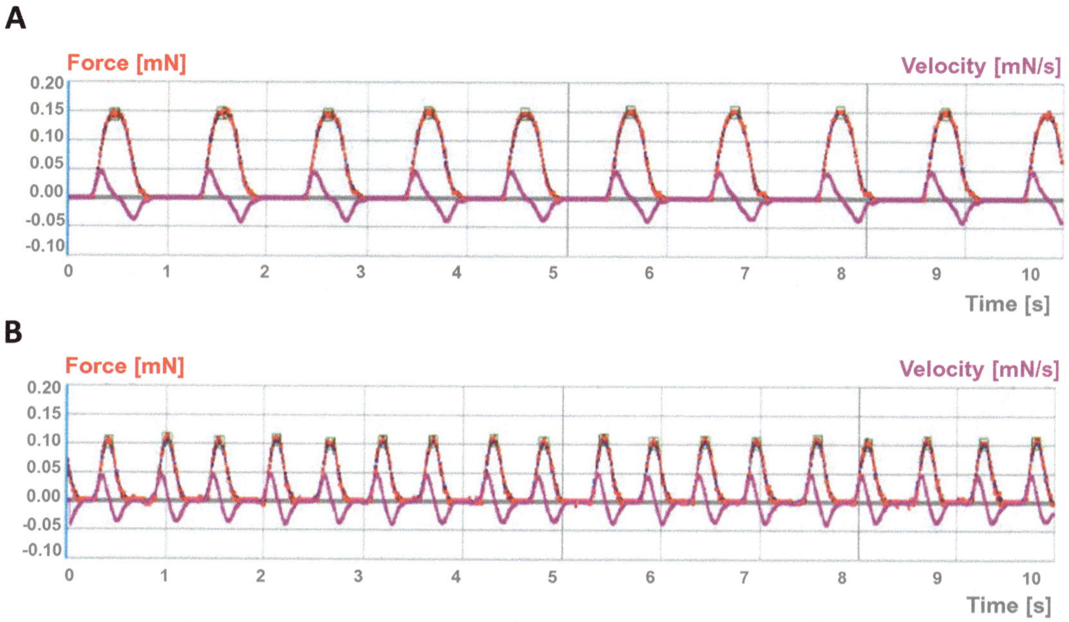

Fig. 2 Original contractility recording graphs of human ventricular and atrial EHT generated by CTMV software. The red line represents the original recording, the pink line shows the velocity. The graphs are representative for typical contractility of hiPSC-derived (**a**) ventricular and (**b**) atrial EHT at day 22

assess the contractility of EHT is by video-optical analysis (*see* Subheading 3.7).

4. After approximately 14 days, the atrial fibrin-based EHTs display longitudinally aligned cells with clearly distinguishable cell strand structure, perform regular coherent contraction, and beat continuously. Atrial and ventricular EHT differ in contractility regarding force, activity, and contraction time which can be measured by, for example, video-optical analysis (Fig. 2, also refer to Subheading 3.7) [12].

3.7 Video-Optical Contractility Analysis Using EHT Technologies Equipment (Fig. 3)

The EHT model can be analyzed in various ways to assess morphology, gene expression, or contractile function. Moreover, 3D tissue as a big advantage, offers the possibility for contractility measurements. To this end, suitable pattern recognition software, such as Musclemotion [28] or software provided by IonOptix can be used. For users working with the video-optical recording and analysis set up by EHT Technologies, we here provide a detailed click-to-click description as an initial guide through the software. We recommend following this procedure and consulting the manual in parallel, which is available from EHT Technologies.

1. Turn on the computer as well as the heating, gas supply, LEDs, and the axis system of the bench-top incubator and measurement device.

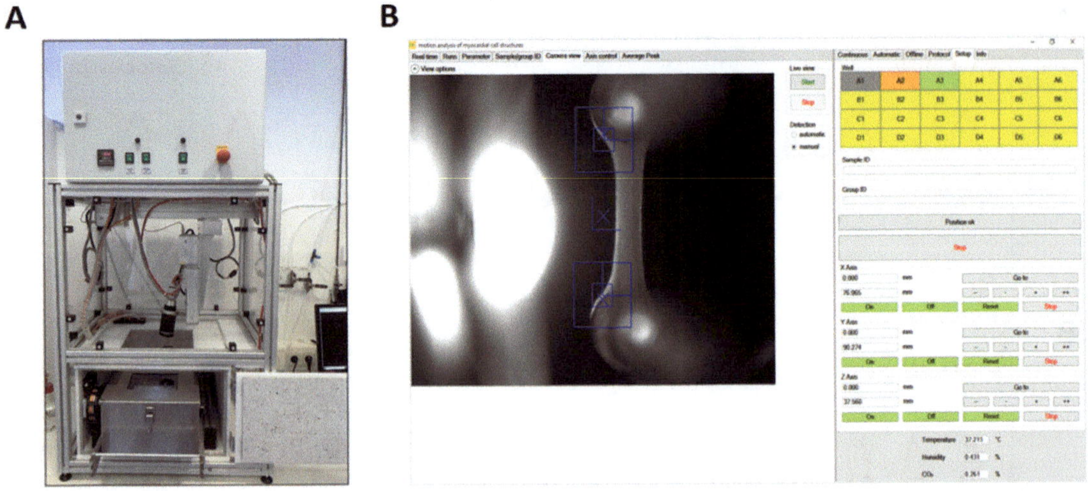

Fig. 3 Video-optical contractility analysis. (**a**) Setup of the EHT Technologies equipment including an incubator (bottom), camera, and electronics compartment (top). (**b**) Screenshot of the custom-designed CTMV software which allows for automatic contractility analysis by figure-recognition. After selecting the EHT and parameters for contractility measurements, the software automatically calculates beating frequency, force, contraction time, relaxation time, and rhythmicity

2. Start the CTMV software.
3. Before starting your measurement, wait for the incubator to reach 37 °C as well as elevated humidity and CO_2, which can all be monitored on the CTMV software start screen. Then, place the 24-well plate with the EHTs in the gas- and temperature-controlled incubator. Take care not to open the incubator for too long to prevent excessive cooling and loss of CO_2.
4. Move to the tab "Protocol." This panel provides different fields, which allow you to add relevant information including species, cell type, age of the EHTs, and also the duration of your measurement per EHT. Only the entered measurement duration will affect the actual measurement. All other meta information will appear in the PDF file report and be stored in this experimental "run" (*see* **step 4**). For future measurements, you can select a previous run by moving to the tab "Runs." The parameters of this measurement run will be loaded by the software and can then be modified for the new experiment.
5. Move to the tab "Setup" and select the wells you want to measure by double-clicking on them. Each well can be displayed in four different colors indicating that the well is either excluded (grey) or included in the measurement (yellow), currently selected (green), or that parameters have just been defined for contractility analysis (orange, *see* **step 10**). All wells you want to include should appear in yellow/orange.

6. Next, open the left side tab "Camera view" and start the "live" view of the EHTs by clicking on the start button below "Live view."
7. Select one well after another (color turns green) to optimize the *X*-, *Y*-, *Z*-coordinates for EHT recognition: The tab "Setup" on the right includes a panel of ± buttons, which allow to move the camera image left/right (*X*-axis), or up/down (*Y*-axis) on the screen and to adjust the camera focus (*Z*-axis). After optimization, the EHT should be focused and in the center of the live view with the illuminated LED light on the left side of the EHT.
8. The field "Sample ID" and "Group ID" allows for including sample descriptions. Each EHT will be analyzed separately and all EHTs with the exact same Group ID will be additionally analyzed as one group. Adding information here is optional.
9. Next, you have to define the EHT ends for pattern recognition and motion analysis by positioning the blue crosses. To this end, choose "Manual detection" under the tab "Camera view" which allows you to click on a region of high contrast between EHT and post to position the blue pattern recognition cross. Monitor EHT contractions to ensure movement of the blue crosses does indeed reflect the exact movement of the EHT and silicone post as force calculation depends on it.
10. When you have finished optimizing the parameters for each EHT, click the button "Position ok" before moving to the next well. When the well appears orange, the parameters were successfully modified and saved.
11. For automated contraction analysis, you need to next set the parameters for all wells by selecting the tab "Parameters" and the subcategory "Peak force" (*see* **Note 10**). We suggest to use the following standard parameters for automated contraction analysis of human EHT [29]:

    Peak force: 4.

    Filter level: 10.

    Baseline level: 0.95.

    Force threshold: 0.02–0.05 mN.

    Minimum factor: 0.2.

    Maximum peak distance: 10 s.

    Contraction velocity/relaxation velocity threshold factor: 0.6.

    These parameters are critical for the software to identify contraction peaks. Once you have selected these parameters for one well, click on the button "use parameters on all wells" to analyze all EHTs with the same settings.

For the subcategory "Force calculation," make sure to use the parameters, which accurately reflect the mechanical characteristics of your silicone posts, for example, elasticity and geometry. Otherwise, the software will not calculate the correct force.

12. Start your measurement by choosing the tab "Automatic" and clicking on the button "Start." To directly monitor the contraction of the EHTs while the device performs the automatic measurement, click on the tab "Real time" on the left side. In addition to displaying the recorded EHT in real-time, the software also presents a contractility graph. Correctly detected contraction peaks are marked by a green square at the top of each peak. Three differently colored lines represent the movement of the EHT (red, blue, pink). The red line represents the original recording, the blue line the filtered contractility graph, and the pink line shows the velocity (first derivative of the position curve).
13. When your measurements are finished, the software automatically generates a PDF report, which comprises different calculated contractile parameters for each individual EHT and group of EHTs including beating frequency, force, contraction, and relaxation time (*see* **Note 11**).

4 Notes

1. EHT culture medium: Always use freshly prepared medium for each media change. The horse serum should be freshly thawed, avoiding excess freeze-thaw cycles, and therefore be initially aliquoted, stored at −20 °C and only thawed once when preparing the EHT culture medium.
2. Human EHT: Make sure your differentiated stem cells contain at least 65% cardiac troponin T-positive cardiomyocytes. A protocol to determine the proportion of cardiac troponin T-positive cells by flow cytometry after dissociation has been described by Breckwoldt et al. [26].
3. Dissociated hiPSC-derived cardiomyocytes: Centrifuge the cells at 100 × *g* and 4 °C for 10 min, carefully resuspend the cell pellet in EHT casting medium and store them on ice to continue with Subheading 3.4. Note that cardiomyocytes are sensitive to high-speed centrifugation: maximally centrifuge at 100 × *g*.
4. Atrial identity: When establishing the atrial differentiation for the first time, we suggest starting with the differentiation of both ventricular and atrial-like cardiomyocytes in parallel to compare the expression of chamber-specific genes. RA

treatment during hiPSC differentiation should result in a markedly higher abundance of atrial-specific genes and much lower expression of ventricular markers. This will be more pronounced if you perform gene expression or histological analyses of atrial and ventricular EHT at the end of your experiments, as described by Lemme et al. [12].

5. Thawing of hiPSC-derived cardiomyocytes: The dropwise addition of the medium to the cells is essential to reduce osmotic shock.
6. Preparation of casting mouds: Maximally prepare 8 wells at a time since the agarose solidifies quickly. Prepare the agarose casting moulds immediately prior to use, as too long storage time might affect the quality of the agarose due to drying, which makes the generation of EHTs more difficult. Only remove the spacers after at least 10–15 min of incubation, otherwise the agarose might not have solidified properly to form optimal casting molds.
7. Preparation of master mix: Fibrinogen needs to be warm (room temperature) for pipetting and added last. To account for the viscosity of fibrinogen, fill your pipet tip very slowly to avoid pipetting errors. Directly add it to your cold master mix. During this step, do not touch the master mix surface to prevent the fibrinogen from becoming too viscous inside the pipette tip. Mix the master mix until the fibrinogen is completely dissolved. Since cardiomyocytes are sensitive to shear stress, only resuspend your master mix when all components are added, using a serological pipette and avoid air bubble formation. Repeat resuspension after 8 EHTs. Keep the master mix on ice at all times.

   In addition, note that the serum used for the EHT casting medium needs to be heat inactivated to avoid premature polymerization processes or even degeneration of the EHT matrix. For the same reason, when freezing CM, use heat-inactivated serum, too.
8. EHT generation: Change the filter tip for each EHT since thrombin remains will otherwise induce polymerization of the fibrinogen-containing master mix in your filter tip or even in the master mix tube. To avoid the formation of air bubbles in your EHT, mix and dispense the master mix for each EHT by only pushing to the first pressure point of your pipet to not empty the filter tip completely. Air bubbles in EHT will usually resolve over time, but may compromise EHT structure.
9. Transfer of EHTs: Covering the EHTs with liquid (e.g., DMEM, PBS) supports their detachment from the surrounding agarose and thus allows for undamaged transfer into EHT culture medium.

10. Video-optical measurements: The video-optical recordings are automatically analysed by custom-designed software (Consulting Team Machine Vision; CTMV). The software uses figure-recognition to track deflection of the EHT mounting posts during contraction. Based on the defined setup parameters and movement of the EHT, different values for beating frequency, force, contraction time, relaxation time, and rhythmicity are automatically calculated. For regular baseline contraction analysis, measure EHT approximately 1 h after medium change to allow for enough time for equilibration to the fresh medium.
11. Offline analysis: If you recognize errors in the selection of the parameters or the recognition of the EHT after contractility measurement, you can still perform additional offline analysis for correction, based on the recorded video files. Consult the manual for instructions on how to optimize figure-recognition.

References

1. Kirchhof P, Benussi S, Kotecha D et al (2016) 2016 ESC Guidelines for the management of atrial fibrillation developed in collaboration with EACTS. Eur Heart J 37:2893–2962
2. Iwasaki YK, Nishida K, Kato T et al (2011) Atrial fibrillation pathophysiology: implications for management. Circulation 124:2264–2274
3. Devalla HD, Schwach V, Ford JW et al (2015) Atrial-like cardiomyocytes from human pluripotent stem cells are a robust preclinical model for assessing atrial-selective pharmacology. EMBO Mol Med 7:394–410
4. Milani-Nejad N, Janssen PML (2014) Small and large animal models in cardiac contraction research: advantages and disadvantages. Pharmacol Ther 141:235–249
5. Blazeski A, Zhu R, Hunter DW et al (2012) Cardiomyocytes derived from human induced pluripotent stem cells as models for normal and diseased cardiac electrophysiology and contractility. Prog Biophys Mol Biol 110:166–177
6. Ma J, Guo L, Fiene SJ et al (2011) High purity human-induced pluripotent stem cell-derived cardiomyocytes: electrophysiological properties of action potentials and ionic currents. Am J Physiol Circ Physiol 301:H2006–H2017
7. van den Berg CW, Elliott DA, Braam SR et al (2016) Differentiation of human pluripotent stem cells to cardiomyocytes under defined conditions. Methods Mol Biol 1353:163–180
8. Marczenke M, Piccini I, Mengarelli I et al (2017) Cardiac subtype-specific modeling of Kv1.5 ion channel deficiency using human pluripotent stem cells. Front Physiol 8:469
9. Lee JH, Protze SI, Laksman Z et al (2017) Human pluripotent stem cell-derived atrial and ventricular cardiomyocytes develop from distinct mesoderm populations. Cell Stem Cell 21:179–194.e4
10. Zhang Q, Jiang J, Han P et al (2011) Direct differentiation of atrial and ventricular myocytes from human embryonic stem cells by alternating retinoid signals. Cell Res 21: 579–587
11. Zaffran S, Robrini N, Bertrand N (2014) Retinoids and cardiac development. J Dev Biol 2: 50–71
12. Lemme M, Ulmer M, Lemoine MD et al (2018) Atrial-like engineered heart tissue: an in vitro model of the human atrium. Stem Cell Reports 11:1378–1390
13. Cyganek L, Tiburcy M, Sekeres K et al (2018) Deep phenotyping of human induced pluripotent stem cell–derived atrial and ventricular cardiomyocytes. J Clin Invest 3(12):e99941
14. Besser RR, Ishahak M, Mayo V et al (2018) Engineered microenvironments for maturation of stem cell derived cardiac myocytes. Theranostics 8:124–140
15. Weinberger F, Mannhardt I, Eschenhagen T (2017) Engineering cardiac muscle tissue. Circ Res 120:1487–1500
16. Kolanowski TJ, Antos CL, Guan K (2017) Making human cardiomyocytes up to date: derivation, maturation state and perspectives. Int J Cardiol 241:379–386
17. Schaaf S, Eder A, Vollert I et al (2014) Generation of strip-format fibrin-based engineered

heart tissue (EHT). Methods Mol Biol 1181: 121–129

18. Eder A, Vollert I, Hansen A et al (2016) Human engineered heart tissue as a model system for drug testing. Adv Drug Deliv Rev 96: 214–224
19. Mannhardt I, Breckwoldt K, Letuffe-Brenière D et al (2015) Human engineered heart tissue: analysis of contractile force. Stem Cell Reports 7:29–42
20. Ulmer BM, Stoehr A, Schulze ML et al (2018) Contractile work contributes to maturation of energy metabolism in hiPSC-derived cardiomyocytes. Stem Cell Reports 10:834–847
21. Lemoine MD, Krause T, Koivumäki JT et al (2018) Human iPSC-derived engineered heart tissue as a sensitive test system for QT prolongation and arrhythmic triggers. Circ Arrhythmia Electrophysiol 11:e006035
22. Lemoine MD, Mannhardt I, Breckwoldt K et al (2017) Human iPSC-derived cardiomyocytes cultured in 3D engineered heart tissue show physiological upstroke velocity and sodium current density. Sci Rep 7:5464
23. Goldfracht I, Protze S, Shiti A et al (2020) Generating ring-shaped engineered heart tissues from ventricular and atrial human pluripotent stem cell-derived cardiomyocytes. Nat Commun 11:1–15
24. Lemme M, Braren I, Prondzynski M et al (2019) Chronic intermittent tachypacing by an optogenetic approach induces arrhythmia vulnerability in human engineered heart tissue. Cardiovasc Res 116(8):1487–1499
25. Lee MO, Jung KB, Jo SJ et al (2019) Modelling cardiac fibrosis using three-dimensional cardiac microtissues derived from human embryonic stem cells. J Biol Eng 13:1–17
26. Breckwoldt K, Letuffe-Brenière D, Mannhardt I et al (2017) Differentiation of cardiomyocytes and generation of human engineered heart tissue. Nat Protoc 12:1177
27. van Meer BJ, de Vries H, Firth KSA et al (2017) Small molecule absorption by PDMS in the context of drug response bioassays. Biochem Biophys Res Commun 482:323–328
28. Sala L, Van MBJ, Tertoolen LGJ et al (2018) Musclemotion: a versatile open software tool to quantify cardiomyocyte and cardiac muscle contraction in vitro and in vivo. Circ Res 122: e5–e16
29. Mannhardt I, Saleem U, Benzin A et al (2017) Automated contraction analysis of human engineered heart tissue for cardiac drug safety screening. J Vis Exp 2017:1–10

Chapter 12

Design and Fabrication of Biological Wires for Cardiac Fibrosis Disease Modeling

Erika Yan Wang, Jacob Smith, and Milica Radisic

Abstract

Extensive progress has been made in developing engineered models for elucidating human cardiac disease. Cardiac fibrosis is often associated with all forms of cardiac disease and has a direct deleterious effect on cardiac function. As currently there is no effective therapeutic strategy specifically designed to target fibrosis, in vitro diagnostic platforms for drug testing have generated significant interest. In this context, we have developed an innovative approach to generate human cardiac fibrotic tissues on Biowire II platform and established a compound screening system. The disease model is constructed to recapitulate contractile, biomechanical, and electrophysiological complexities of fibrotic myocardium. Additionally, an integrated model with fibrotic and healthy cardiac tissues coupled together can be created to mimic focal fibrosis. The methods for constructing the Biowire fibrotic model will be described here.

Key words Cardiac fibrosis, Disease modeling, Biowire, Microfabrication, Soft lithography, Replica molding, Hot embossing, Electrical stimulation, Functional assessment, Force sensor

1 Introduction

Cardiac fibrosis is often associated with all forms of cardiac conditions and is believed to have a pivotal role in causing heart failure [1]. It is characterized by activation of myofibroblasts and excessive deposition of extracellular matrix in the fibrotic loci. To date, there is no efficient anti-fibrotic therapeutic approach available. A high-fidelity in vitro model of cardiac fibrosis can potentially elucidate the disease mechanisms and facilitate the development of therapeutic strategies. However, commonly used monolayer cell cultures cannot fully capture the structural and functional properties of fibrotic myocardium.

Recent advances in organ-on-chip (OOC) technologies have led to the development of powerful biomimetic systems for elucidating human disease [2, 3]. Several OOC models of myocardial fibrosis have been reported [4, 5]. These engineered systems

Kareen L. K. Coulombe and Lauren D. Black III (eds.), *Cardiac Tissue Engineering: Methods and Protocols*, Methods in Molecular Biology, vol. 2485, https://doi.org/10.1007/978-1-0716-2261-2_12,

contain more sophisticated topographic and biomechanical cues for disease modeling than cells growing on a flat petri dish. The most common limitation of the pre-existing OOC models lies in the ability to capture key hallmarks of adult-like human myocardium and enable high-content functional readouts predictive of cardiac function. Moreover, most platforms require complex fabrication procedures that are difficult to scale up for high-content testing.

To address this issue, our group developed Biowire fibrotic model, a disease-on-a-chip system exhibiting biomechanical and electrophysiological features of fibrotic cardiomyopathy in the adult human heart [6]. The design of the device was geared towards a streamlined manufacturing process based on standard microfabrication techniques including soft lithography, replica molding, and hot embossing. In the Biowire II platform, cylindrical cardiac tissues are cultured in an array of micro-chambers on a polystyrene chip, and fitted between two poly (octamethylene maleate (anhydride) citrate) (POMaC) wires [6, 7]. Application of the inert polystyrene as the base material reduces absorption of small hydrophobic molecules compared to commonly used PDMS devices [8].

The cellular composition and mechanical properties of the tissues are tailored by systematically manipulating the cell populations in a tuned hydrogel. Pacing cardiac tissues with electrical stimulation has been shown to promote tissue maturation over time, as well as standardize beating frequency during functional assessment to enhance testing reliability [9]. This platform design enables structural and functional maturation of three-dimensional constructs with electrical conditioning. In addition, a heteropolar Biowire tissue can be created by spatially patterning the fibrotic and normal compartments in a same construct. This scar-myocardium integrated model can be used to model focal fibrosis, and mimic the interaction between scar lesion and adjacent healthy myocardium [6].

A key feature of the Biowire II platform is the ability to noninvasively assess cardiac function on the basis of tissue contractility and compaction without disturbing long-term tissue cultivation. As POMaC is intrinsically fluorescent, the deflection of the polymer wire due to tissue contraction can be isolated from tissue movement and recorded under blue fluorescent light. Thus, this platform uses the deflection of the polymer wires to continuously measure absolute contractile systolic and diastolic properties, thereby allowing real-time functional readouts of active force, passive tension, and force dynamics. The high-fidelity in vitro model system combined with convenient functional readouts enables the use of the system in the evaluation of anti-fibrotic compounds. In this chapter, we will describe the design, fabrication as well as functional assessment methods of the Biowire fibrotic model in detail.

2 Materials

2.1 SU-8 Master Molds

Photomasks for fabricating the master molds are designed using AutoCAD and printed using laser pattern generator (Heidelberg Instruments Mikrotechnik GmbH) onto negative photoresist chrome-glass masks, or onto film masks (CADART).

The master molds derived from the photomasks are fabricated on 6-inch silicon wafers by soft lithography using negative photoresist SU-8 (MICRO CHEM).

2.2 PDMS Molds

Polydimethylsiloxane (PDMS) molds are made by replica molding from the SU-8 master molds by mixing 184 silicone elastomer base and curing agent (Sylgard 184 silicone elastomer kit, Dow Corning).

2.3 POMaC Prepolymer Solution

POMaC prepolymer solution for wire fabrication is synthesized from citric acid (Caledon), maleic anhydride (Sigma), and 1,8-octanediol (Sigma). Purified prepolymer is mixed with poly (ethylene glycol) dimethyl ether (PEGDM) (Sigma) and 2-hydroxy-1-[4(hydroxyethoxy)phenyl]-2-methyl-1-propane (Irgacure 2959) (Sigma) to form the final solution (*see* Subheading 3.4, **step 1**).

2.4 Polystyrene Chips

The PDMS mold for the chip fabrication is fixed to a silicon wafer by plasma bonding using a plasma etcher.

The features are imprinted onto a semi-clear white polystyrene sheet (McMaster-Carr) using a semi-automated hot embossing system (EVG520).

Clear polyurethane two-part adhesive (GS Polymers Inc.) is used to fix the polymer wires on the chip.

2.5 Cells and Cell Culture Media

Predominantly ventricular cardiomyocytes (CMs) are derived from the human-induced pluripotent stem cell (hiPSC) line BJ1D. Commercially available iCell human cardiomyocytes (CDI) can be used as alternative cell source.

Human ventricular cardiac fibroblasts (cFBs) are obtained from LONZA (Clonetics NHCF-V). FGM-3 Cardiac Fibroblast Growth Medium-3 (Lonza) is used for fibroblast culture.

Induction 3 Medium (I3M) (StemPro-34 complete media, 20 mM HEPES, 1% GlutaMAX, 1% penicillin–streptomycin, Life Technologies; 213 μg/mL 2-phosphate ascorbic acid, Sigma-Aldrich; 150 μg/mL transferrin, Sigma-Aldrich) is used for tissue cultivation.

2.6 Fibrin-based hydrogel

The hydrogel for cell seeding is prepared by combining 33 mg/mL fibrinogen (Sigma-Aldrich) in Hanks balanced salt solution (Sigma-Aldrich) with 15% (v/v) growth factor reduced Matrigel

(Corning). 25 IU/mL thrombin (Sigma-Aldrich) is used for converting fibrinogen to fibrin (*see* Subheading 3.5, **step 4**).

2.7 Electrical Stimulation Chamber and External Electrical Stimulator

Electrical stimulation chamber is made using 1/8 inch-diameter carbon rods (Ladd Research Industries), 10 cm tissue culture dish, polyurethane two-part adhesive (GS Polymers), and platinum wires (Ladd Research). Field electrical stimulation is applied by an external electrical stimulator (Grass Technology S88X Square Pulse Stimulator) (*see* Subheading 3.7, **step 1**).

3 Methods

Microfabrication is carried out in the cleanroom; reagent preparation and cell culture handling are carried out in a biological safety cabinet under sterile conditions in a biosafety level 2 laboratory.

3.1 Part I. Platform Design and Fabrication: Photomask Design and Preparation

1. Photomasks for fabricating the tissue chip and polymer wires are designed using AutoCAD. The photomasks for chip fabrication consist of a repeating pattern consisting with rectangular micro-chambers (5 mm × 1 mm × 300 μm, L × W × H), spaced at 4 mm apart horizontally and 8 mm vertically. The micro-chambers are interconnected by two parallel grooves (200 μm × 100 μm, W × H) (*see* Fig. 1a). The photomask for wire fabrication consists of an array of channels (100 μm × 100 μm, W × H) (*see* Fig. 2a).
2. The designs are patterned on photomasks ordered from CADART, setting the device features as transparent and the surrounding regions dark. Alternatively, Heidelberg® μPG101 laser pattern generator can be used to print the AutoCAD design onto photoresist chrome-glass masks.

3.2 SU-8 Master Mold and PDMS Mold Fabrication

1. The SU-8 master molds are fabricated on silicon wafers using standard soft lithography technique. Negative photoresist SU-82050 is used based on the protocol provided by MicroChem®. For greater clarity, a standard protocol for this technique can be followed [10].
2. PDMS molds used to produce polymer wires and to enable hot embossing of the polystyrene base are made by replica molding from the SU-8 master molds. 184 silicone elastomer base is mixed with curing agent at a 15:1 ratio, and cast onto the SU-8 molds. Leave at room temperature for 48 h to allow cross-linking (*see* **Note 1**).

3.3 Polystyrene Chip Fabrication

1. Fix the back of PDMS mold for the chip fabrication on a 6-inch silicon wafer by plasma bonding or corona etching.

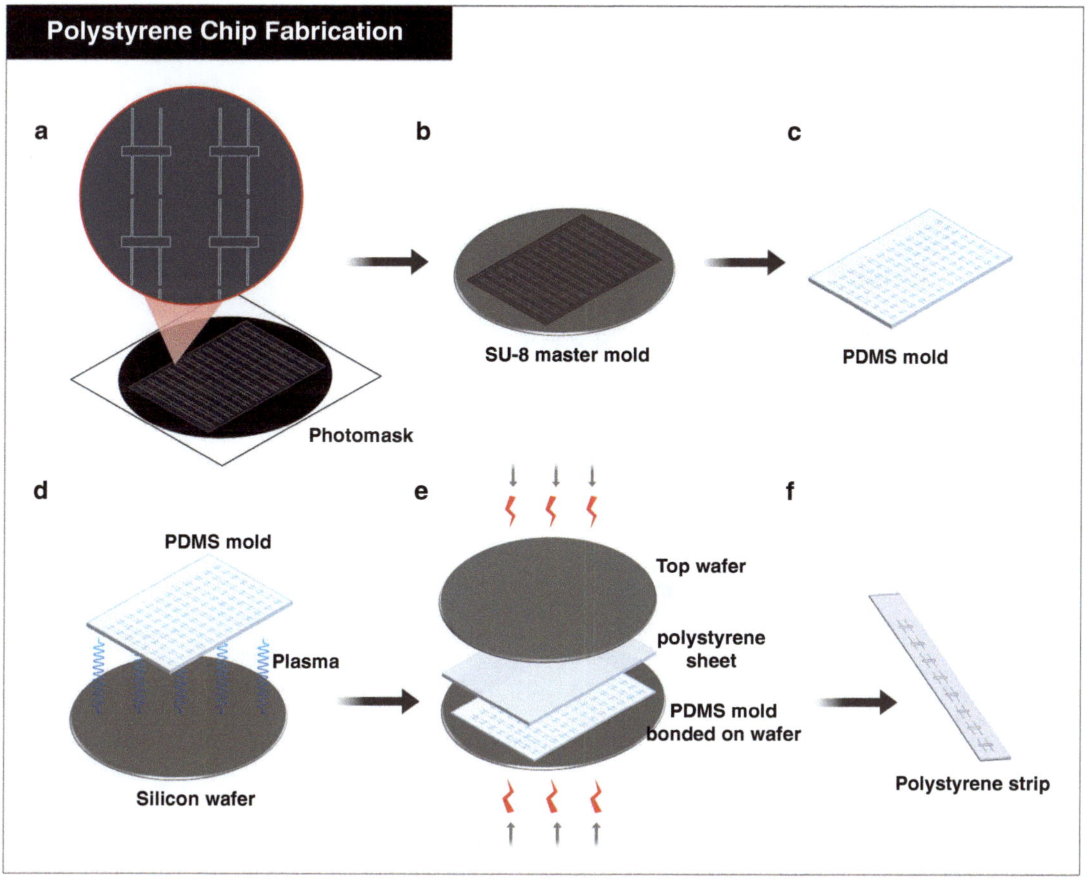

Fig. 1 Schematic illustrations of polystyrene chip fabrication. (**a**) The design of photomasks for chip fabrication. The pattern of rectangular micro-chambers and parallel grooves is shown in the zoomed-in window. (**b**) SU-8 master mold fabricated on a silicon wafer using soft lithography. (**c**) PDMS mold for hot embossing made by replica molding from the SU-8 master mold. (**d**) Plasma bonding of the PDMS mold and a 6-inch silicon wafer. (**e**) Hot embossing procedure to produce the polystyrene base. (**f**) Polystyrene strip containing eight micro-chambers.

2. In a hot embosser, the PDMS master is exposed to 150 °C and 3500 N for 30 min. A polystyrene sheet is molded against the PDMS master to yield a customized polystyrene base. Specifically, in the hot embosser the polystyrene sheet is placed on top of the silicon wafer with the PDMS master mold (*see* **Note 2**). Then, a top silicon wafers and a graphite disc are added on the polystyrene sheet to even out the pressure (*see* Fig. 1e).
3. Cut the polystyrene sheet into strips containing eight micro-chambers with a razor blade. Safety goggles must be worn.

3.4 POMaC Wire Fabrication and Assembly

1. First, polymer is synthesized as previously reported [11, 12]. Briefly, the pre-POMaC is formed by mixing and melting 1,8-octanediol, maleic anhydride, and citric acid

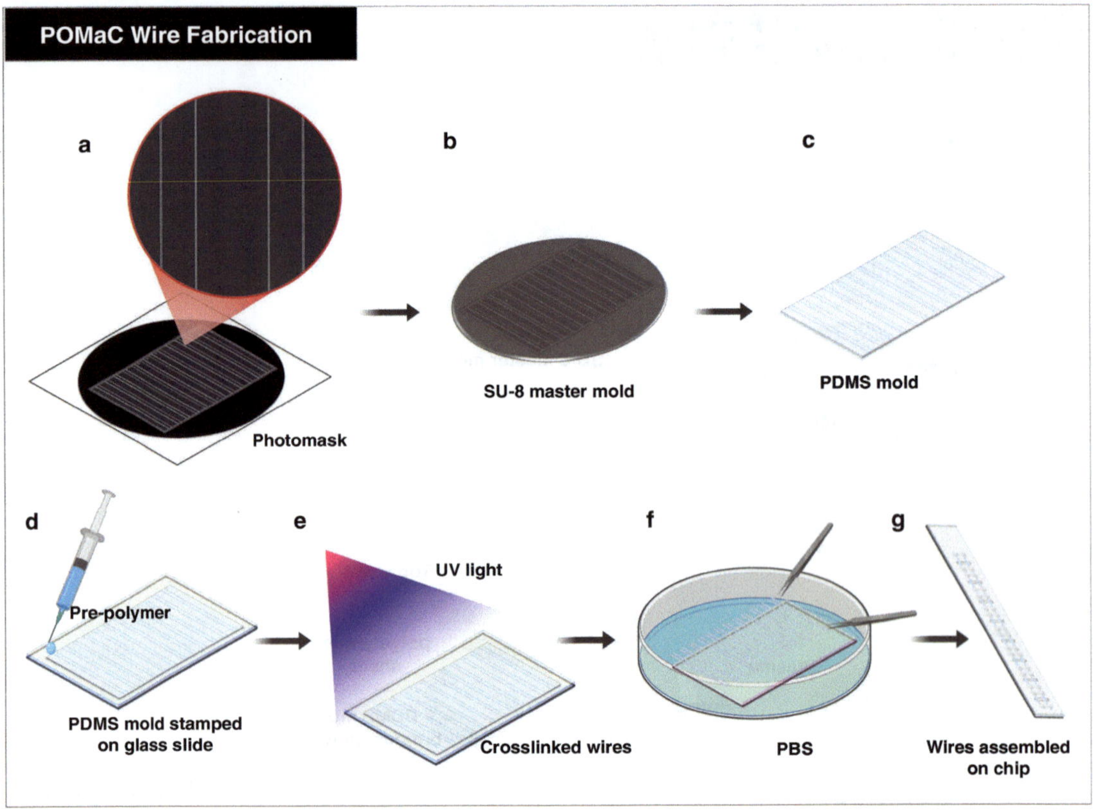

Fig. 2 Schematic illustrations of the polymer wires preparation and assembly. (**a**) The design of photomasks for wire fabrication. The pattern of channels is shown in the zoomed-in window. (**b**) SU-8 master mold fabricated on a silicon wafer using soft lithography. (**c**) The PDMS mold for wire fabrication made by replica molding. (**d**) POMaC prepolymer perfusion on a glass slide. (**e**) UV exposure to generate crosslinked elastomeric wires. (**f**) Wire release from the glass slide in PBS. (**g**) Assembled wires on polystyrene strip.

under nitrogen flow at 160 °C. Polycondensation is carried out to obtain a purified concentrated prepolymer. POMaC polymer solution is prepared by mixing the prepolymer with 5% (w/w) UV initiator Irgacure 2959 and 40% (w/w) poly (ethylene glycol) dimethyl ether (PEGDM). The solution can be stored at 4 °C until wire fabrication.

2. Press the PDMS mold for wire production onto a 178 × 127 mm large microscope slide (Agar Scientific) (*see* **Note 3**). Dispense a drop of POMaC prepolymer solution to one end of the wire channel with syringe. Leave it for 48 h and protect from light at room temperature, allowing the solution to perfuse through the micro-channels by capillary action (*see* Fig. 2d).

3. Following full perfusion, expose the POMaC prepolymer to UV light under a mask aligner (OAI) to generate an array of crosslinked elastomeric wires. Total energy of 5100 mJ is

applied at an intensity of 10 mJ/s for 8.5 min to crosslink the polymer (*see* **Note 4**).

4. Gently remove the PDMS layer, crosslinked wires should remain on the glass slide due to the stronger adhesion of the POMaC wires to glass than PDMS.
5. Soak the POMaC wires in phosphate-buffered saline (PBS) to release them from the glass slide.
6. Manually place the wires into the two parallel grooves patterned into the polystyrene strip. To do so, pick up the wires from both ends using fine tweezers to lift the wires form PBS, align the wire with the groove and place it in the groove. Do not overstretch the wire during the process (*see* **Note 5**).
7. Clear polyurethane two-part adhesive is used in a minimum quantity to fix the POMaC wires in place in between each micro-chamber (*see* **Note 6**). Once solidified, seal the strips in plastic packages and sterilize them by gamma irradiation.
8. Store the strips under dry condition before seeding (*see* **Note 7**).

3.5 Part II. Tissue Construction and Assessment: Generation of Interstitial Fibrotic Cardiac Tissues and Healthy Controls

1. Polystyrene strips are transferred to a sterile 10 cm tissue culture dish in a biosafety cabinet. Rinse the strip surface with 5% (w/v) Pluronic Acid (filter sterilized using 0.22 μm syringe filter) for around 5 min. Aspirate the solution from the dish and air dry the strips in the biosafety cabinet.
2. Ventricular cardiomyocytes are derived from the human-induced pluripotent stem cell (hiPSC) line BJ1D using the monolayer differentiation protocol previously described by the Palecek group [13]. At day 21 of differentiation, cardiomyocytes are disassociated into single cells. iCell cardiomyocytes can be freshly thawed according to the manufacturer's instructions before seeding (*see* **Note 8**). Ventricular cardiac fibroblasts are passaged on T75 cell culture flasks and disassociated into single cells before seeding (*see* **Note 9**).
3. Dissociated CMs and cFBs are mixed in 1:3 (fibrotic) and 3:1 (control) cell number ratios for constructing interstitial fibrotic tissues and healthy control tissues. Centrifuge CM and cFB cell mixture and aspirate remaining media (*see* **Note 10**). Place the pellets on ice while preparing hydrogel (*see* Fig. 3).
4. Prepare the Fibrin/Matrigel hydrogel by combining 33 mg/mL fibrinogen with 15% (v/v) growth factor reduced Matrigel in a 3:1 ratio. Add a volume of 0.5 μL of 25 IU/mL thrombin to each micro-chamber prior to seeding (*see* **Note 11**).
5. Resuspend the cell pellets at a concentration of 5.5×10^7 cells/mL in the hydrogel. Pipette several times to get homogenous mixture.

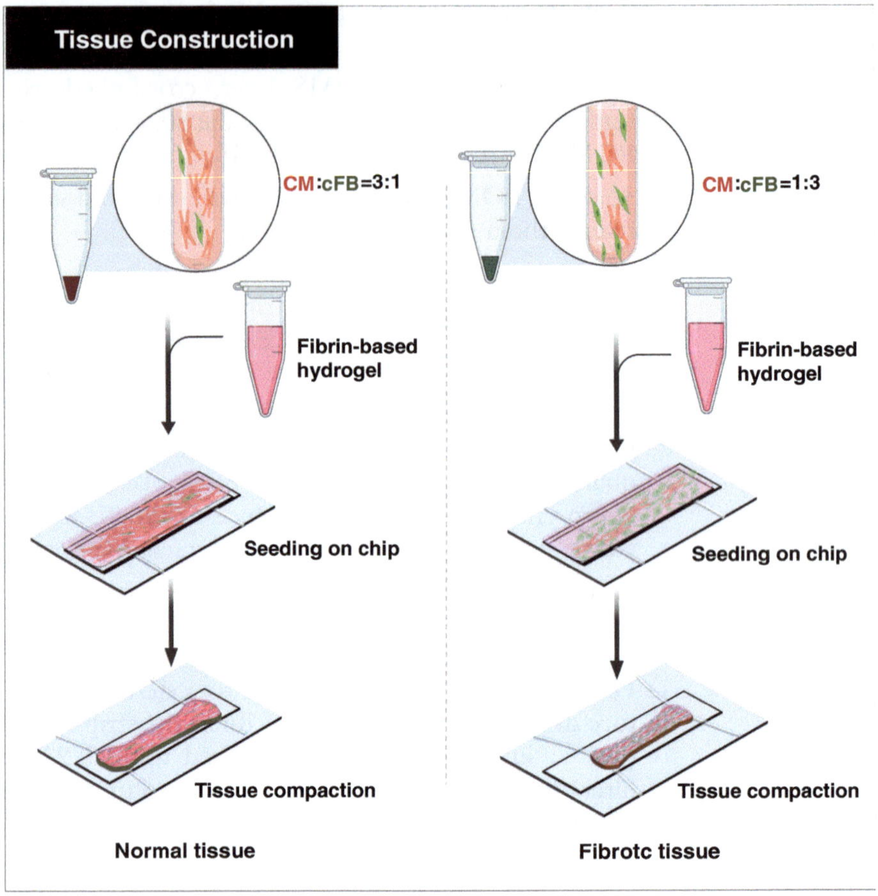

Fig. 3 A general scheme for construction of the normal and fibrotic tissues. Dissociated cardiomyocytes (CM) and cardiac fibroblasts (cFB) are mixed in and 3:1 (normal) and 1:3 (fibrotic) cell number ratios for constructing healthy control and fibrotic tissues. Pelleted cells are resuspended in a fibrin-based hydrogel and seeded into micro-chambers on the polystyrene strip. Tissues are cultured for 7 days to allow compaction before initiating electrical stimulation.

6. Gently add 2 μL cell-hydrogel suspension into each micro-chamber on the polystyrene strip to give a final seeding concentration of 1.1×10^5 cells/micro-chamber (*see* **Note 12**).
7. The plate is incubated at 37 °C and 5% CO_2 for 10 min to allow gelation to complete (*see* **Note 13**).
8. A volume of 10 mL of I3M culture media is then added to the tissue culture dish in the biosafety cabinet, ensuring that the media is not directly applied to the tissue but to the side of the dish. Supplement the culture media with 10 μM Aprotinin for the first 7 days of culture (*see* **Note 14**).

3.6 Generation of Focal Fibrotic Cardiac Tissues

1. To construct the focal fibrotic tissue with the two opposing sides, dissociated CM and cFB are mixed in 3:1 and 1:3 cell number ratio in two sterile microcentrifuge tubes for generating the normal and fibrotic sides of the integrated tissue.
2. The mixed cells are pelleted and resuspended at a concentration of 5.5×10^7 cells/mL in hydrogel in separate tubes.
3. Seed 1 μL of cell-hydrogel mixture containing 25% cFB and 75% CMs into one side of the micro-chamber. Tilt the tissue culture dish at an angle of 20°–30° immediately following seeding to prevent migration of the mixture in the micro-chamber. Leave the culture dish in the biosafety cabinet at room temperature for 10 min for gelation.
4. Seed 1 μL of cell-hydrogel mixture containing 75% cFBs and 25% CMs on the other side of the same micro-chamber. Repeat **step 3** to prevent the mixture from spreading to the opposing side.
5. Add a volume of 10 mL of culture media to the tissue culture dish after gelation. Label the tissues with corresponding sides on the lid.

3.7 Fabrication of Electrical Stimulation Chambers

1. A pair of 1/8 inch-diameter carbon rods are fixed with 1 cm distance (inner edge-to-inner edge) onto the bottom of a 10 cm petri dish using polyurethane two-part adhesive.
2. Two 5 cm-long platinum wires are affixed to opposing ends of each carbon rod by wrapping it around the rod (*see* **Note 15**) and leading the wires out of the chamber under the lid of the petri dish (*see* Fig. 4a). Connect the platinum wires to alligator clip lead wire of an external electrical stimulator (*see* Fig. 4b).
3. The electrical stimulation chamber is sterilized with the lid on by gamma irradiation (*see* **Note 16**).

3.8 Electrical Stimulation

1. After seeding, tissues are first cultured for 7 days to allow for remodeling and compaction around the POMaC wires.
2. After 7 days of culture, transfer each strip of eight tissues to an electrical stimulation chamber and place the strip right in between the electrodes (*see* Fig. 4a).
3. Start electrical stimulation at 1 Hz on day 7, followed by 1 Hz weekly step-up until the frequency reached 6 Hz (*see* Fig. 4c).
4. The stimulation voltage is adjusted weekly to 1.5 times excitation threshold (ET) down to a minimum voltage of 3.5 V/cm.
5. End point assessments are performed when a positive force-frequency relationship (FFR) (at least 1–3 Hz) is achieved. If a positive FFR is not observed once the tissues reached 6 Hz, you may continue stimulation at 6 Hz until a positive FFR is observed (*see* **Note 17**).

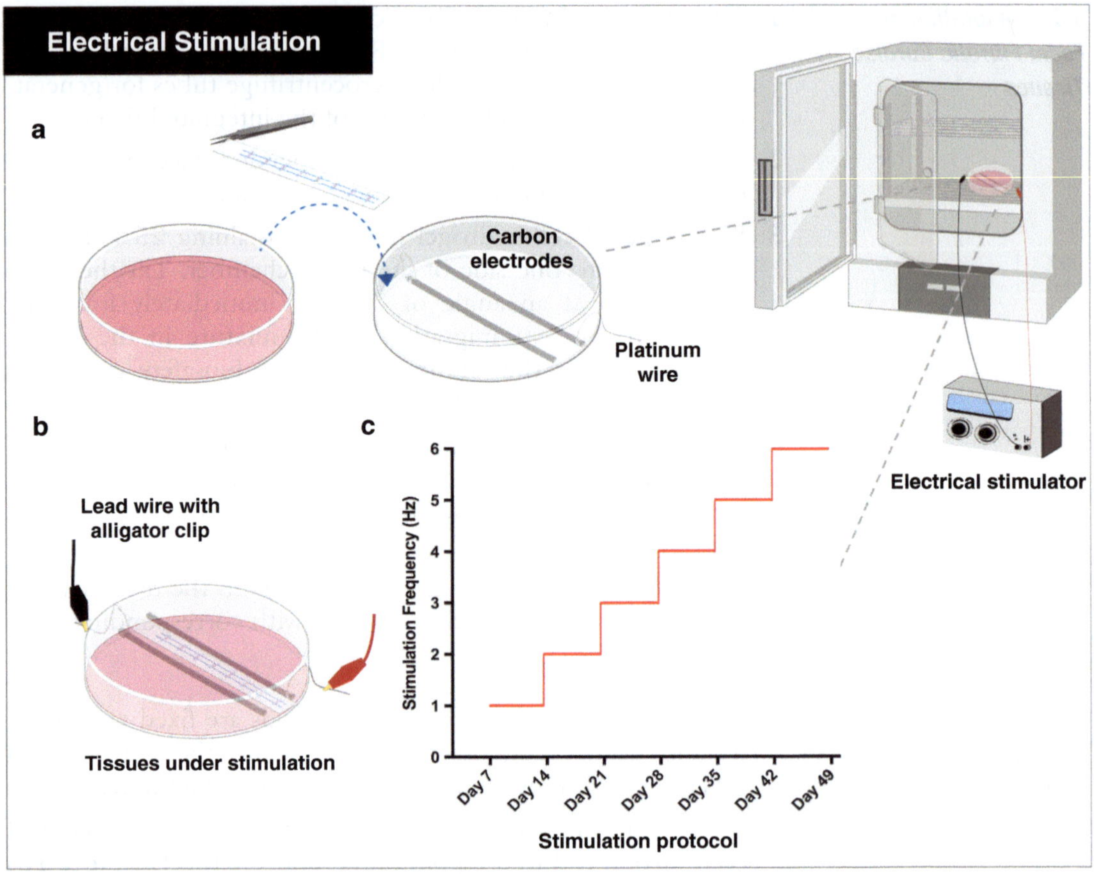

Fig. 4 Electrical stimulation setup and protocol. (**a**) After 7 days of culture, a strip of eight tissues is transferred to an electrical stimulation chamber in between the electrodes. (**b**) The platinum wires are connected to alligator clip lead wires of an external electrical stimulator to initiate stimulation. (**c**) Electrical stimulation is initiated at 1 Hz on day 7, followed by 1 Hz weekly step-up until the frequency reached 6 Hz.

3.9 Tissue Compaction Assessment

1. The tissues undergo a process of gel compaction over the first 7 days of culture, wherein the ECM becomes denser in the fibrotic tissues compared to control. This process is used to monitor ECM remodeling noninvasively without disturbing tissue cultivation [6].
2. During this time course, daily brightfield images of the tissues are taken using an Olympus CKX41-inverted microscope and CellSens software (Olympus Corporation).
3. The average tissue length and width are measured to generate the tissue compaction curve.

3.10 Polymer Wire Force-Displacement Curves

1. The force required to displace the POMaC wire is determined by mechanical testing using a microscale mechanical tester MicroSquisher (Cell Scale).

2. Custom tips of three different diameters (0.5 mm, 0.7 mm, and 0.8 mm) are designed to model different tissue widths on the POMaC wire. The tips are fabricated from an SU-8 master by soft lithography, and attached to a 0.1524 mm diameter tungsten probe using T-GSG-01 Titan Gel.
3. The polymer wires assembled on the strip are soaked in tissue culture media for 7 days. Prior to testing, the polystyrene strip is affixed to a 10 cm petri dish and topped up with 10 mL media.
4. Place the probe tip at one end of micro-chamber and move it towards the wire at a velocity of 2.5 mm/s. Upon reaching the wire at midsection, apply force perpendicular to the long axis of the wire. Record force over the probe displacement range of 0–150 mm.
5. The experimental data for each custom tip are fit to a third-degree polynomial equation, generating a force-displacement calibration curve for corresponding tissue width.

3.11 Functional Assessment and Force Measurement

1. To record cardiac tissue contractile pattern and force displacement for the force calculation, 4× brightfield movies of tissue beating spontaneously and under stimulation at 1 Hz are taken on day 7 and weekly thereafter.
2. Excitation threshold (ET) indicating the minimum voltage/cm required to stimulate the synchronized contraction of the tissue, and maximum capture rate (MCR) indicating highest frequency the tissue could contract in response to the stimulation pulse at two times ET are measured by changing pacing voltage and frequency, respectively.
3. 10× blue channel movies are taken using a fluorescence microscope to record the bending movement of the POMaC wire during tissue contraction from 1 to 6 Hz stimulation to measure contractile force and the force-frequency relationship (FFR) (*see* **Note 18**).
4. To measure post-rest potentiation (PRP) of the tissue, after being stimulated at 6 Hz for 20 s, stimulation is stopped for 10 s (rest period) and reinitiated at 1 Hz.
5. In a still frame of the 4× brightfield videos in the relaxed position, the width of the Biowire tissues is measured at three points along the central length to determine the average width of the tissue. The width of the tissue on the polymer wire (Tw) is also measured (*see* Fig. 5).
6. In the 10× blue channel videos, passive wire deflection is measured at the tissue resting state (*see* **Note 19**). Blue channel image sequences are analyzed using a custom MATLAB code that traces the maximum deflection of the POMaC wire (*see* Fig. 5).

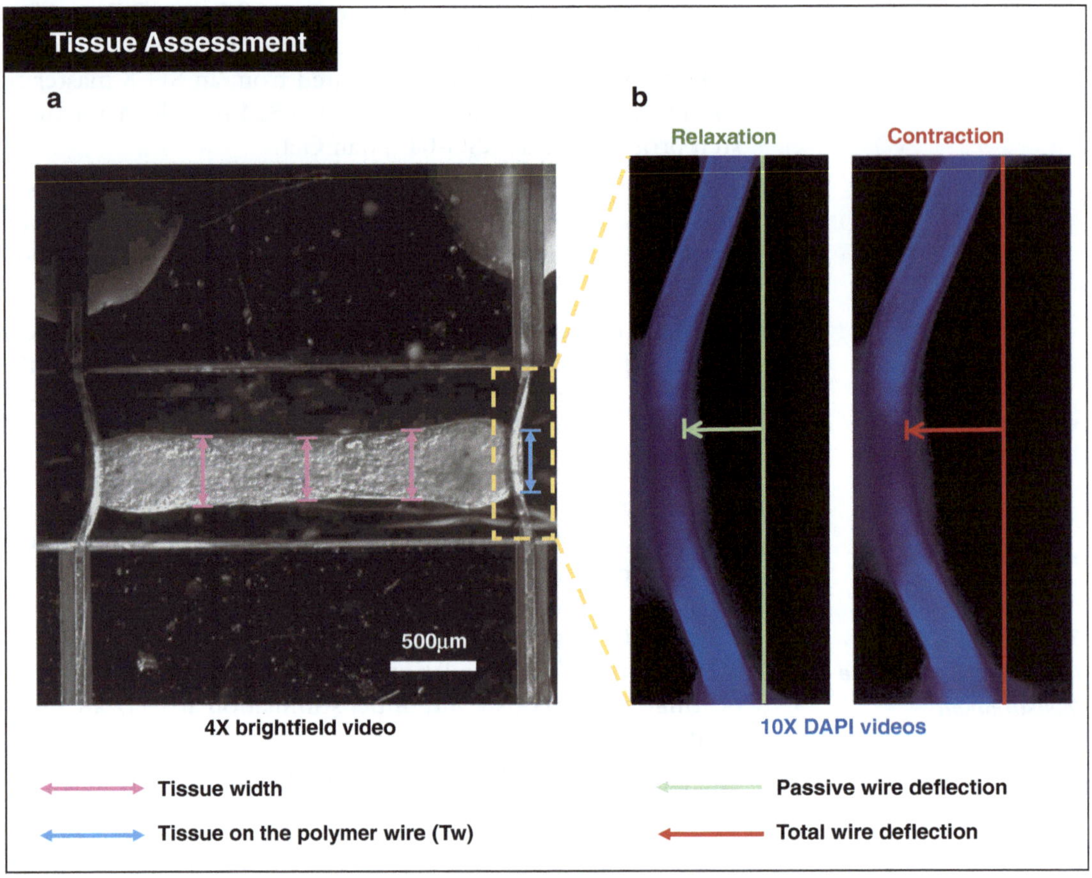

Fig. 5 Tissue assessment based on polymer wire deflection. (**a**) The tissues are imaged in a stimulation chamber using an inverted fluorescent microscope. The average tissue width (pink lines) and width of the tissue on the polymer wire (Tw) (blue line) are measured from still frames of the 4× brightfield videos in the relaxed position. (**b**) 10× blue channel movies are taken to record the bending movement of the POMaC wire. Total (red line) and passive (green line) wire deflection are recorded and converted to force measurements based on the standard forces curves using MATLAB.

3.12 MATLAB Data Analysis

1. Before being loaded into MATLAB, each video is cropped to a height of 50 pixels and a width of 160 pixels from the center of the fluorescent wire.
2. Two MATLAB scripts, available by email, are used for the analysis. First, run the MATLAB script for wire tracking. Select the cropped video to be analyzed. Once MATLAB has finished tracking the position of the wire, an Excel spreadsheet file is created containing the tracked position of the wire.
3. Then, run the MATLAB script for force calculation. Select the Excel file containing the tracked position of the wire calculated by the first MATLAB script in **step 2**. The user must specify the start and end frequency of the experiment, the Tw, and the passive wire deflection.

4. The readouts for the total and passive force of tissue are then interpolated according to the Tw and corresponding force-displacement calibration curve previously generated using the method described in Subheading 3.10. Total (at peak contraction) and passive (at rest) POMaC wire deflection are converted to force measurements (μN) based on the standard curves. The active force is calculated as the difference between the total and passive force.
5. The average cross-sectional area, determined from the average tissue width measurement assuming a cylindrical cross-section, is used in the conversion of total active force (μN) produced on the wire to the average stress (μN/mm^2) exhibited by the tissue.

4 Notes

1. We opt for crosslinking PDMS mold at room temperature rather than in an oven to minimize the effect of thermal expansion. Thus, the resultant PDMS mold will retain the designed dimensions across the large surface area.
2. The surface of the polystyrene chip should have minimal light scattering to ensure good imaging quality. Overall thickness of the polystyrene chip should be within 1 mm to allow high magnification microscopy imaging.
3. Before stamping the PDMS mold onto the glass slide for POMAC wire fabrication, clean the PDMS surface thoroughly with scotch tape to get rid of PDMS residuals and dirt in the channels.
4. To mitigate the effects of batch-to-batch variation of the bulk polymer solution, the UV crosslinking energy should be adjusted for each batch of polymer by tensile testing to reach the desired Young's modulus (~250 kPa). This can be conducted with a Myograph (Kent Scientific) using a modified ASTM D638-10 standard test method for tensile properties of plastics as previously described [14]. For assessment of cardiac tissues, the force sensor should allow force measurement within the range of 1–250 μN with 1 μN sensitivity.
5. It is important to place the polymer wire on chip in a controllable and consistent manner. Ensure that the total wire length and stretching force across the eight micro-chambers are consistent. Always check the wire placement under the optical microscope after assembly. The wires should be aligned within the grooves uniformly and no obvious defects should be present on chip.

6. When fixing the POMaC wires on the chip, make sure that the polyurethane adhesive is properly mixed in a 35 mm petri dish and leave the mixture at room temperature for 10 min to allow partial gelation. Transfer a small amount using a 10 μL pipette tip to fix the wire. This will prevent the adhesive from leaking into wire groove and micro-chamber.
7. Although the complete microfabrication workflow can take up to 6 days, most of the different steps can be carried out in parallel and concurrently to achieve higher throughput.
8. Always characterize the cell population of ventricular cardiomyocytes from BJ1D-derived CMs based on cardiac troponin T expression with flow cytometry to ensure the accuracy of the final CM: cFB ratio. Similarly, check cardiomyocytes purity of purchased iCells before seeding.
9. Use human cardiac fibroblasts of early passages, specifically under 5 from the time they are received from the distributor, to reduce the population of myofibroblasts activated by the stiff flask surface over the course of culture. Fibroblasts from the same passage number should be used throughout the experiment to ensure consistency.
10. After the cell mixtures are pelleted, it is important to aspirate all traces of media away from the cell pellet by tipping the tube towards the pipet without being too close to the cells. This step is essential for the final concentration of cells in hydrogel.
11. Keep the cell pellets, hydrogel and pipette tips on ice during gel preparation and cell seeding.
12. Using 0.2–10 μL gel loading pipette tips can facilitate precise seeding into the tiny micro-chambers with easier control, compared to regular pipette tips.
13. Immediately after seeding, place a small volume of media besides the strip in the tissue culture petri dish to retain moisture until gelation is complete.
14. Aprotinin should be added to the media in the first 7 days of culture to maintain the integrity of fibrin.
15. A small notch can be carved on each carbon rod with a dremel, and then wrap the platinum wire around it to prevent the wires from slipping off the electrodes.
16. Upon electrical stimulation chamber assembly, it is critical to ensure with an oscilloscope that the prescribed voltage is delivered by the electrodes immersed in the culture media.
17. During electrical stimulation for tissue cultivation and assessment, ensure that the platinum wires are not shorting or making contact with one another.

18. For analysis consistency, record videos of the wire on the same side of each tissue. Ensure that each recording uses the same exposure time, framerate and lamp intensity.
19. Tissue width measurements can be completed using CellSens's builtin digital ruler, or the open-source image processing program ImageJ.

Acknowledgments

We thank Drs. Yimu Zhao, Locke Davenport Huyer, and Boyang Zhang who established the initial methods of Biowire II platform design and polymer synthesis on which this chapter was based.

Some components of schematics were created with BioRender.com.

Radisic lab is funded by the Heart and Stroke Foundation Grant-in-Aid (G-16-00012711), Natural Sciences and Engineering Research Council of Canada (NSERC) Discovery Grant (RGPIN 326982-10), NSERC-CIHR Collaborative Health Research Grant (CHRP 493737-16), and National Institutes of Health Grant 2R01 HL076485. MR was supported by NSERC Steacie Fellowship and Canada Research Chair, LDH was supported by CIHR Vanier Scholarship. The authors acknowledge the Canada Foundation for Innovation, Project 19119, and the Ontario Research Fund for funding of the Centre for Spectroscopic Investigation of Complex Organic Molecules and Polymers.

References

1. Leask A (2015) Getting to the heart of the matter: new insights into cardiac fibrosis. Circ Res 116:1269–1276
2. Ahadian S, Civitarese R, Bannerman D, Mohammadi MH, Lu R, Wang E, Davenport-Huyer L, Lai B, Zhang B, Zhao Y, Mandla S, Korolj A, Radisic M (2018) Organ-on-a-chip platforms: a convergence of advanced materials, cells, and microscale technologies. Adv Healthc Mater 7(2):1700506. https://doi.org/10.1002/adhm.201700506
3. Savoji H, Mohammadi MH, Rafatian N, Toroghi MK, Wang EY, Zhao Y, Korolj A, Ahadian S, Radisic M (2019) Cardiovascular disease models: a game changing paradigm in drug discovery and screening. Biomaterials 198:3–26. https://doi.org/10.1016/j.biomaterials.2018.09.036
4. Asmani M, Velumani S, Li Y, Wawrzyniak N, Hsia I, Chen Z, Hinz B, Zhao R (2018) Fibrotic microtissue array to predict antifibrosis drug efficacy. Nat Commun 9(1):2066. https://doi.org/10.1038/s41467-018-04336-z
5. Sadeghi AH, Shin SR, Deddens JC, Fratta G, Mandla S, Yazdi IK, Prakash G, Antona S, Demarchi D, Buijsrogge MP, Sluijter JPG, Hjortnaes J, Khademhosseini A (2017) Engineered 3D cardiac fibrotic tissue to study fibrotic remodeling. Adv Healthc Mater 6(11). https://doi.org/10.1002/adhm.201601434
6. Wang EY, Rafatian N, Zhao Y, Lee A, Lai BFL, Lu RX, Jekic D, Davenport Huyer L, Knee-Walden EJ, Bhattacharya S, Backx PH, Radisic M (2019) Biowire model of interstitial and focal cardiac fibrosis. ACS Central Sci 5(7):1146–1158. https://doi.org/10.1021/acscentsci.9b00052
7. Zhao Y, Rafatian N, Feric NT, Cox BJ, Aschar-Sobbi R, Wang EY, Aggarwal P, Zhang B, Conant G, Ronaldson-Bouchard K, Pahnke A, Protze S, Lee JH, Davenport Huyer L, Jekic D, Wickeler A, Naguib HE,

Keller GM, Vunjak-Novakovic G, Broeckel U, Backx PH, Radisic M (2019) A platform for generation of chamber-specific cardiac tissues and disease modeling. Cell 176(4):913–927. e918. https://doi.org/10.1016/j.cell.2018.11.042

8. Zhao Y, Wang EY, Davenport LH, Liao Y, Yeager K, Vunjak-Novakovic G, Radisic M, Zhang B (2019) A multimaterial microphysiological platform enabled by rapid casting of elastic microwires. Adv Healthc Mater 8(5): 1801187. https://doi.org/10.1002/adhm.201801187
9. Tandon N, Cannizzaro C, Chao P-HG, Maidhof R, Marsano A, Au HTH, Radisic M, Vunjak-Novakovic G (2009) Electrical stimulation systems for cardiac tissue engineering. Nat Protoc 4(2):155–173. https://doi.org/10.1038/nprot.2008.183
10. Sahin O, Ashokkumar M, Ajayan PM (2018) 3-Micro- and nanopatterning of biomaterial surfaces. In: Balakrishnan SMS, Thomas S (eds) Fundamental biomaterials: metals. Woodhead Publishing, pp 67–78. https://doi.org/10.1016/B978-0-08-102205-4.00003-9
11. Montgomery M, Davenport Huyer L, Bannerman D, Mohammadi MH, Conant G, Radisic M (2018) Method for the fabrication of elastomeric polyester scaffolds for tissue engineering and minimally invasive delivery. ACS Biomater Sci Eng 4(11):3691–3703. https://doi.org/10.1021/acsbiomaterials.7b01017
12. Zhang B, Lai BFL, Xie R, Davenport Huyer L, Montgomery M, Radisic M (2018) Microfabrication of AngioChip, a biodegradable polymer scaffold with microfluidic vasculature. Nat Protoc 13(8):1793–1813. https://doi.org/10.1038/s41596-018-0015-8
13. Lian X, Hsiao C, Wilson G, Zhu K, Hazeltine LB, Azarin SM, Raval KK, Zhang J, Kamp TJ, Palecek SP (2012) Robust cardiomyocyte differentiation from human pluripotent stem cells via temporal modulation of canonical Wnt signaling. Proc Natl Acad Sci U S A 109(27): E1848–E1857. https://doi.org/10.1073/pnas.1200250109
14. Davenport Huyer L, Zhang B, Korolj A, Montgomery M, Drecun S, Conant G, Zhao Y, Reis L, Radisic M (2016) Highly elastic and moldable polyester biomaterial for cardiac tissue engineering applications. ACS Biomater Sci Eng 2(5):780–788. https://doi.org/10.1021/acsbiomaterials.5b00525

Chapter 13

Methods for Transepicardial Cell Transplantation in a Swine Myocardial Infarction Model

Fanny Wulkan, Rocco Romagnuolo, Beiping Qiang, and Michael A. Laflamme

Abstract

The transplantation of human pluripotent stem cell-derived cardiomyocytes (hPSC-CMs) has garnered significant attention as a potential means of restoring lost muscle mass and contractile function in injured hearts. Early preclinical work with hPSC-CMs employed rodent models, but the field has recently advanced to transplantation studies in more translationally relevant large animal models including non-human primates and swine. The pig is a particularly attractive model for such studies because the size, structure, and physiology of the porcine heart is very similar to that of humans. The pig model has reasonably high throughput, is readily amenable to clinically available cell delivery methods and imaging modalities and has been used frequently to test the safety and efficacy of new cardiac therapies. Here, we describe methods that were established in our laboratory for the specific purpose of testing hPSC-CM transplantation in a pig model of subacute myocardial infarction, but these same techniques should be broadly applicable to the transepicardial delivery of other biologicals including other candidate cell populations, biomaterials, and/or viral vectors.

Key words Myocardial infarction, Cell transplantation, Porcine model, Pluripotent stem cell-derived cardiomyocytes, Transepicardial delivery, Cardiac regeneration

1 Introduction

Ischemic heart disease remains a leading cause of morbidity and mortality worldwide. After a myocardial infarction (MI), the damaged muscle is replaced by non-contractile scar tissue, and both the direct loss of force-generating units and subsequent adverse ventricular remodeling often lead to progressive heart failure [1]. Currently, whole-organ transplantation is the only available therapeutic option for replacing lost myocardium, but this intervention is limited by the restricted supply of donor hearts and a requirement for life-long immunosuppression. Given this situation, stem cell-based therapies have attracted significant attention as a potential alternative means of restoring lost muscle mass and contractile

Kareen L. K. Coulombe and Lauren D. Black III (eds.), *Cardiac Tissue Engineering: Methods and Protocols*, Methods in Molecular Biology, vol. 2485, https://doi.org/10.1007/978-1-0716-2261-2_13,

function [2]. A wide variety of candidate cell sources have been considered for this application including skeletal myoblasts, bone marrow derivatives, as well as both adult and pluripotent stem cells and their derivatives [2, 3].

Our own group has been particularly focused on the preclinical testing of a regenerative cardiac therapy based on human pluripotent stem cells (hPSCs) because the latter cell type represents an abundant source of bona-fide human cardiomyocytes. In the past work, we and others in the field have shown that transepicardial injection of hPSC-derived cardiomyocytes (hPSC-CMs) results in the partial remuscularization of injured hearts with electromechanically integrated implants of human myocardium and mediates beneficial effects on left ventricular contractile function [4–8]. While early transplantation studies with hPSC-CMs employed rodent models, the field has recently advanced to testing in more translationally relevant large animal models [9–12]. Rodent models offer advantages in terms of higher throughput and lower costs, but these species obviously differ greatly from humans in terms of cardiovascular physiology. For example, the typical adult rat heart weighs 1 g and has a sinus rate of >300 beats per minute (bpm) versus corresponding values of approximately 250 g and 60–100 bpm for an adult human heart [13]. Likely reflecting these species differences, we have observed frequent graft-related tachyarrhythmias following hPSC-CM transplantation in infarcted swine and non-human primates [9, 12] that were not predicted by earlier work in smaller rodents with faster heart rates [4, 6]. Rodent hearts also have limited utility for investigating cell dose and routes of administration in eventual human patients, plus many imaging modalities (e.g., electroanatomic mapping) are challenging or unavailable in smaller hearts.

Moreover, having worked in both swine and non-human primates, we have concluded that the former is both more practical and better suited for modeling outcomes in human patients. Both models have their advantages and disadvantages, but primates are more challenging to work with and prohibitively expensive. Relative to small macaques, the pig is comparatively higher throughput, its larger size facilitates better imaging, and its heart size and sinus rate (~80–90 bpm vs ~120–160 bpm in small macaques [9, 10, 14–16]) are closer to those in humans. Given these factors, the pig has been much more frequently used most to validate novel cardiac interventions, and regulatory agencies are accustomed to receiving preclinical safety and efficacy data from this species.

We describe below a protocol for the reliable induction of MI via balloon coronary occlusion in this species, followed later by transepicardial (intramyocardial) cell transplantation and post-transplant physiological monitoring. Figure 1 depicts the sequence and timing of these procedures as they were employed in our recently reported study describing structural and functional

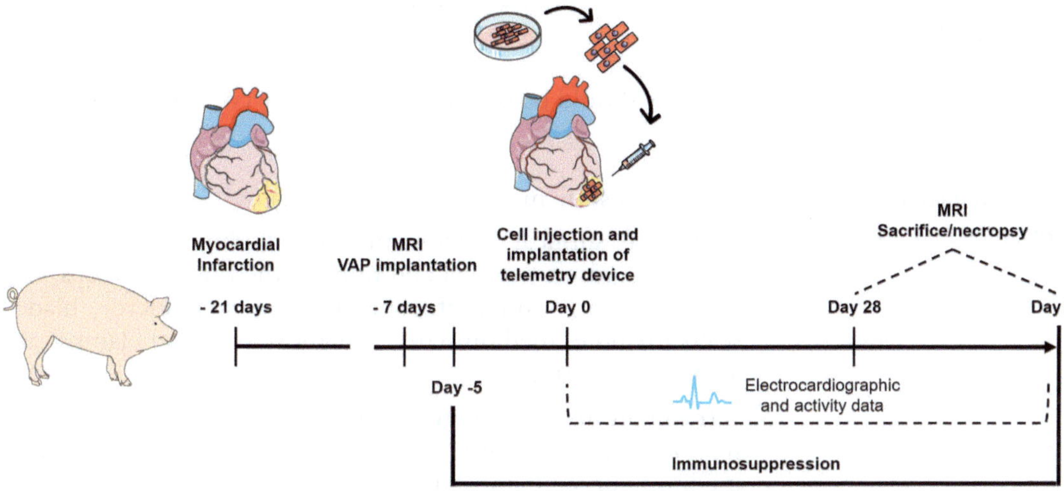

Fig. 1 Flowchart for typical cell transplantation experiments in porcine model of subacute MI. This precise sequence and timing of procedures was recently used by our group to investigate structural and functional outcomes after hPSC-CM transplantation in infarcted pigs [12]. MRI scans are acquired 7 days prior to cell transplantation and at 28- and 56-days post-transplantation, while telemetric ECG recordings are continuously acquired from the time of cell transplantation to euthanasia

outcomes after hPSC-CM transplantation in a porcine model of subacute MI [12]. In brief, MIs were induced in adult swine via a closed-chest approach and inflation of a percutaneous balloon dilation catheter, followed by reperfusion. At 14-days post-MI, infarcted animals underwent implantation of an indwelling vascular access port (VAP) to facilitate the intravenous administration of immunosuppressive drugs and routine blood sampling. At 21-days post-MI, we performed a mini-thoracotomy and directly injected 1×10^9 hPSC-CMs via the transepicardial route. During this cell implantation procedure, we typically also implanted a subcutaneous telemetry device for the subsequent noninvasive collection of physiological data (e.g., electrocardiogram, body temperature, and activity). While we chose to deliver hPSC-CMs at 21-days post-MI to determine the effects of cell therapy during the subacute phase of infarct healing, the timing of this procedure can obviously be adjusted as required experimentally. (Ghugre and colleagues have shown that scar size and ventricular remodeling are essentially stabilized in infarcted pigs by 6–8 weeks post-MI, so intervention at this time-point suffices to model outcomes in chronic MIs [17]). Finally, while our laboratory developed these methods for the specific purpose of testing hPSC-CM transplantation in infarcted swine, they should be readily adaptable for testing other candidate cell types and biological therapies (e.g., viral vectors).

2 Materials

Sterile packs should be prepared and autoclaved at least 1 day prior to surgery.

2.1 Induction of Myocardial Infarction

1. Adult Yorkshire male swine (20–30 kg).
2. A general surgical instrument set including needle holder, tweezers, Adson toothed and non-toothed tissue forceps, kidney bowls, iris and Metzenbaum scissors, retractor, blades, scalpel handle, medium-curved hemostats, towel clamps, and 4-in. × 4-in. gauzes.
3. A tracheotomy instrument set including small Weitlaner retractor, toothed standard forceps, surgical scissors, and four medium-curved hemostats.
4. Anesthesia induction supplies including laryngoscope, stylet, lidocaine spray, ophthalmic lubricating ointment, endotracheal (ET) tubes (6.0–7.0 mm inner diameter (ID)), and cuff inflator.
5. Gown pack, sterile drapes, and surgical gloves.
6. Programmable volumetric infusion pump (set at 10 drops/mL) and stopcocks.
7. 22G angiocatheter.
8. Sutures: 2-0 silk taper needle, 2-0 PDS (polydioxanone) II taper needle, and 2-0 Monocryl (poliglecaprone).
9. Syringes: 1, 3, 5, 10, 20, and 30 mL.
10. Needles: 25G, 22G, and 18G.
11. Monitoring equipment for oxygen saturation level (SpO_2), end-tidal CO_2 ($EtCO_2$), and electrocardiogram (ECG).
12. Ventilator and anesthetic vaporizer.
13. Defibrillator and conductive gel.
14. 6F arterial sheath.
15. 6F Judkins left guiding catheter.
16. Monorail PTCA balloon catheter (2.5–3.0 mm × 10–20 mm).
17. 0.014-in. × 180 cm PTCA wire and 0.035-in. J-tip guiding wire.
18. PCI inflation pump kit (including inflation pump, needle introducer, screw type Y connector, and torque).
19. Suitable fluoroscopy system (e.g., General Electric OEC 9900 Elite mobile C-arm angiographic system).
20. Lead aprons and dosimeters.
21. Circulating water blanket.

22. Non-ionic radiocontrast (Isovue or Hexabrix), 20 cc.
23. Normal saline (0.9% wt/vol sodium chloride), two 500 cc bags.
24. Skin cleaning agents: 70% ethanol, chlorhexidine gluconate, and Betadine.
25. Anesthetic agents: ketamine (20 mg/kg), midazolam (0.3 mg/kg), and isoflurane (2–3% inhalational).
26. Analgesics: meloxicam (0.3 mg/kg) and extended-release buprenorphine (0.24 mg/kg).
27. Antibiotics: ceftiofur (5 mg/kg) and cefazolin (20 mg/kg).
28. Other drugs: heparin (1000 units/mL), lidocaine (2% wt/vol, delivered at 0.025–0.030 mg/kg/min), epinephrine (0.01 mg/kg, prepared by diluting stock epinephrine (at 1 mg/mL) with normal saline in a 10 mL syringe), amiodarone (4 mg/kg), nitroglycerin (0.2 mg, prepared by diluting stock nitroglycerin (at 5 mg/mL) with saline in a 10 mL syringe), and atropine (0.04 mg/kg).

2.2 Implantation of Indwelling VAP

1. A general surgical instrument set including needle driver, tweezers, Adson toothed and non-toothed tissue forceps, kidney bowls, Metzenbaum and iris scissors, Weitlaner retractor, blades, scalpel handle, small- and medium-curved hemostats, medium straight hemostat, trocar (10G), 4-in. × 4-in. gauzes, towel clamps, and sterile swabs.
2. A tracheotomy instrument set including small Weitlaner retractor, toothed standard forceps, surgical scissors, and four medium-curved hemostats.
3. Anesthesia induction supplies including laryngoscope, stylet, lidocaine spray, ophthalmic lubricating ointment, ET tubes (7.0–7.5 mm ID), and cuff inflator.
4. Gown pack, sterile drapes, and surgical gloves.
5. Programmable volumetric infusion pump (set at 10 drops/mL) and stopcocks.
6. 22G angiocatheter.
7. Sutures: 2-0 silk taper needle, 2–0 PDS*II taper needle, 0 PDS*II, and 2-0 Monocryl.
8. Syringes: 1, 3, 5, and 30 mL.
9. Needles: 25G, 22G, and Huber point needle.
10. Film dressing with nonadherent pad (e.g., 3 M Tegaderm transparent dressing).
11. Monitoring equipment for oxygen saturation level (SpO_2), end-tidal CO_2 ($EtCO_2$), and electrocardiogram (ECG).
12. Ventilator and anesthetic vaporizer.

13. Defibrillator and conductive gel.
14. Electrocautery pen and grounding pad.
15. Indwelling vascular access port (VAP) (e.g., Access Technologies, Skokie, IL. Cat Number: INLINEPORT-AC-91S).
16. Circulating water blanket.
17. Normal saline (0.9% wt/vol sodium chloride), two 500 cc bags.
18. Skin cleaning agents: 70% ethanol, chlorhexidine gluconate, and Betadine.
19. Anesthetics: ketamine (10 mg/kg), midazolam (0.3 mg/kg), and isoflurane (2–3% inhalational).
20. Analgesics: meloxicam (0.3 mg/kg) and extended-release buprenorphine (0.24 mg/kg).
21. Antibiotics: ceftiofur (5 mg/kg) and cefazolin (20 mg/kg).
22. Other drugs: heparin (1000 units/mL), lidocaine (2% wt/vol, delivered at 0.025–0.030 mg/kg/min), epinephrine (0.01 mg/kg, prepared by diluting stock epinephrine (at 1 mg/mL) with normal saline in a 10 mL syringe), amiodarone (4 mg/kg), nitroglycerin (0.2 mg, prepared by diluting stock nitroglycerin (at 5 mg/mL) with saline in a 10 mL syringe), and atropine (0.04 mg/kg).

2.3 Thoracotomy, Transepicardial Cell Injection, and Implantation of Subcutaneous Telemetry Device

1. Cell product to be implanted (stored as appropriate until the time of delivery, typically at 4 °C for hPSC-CMs).
2. A general surgical instrument set including needle driver, tweezers, forceps (Adson toothed, non-toothed tissue, and DeBakey atraumatic), kidney bowls, scissors (iris and Metzenbaum), Weitlaner retractor, blades, scalpel handle, self-retaining chest retractor (single blade style), malleable retractor, rib approximator, large- and medium-curved hemostats, medium straight hemostat, trocar (9G), large suction tip, 4-in. × 4-in. gauzes, towel clamps, and sterile swabs.
3. A tracheotomy instrument set including small Weitlaner retractor, toothed standard forceps, surgical scissors, and four medium-curved hemostats.
4. Anesthesia induction supplies: laryngoscope, stylet, lidocaine spray, ophthalmic lubricating ointment, ET tubes (7.0–8.0 mm ID), and cuff inflator.
5. Double gown pack, sterile drapes, and surgical gloves.
6. Programmable volumetric infusion pump (set at 10 drops/mL) and stopcocks.
7. 22G angiocatheter.

8. Sutures: 2-0 silk taper needle, 2-0 PDS*II taper needle, 2.0 vicryl, 1.0 prolene suture with long curve needle, 2-0 Monocryl, and 5-0 Ethibond.
9. Syringes: 1 mL tuberculin, 3, 5, and 30 mL.
10. Needles: 18G, 19G, and 27G.
11. Film dressing with nonadherent pad (e.g., 3 M Tegaderm transparent dressing).
12. Monitoring equipment for oxygen saturation level (SpO_2), end-tidal CO_2 ($EtCO_2$), and electrocardiogram (ECG).
13. Ventilator and anesthesia vaporizer.
14. Defibrillator, internal and external defibrillator paddles, and conductive gel.
15. Electrocautery pen and grounding pad.
16. Antimicrobial incise drape (e.g., 3 M Steri-Drape 2, 60×85 cm).
17. Chest tube drainage system.
18. 20F chest tube.
19. Implantable telemetry devices (e.g., Data Sciences International (DSI) PhysioTel Digital large animal telemetry system and M01 implants).
20. Circulating water blanket.
21. Normal saline (0.9% wt/vol sodium chloride), two 500 cc bags.
22. Skin cleaning agents: 70% ethanol, chlorhexidine gluconate, and Betadine.
23. Ringer's lactate solution, 500 cc bag (infused at 5–10 mL/kg/h).
24. Anesthetics: Anesthetics: ketamine (10 mg/kg), midazolam (0.3 mg/kg), and isoflurane (2–3% inhalational).
25. Analgesics: meloxicam (0.3 mg/kg), buprenorphine (30 μg/kg), and extended-release buprenorphine (0.18 mg/kg).
26. Antibiotics: ceftiofur (5 mg/kg) and cefazolin (20 mg/kg).
27. Other drugs: heparin (1000 units/mL), lidocaine (2% wt/vol, delivered at 0.025–0.030 mg/kg/min), epinephrine (0.01 mg/kg, prepared by diluting stock epinephrine (at 1 mg/mL) with normal saline in a 10 mL syringe), amiodarone (4 mg/kg), nitroglycerin (0.2 mg, prepared by diluting stock nitroglycerin (at 5 mg/mL) with saline in a 10 mL syringe), and atropine (0.04 mg/kg).

3 Methods

3.1 Induction of Myocardial Infarction

MI induction and all subsequently described animal procedures should be performed in strict compliance with institutional, local, and national regulations. Our own experiments were approved by the University Health Network (UHN) Animal Care Committee and performed with the supervision and assistance of UHN veterinary staff. The protocol below describes methods for the induction of highly reproducible MIs using a closed-chest approach and standard cardiac catheterization techniques. For this, a percutaneous balloon dilation catheter is maneuvered into the mid left anterior descending (LAD) coronary artery under fluoroscopic guidance, then the balloon is transiently inflated at a site distal to the first diagonal branch [17]. In our hands, when these methods (including 90 min of LAD occlusion) were applied to 20–30 kg adult male Yorkshire swine, they resulted in infarct scars that approximated 16% of total LV mass [12]. The duration of occlusion or placement of the balloon can obviously be varied if infarcts of different sizes are preferred given one's experimental design (e.g., a short duration of occlusion will result in smaller, non-transmural infarct scars). These same techniques can also be readily adapted to other porcine strains although the precise site and duration of occlusion may need to be modified to account for strain-dependent differences in coronary anatomy and/or the extent of collaterals [16, 18].

To ensure correct positioning of both the guiding catheters and the balloon, one must employ a fluoroscopy system that is suitable for angiographic procedures of this nature. In our case, we have used a General Electric OEC 9900 mobile C-arm to generate real-time high-resolution angiograms to evaluate the coronary anatomy and extent of collaterals and make decisions about balloon positioning and LAD occlusion. Moreover, if other noninvasive cardiac imaging modalities are available (e.g., echocardiography or MRI), it is strongly recommended that they be performed at baseline to confirm normal cardiac structure and function before MI model creation. Post-infarct imaging studies are obviously also required to follow structure and function over time, including before and after cell transplantation. Such imaging methods are beyond the scope of this protocol, but they have been well described elsewhere [19].

All experimental procedures in the pig, including MI induction as described here, should be performed using good sterile techniques along with inhalational anesthesia and fluid support. Anesthetized animals are particularly susceptible to hypothermia, so supplemental heat sources are also important to maintain body temperature. The animal's electrocardiogram (ECG), respiratory rate, end-tidal carbon dioxide saturation level ($EtCO_2$), peripheral oxygen saturation (SpO_2), and blood pressure should be closely monitored throughout the procedure. Finally, the protocol below assumes that early steps including intubation of the animal, establishment of intravenous (IV) access, and initial preparation of the incision site will occur in a separate surgical prep area. Only after the latter steps are completed (**steps 1–7** in Subheading 3.1.1 below) is the animal transferred to the operating room (OR) suite and secured to the OR table. We have found this staging to work well for MI induction and all subsequently described surgical procedures.

3.1.1 Pre-anesthesia, General Anesthesia, and Preparation for MI Procedure

1. Acclimatize the pig for a least 5–7 days between entry into the vivarium and commencing invasive procedures such as MI induction.
2. Fast the animal overnight prior to the procedure, but do not restrict access to water to avoid dehydration.
3. Sedate the animal with an intramuscular (IM) injection of a pre-anesthetic cocktail comprised of atropine (0.04 mg/kg body weight (BW)), ketamine (20 mg/kg BW), and midazolam (0.3 mg/kg BW).
4. Once the pig is sedated, transfer the animal from the vivarium to the surgical prep area, place it in a supine position, and secure it with sandbags.
5. Intubate the animal with an appropriately sized ET tube (typically 6.5 mm ID. for an approximately 25 kg pig), then deliver ongoing inhalational maintenance anesthesia with isoflurane (2–3%) mixed with 100% oxygen via a mechanically ventilated closed-loop anesthesia machine. Ventilate at a rate of 20 breaths per minute and a tidal volume of 10–15 mL/kg BW. Concentrations of isoflurane higher than 3% can be used in special circumstances (e.g., if pig does not respond well to pre-medication).
6. Once a surgical plane of anesthesia is achieved, insert a 22G angiocatheter into the lateral auricular vein to administer IV maintenance fluids.
7. Shave the incision site (right anterior cervical region) and clean the area thoroughly with chlorhexidine cleanser.
8. Transfer the pig from the surgical prep area to the OR suite, place it in a dorsal recumbent position on the OR table, and secure with sandbags. Place ECG leads in the left and right axilla, as well as on the left hind limb, and commence ECG monitoring. Continue to deliver maintenance anesthesia as detailed in **step 5**.
9. Administer lidocaine at constant rate infusion (CRI) of 0.025–0.030 mg/kg/min BW. (To prepare this lidocaine solution, withdraw 10 mL of normal saline from a 500 mL bag and

discard. Then, withdraw 10 mL of lidocaine hydrochloride (2% w/v) solution, and mix into the remaining 490 mL of normal saline).

10. 20 min prior to first incision, slowly administer cefazolin (20 mg/kg BW, IV) over 10 min to reduce the risk of surgical site infection.
11. Administer extended-release buprenorphine (0.24 mg/kg BW subcutaneous (SQ)) and meloxicam (0.3 mg/kg BW SQ).
12. Prior to first incision, prepare and drape the surgical site(s) in an aseptic manner. Clean the surgical site with chlorhexidine scrub and water on sterile gauze. Next, apply isopropyl alcohol to the entire surgical site, and allow it to dry fully.

3.1.2 MI Induction via Balloon Occlusion of the Mid-LAD

1. Commence close monitoring of ECG, respiratory rate, $EtCO_2$, SpO_2, and blood pressure.
2. Make a 5 cm incision in the jugular farrow of the right anterior cervical region. Carefully isolate the right common carotid artery using blunt dissection to avoid damage to the vagus nerve. The vessel is then bathed with 2% lidocaine (w/v) to dilate it and facilitate access.
3. Ligate the distal end of the right carotid artery with a 2-0 silk suture. Temporarily occlude the proximal end with another 2-0 silk suture. Make a small incision between the two sutures and insert a 6F sheath into the carotid artery.
4. Anticoagulate by administering heparin (1000 IU/kg BW IV) via the arterial sheath. Continue to deliver heparin (1000 IU/kg BW IV) every 30 min during the procedure.
5. Under the guidance of the C-arm, advance a 6F 3.5 Judkins Left guiding catheter into the aortic arch, followed by an 0.035-in. J-tip wire. Engage the tip of the guiding catheter in the ostium of left coronary artery.
6. Inject nitroglycerin (0.2 mg) into the left coronary artery to prevent vasospasm, then acquire an angiogram of the left coronary artery including the LAD (Fig. 2a).
7. Using fluoroscopy guidance, advance a 0.014-in. guidewire into the LAD, then advance the PTCA dilation catheter over the wire until it is located just distal to the first diagonal branch.
8. Perform an angiogram to confirm that the balloon is indeed located immediately distal to the first diagonal branch. Inflate the PTCA balloon to its nominal pressure, confirm total occlusion of the LAD by angiography (Fig. 2b), then maintain occlusion for a total of 90 min. *See* **Note 1** regarding interventions that may be required in the event of ECG changes during this step.

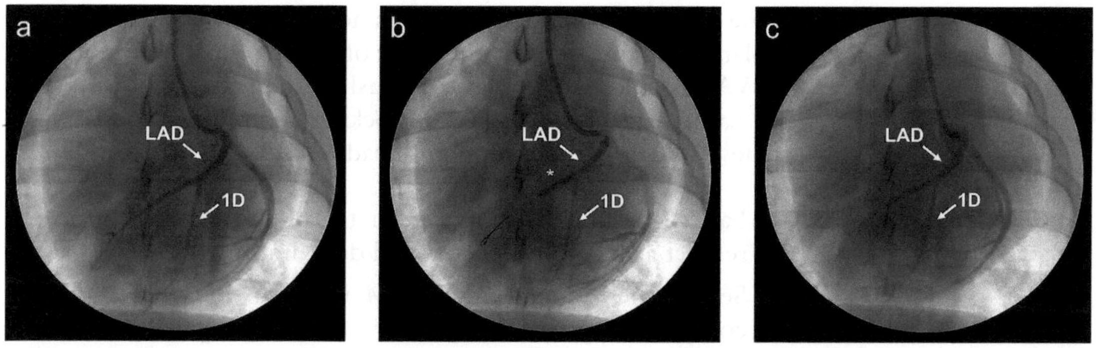

Fig. 2 Angiography before, during, and after LAD occlusion. (**a**) Left coronary angiogram prior to occlusion showing the left anterior descending coronary artery (LAD) and the first diagonal branch (1D). (**b**) Repeat angiogram after occlusion by the PTCA balloon catheter (indicated by asterisk), which is positioned just distal to the first diagonal branch. **(c)** Vessel after deflation of the balloon and reperfusion

9. After 90 min, deflate the PTCA balloon, and confirm successful reperfusion by repeat angiography (Fig. 2c).
10. Remove the PTCA balloon, withdraw the guide catheter and introducers, then ligate the carotid artery unilaterally.
11. Close the neck incision in layers using 2-0 Monocryl sutures.
12. Clean the incision site in a sterile manner and cover with a nonadherent and waterproof dressing pad.
13. Administer ceftiofur (5 mg/kg BW, IM).
14. Allow the animal to slowly recover from anesthesia, leaving it intubated until the recovery of spontaneous jaw movements. When jaw movements return, extubate, and gently transfer the animal from the OR suite back to the vivarium. Have face mask oxygen readily available both in the OR and the animal pen and administer oxygen as necessary. Monitor the animal closely during recovery from MI but try to leave it as undisturbed as possible.

3.2 Implantation of Indwelling VAP

For our experiments, we found it essential to implant a VAP in the external jugular vein that could then be used for the intravenous administration of drugs as well as routine blood sampling. In our case, the VAP was placed at 2-weeks post-MI and 1-week prior to thoracotomy and cell implantation [12]. This timing allowed the pig to have fully recovered from the MI procedure but also facilitated the start of immunosuppression prior to cell transplantation. (*See* **Note 2** for details of the specific immunosuppression protocol that we found efficacious in preventing the rejection of hPSC-CM xenografts in Yorkshire swine [12]. This regimen would likely need to be modified for other cell types or recipient pig strains.)

While very convenient for drug administration and blood sampling, the VAP is nonetheless a potential source of serious infections in the pig. Strict aseptic techniques need to be followed during its implantation and all subsequent use of the system. Once implanted, the VAP should also be regularly flushed to prevent clot formation and catheter occlusion using a "locking" solution of heparinized saline (as described below in Subheading 3.2.2).

3.2.1 VAP Implantation Procedure

1. Fast the animal overnight prior to the procedure, but do not restrict access to water to avoid dehydration.
2. Sedate the animal with an IM injection of a pre-anesthetic cocktail comprised of atropine (0.04 mg/kg BW), ketamine (10 mg/kg BW), and midazolam (0.3 mg/kg BW). (Note the lower ketamine dosage here versus that employed during the MI induction procedure.)
3. Once the pig is sedated, transfer the animal from the vivarium to the surgical prep area, place it in a supine position, and secure it with sandbags.
4. Intubate the animal with an appropriately sized ET tube, then deliver ongoing inhalational maintenance anesthesia with isoflurane (2–3%) mixed with 100% oxygen via a mechanically ventilated closed-loop anesthesia machine. Ventilate at a rate of 20 breaths per minute and a tidal volume of 10–15 mL/kg BW.
5. Once a surgical plane of anesthesia is achieved, insert a 22G angiocatheter into the lateral auricular vein to administer IV maintenance fluids.
6. Shave behind the left ear caudally until the back of the scapula and the left flank of the pig from spine to sternum. Clean this area thoroughly with chlorhexidine cleanser.
7. Transfer the pig from the surgical prep area to the OR suite. Place the animal on the OR table in a right lateral recumbent position. Place ECG leads in the left and right axilla, as well as on the left hindlimb, and commence ECG monitoring. Continue to deliver maintenance anesthesia as detailed in **step 4**.
8. Follow **steps 10–12** of Subheading 3.1.1 for antibiotic prophylaxis, analgesia administration, and preparation of the surgical site.
9. Make a ~5 cm long incision on the left side of neck, at a point approximately 2 cm from the midline. Isolate the left jugular vein with blunt dissection.
10. Put two sutures around the left jugular vein, ligate the sutures, and make a small incision in the jugular vein between them.
11. Insert the VAP catheter into the left jugular vein, and temporarily cap the other end of the catheter to prevent bleeding. (Do not over-tighten the cap at this point.)

12. Using a Huber needle, inject 5 mL of 0.9% heparinized saline into the catheter to flush.
13. Under fluoroscopic guidance, advance the catheter until its tip is located in the superior vena cava just proximal to the right atrium. Alternatively, in the absence of fluoroscopy, advance the catheter into the external jugular vein for a distance of approximately 20 cm. If it is no longer possible to withdraw blood from the catheter, this implies that the tip is pressed against the wall of the right atrium. Retract the catheter approximately 1–2 cm from that point, then secure the proximal end in place with a 2-0 silk suture.
14. Use a 10F trocar to make a tunnel to pass the VAP catheter through the neck incision to a point in the dorsal midline just caudal to the scapula. Make a small incision (approximately 2 cm) at this location.
15. Before inserting the catheter through the trocar, uncap the VAP. Insert and pass the catheter through the trocar up to the incision in the midline.
16. Once the catheter is in place, remove the trocar, and again temporarily cap the end of the catheter to avoid bleeding.
17. Flush the VAP port and catheter with 5 mL of 0.9% heparinized saline via a Huber needle.
18. Before closing the jugular incision site, make a 3–4 cm loop in the VAP catheter close to the neck incision. This loop provides flexibility to accommodate animal growth over time.
19. Close the jugular incision site in layers using 2-0 Monocryl sutures.
20. Return to the midline incision and uncap the VAP. Adjust and cut the catheter length adjacent to the skin surface, and tightly close the end of the catheter with the VAP cap.
21. Secure the VAP port in the midline aperture with 0 PDS*II sutures.
22. To ensure the VAP is working properly, withdraw blood with a syringe. Flush the VAP port and catheter with 5 mL of 0.9% heparinized saline via a Huber needle. Make sure to never leave blood in the catheter or VAP port, as it may clot and clog the VAP line and port.
23. Clean both incision sites in a sterile manner, and cover the VAP with a nonadherent, waterproof dressing pad.
24. Administer ceftiofur (5 mg/kg BW IM), and meloxicam (0.3 mg/kg SQ) and allow the animal to recover from anesthesia. Give an additional dose of ceftiofur and meloxicam on the following day.

3.2.2 VAP Maintenance and Use

1. Using sterile gloves, clean the VAP and surrounding area with chlorhexidine soap. Swab the area four times. Use a new swab each time, starting at the center of the port and working outwards in concentric circles. After finishing, change to a new pair of sterile gloves.
2. Using the thumb and index finger of a gloved hand stabilize the VAP port.
3. Insert a non-coring Huber point needle with firm consistent pressure through the port septum at a 90° angle relative to the port dome. The needle is in the correct position when the tip touches the bottom of the port and one can see it within the solution.
4. Attach a syringe (5 mL or larger) to the Huber point needle hub and aspirate the "locking" solution of heparinized saline. Concurrently, blood is drawn into the hub as you remove the "locking" solution, indicating correct catheter patency and positioning. Remove the syringe and discard it and any fluid.
5. After thus confirming patency, attach a new syringe to the same Huber point needle, then flush the port and catheter with sterile saline (with at least twice the volume of the VAP).
6. Perform blood sampling or drug administration, as necessary.
7. Once sampling is completed, inject fresh heparinized "locking" solution (saline with 100 IU/mL heparin) to maintain catheter patency. The amount depends on the volume of the port and catheter (typically approximately 20 mL).
8. Remove the Huber needle from the VAP while applying gentle positive pressure to the syringe plunger (to prevent reflux of blood into the catheter tip and potential occlusion).
9. Flush the catheter every 1–2 days following implantation and "lock" it with saline with 100 IU/mL heparin each time that it is accessed. Do not try to clear an occluded VAP by over-pressurizing the system, as this could cause catheter rupture and/or embolization.
10. Cover the VAP port site with a nonadherent and waterproof dressing pad.

3.3 Thoracotomy, Transepicardial Cell Injection, and Implantation of Subcutaneous Telemetry Device

Our laboratory has used the methods described in this section for the transepicardial (intramyocardial) implantation of hPSC-CMs into infarcted swine and experienced a perioperative mortality rate of <10% [12]. Because we are aiming to remuscularize the infarct scar, we have typically injected cardiomyocytes directly into the infarct zone at pre-planned injection sites (12–18 sites per heart) that were informed by preoperative imaging (e.g., late gadolinium enhancement MRI scans). That said, these same methods can be readily modified for the delivery of other cell types or biologics to any locations on the anterior and lateral LV wall, including both the infarct and border zones. To minimize the duration of the open-chest procedure, the cell product to be injected should be harvested, thawed, or prepared prior to the start of the thoracotomy. In our case, we have found that it typically works well to commence thawing vials of cryopreserved hPSC-CMs concomitantly with the first steps of preparing the pig for surgery. Our cell product and any syringes or needles used to deliver it are kept chilled at 4 °C until use.

We have also found the thoracotomy and cell implantation procedure to be a convenient time to also implant a subcutaneous telemetry device for the noninvasive acquisition of post-transplant physiological data, especially ECG. Once implanted, this device can be used to monitor the animal's ECG either continuously or at pre-programmed intervals postoperatively depending on the duration of the experiment and the sensor's battery life. The implantation methods described below (Subheading 3.3.3) were specifically developed for M01 telemetry devices (DSI), which are compatible with telemetric ECG system (Ponemah Physiology Platform, Data Sciences International (DSI), St. Paul, MN, USA), but these same techniques should be readily adaptable for similar devices from other vendors. The M01 sensors are small (13.7 g, 11 cc volume) and are contained within a biocompatible silicone elastomer. The leads extending from the main body of each device (helix of medical grade stainless steel wires covered with insulating silicone tubing) are flexible and like those used in cardiac pacemakers in humans. They have a battery-on time of approximately 40 days. In our case, their use has allowed us to closely monitor the incidence and time-course of graft-related tachyarrhythmias in infarcted hPSC-CM recipients for up to 8 weeks post-transplantation [12]. We have not encountered any compatibility issues or imaging artifacts when performing cardiac MRI in implanted animals.

Finally, we strongly recommend performing frequent blood chemistries, full blood counts, and microbiological testing in transplanted animals, particularly if immunosuppression is being used. Cell recipients should also be closely followed by clinical examinations and telemetric ECG monitoring to monitor for adverse effects that may be related to cell transplantation, prior surgical procedures, or immunosuppression.

3.3.1 Pre-anesthesia, General Anesthesia, and Preparation of Surgical Site for Thoracotomy Procedure

1. Fast the animal overnight prior to the procedure, but do not restrict access to water to avoid dehydration.
2. Load cell product to be delivered into multiple 1 cc tuberculin syringes with 27G needles attached. Store at 4 °C until immediately before delivery.

3. Sedate the animal with an intramuscular (IM) injection of a pre-anesthetic cocktail comprised of ketamine (10 mg/kg BW) and midazolam (0.3 mg/kg BW).
4. Once the pig is sedated, transfer it from the vivarium to the surgical prep area. Place the animal in a supine position, and secure with sandbags.
5. Intubate the animal with the appropriately sized ET tube, then deliver ongoing inhalational maintenance anesthesia with isoflurane (2–3%) mixed with 100% oxygen via a mechanically ventilated closed-loop anesthesia machine. Ventilate at a rate of 20 breaths per minute and a tidal volume of 10–15 mL/kg BW. Concentrations of isoflurane higher than 3% can be used in special circumstances (e.g., if pig does not respond well to pre-medication).
6. Once a surgical plane of anesthesia is achieved, insert a 22G angiocatheter into the lateral auricular vein to administer IV maintenance fluids.
7. Position the pig in right lateral recumbency and secure its limbs to the table with adhesive tape. Shave the left half of the torso (from the left pinna to the front of the pelvis and from spine to sternum), as well as the proximal half of the left forelimb. Clean this area thoroughly with chlorhexidine cleanser.
8. Transfer the pig to the OR suite and place it on the procedure table. Place ECG leads in the left and right axilla, as well as on the left hindlimb, and commence ECG monitoring. Continue to deliver maintenance anesthesia as detailed in **step 5**.
9. 20 min prior to first incision, slowly administer cefazolin (20 mg/kg BW diluted in 10 mL saline, IV) over 10 min.
10. Administer buprenorphine (22.5 μg/kg BW IM).
11. Prior to first incision, prepare and drape the surgical sites in an aseptic manner. Thoroughly clean the skin of the left torso and proximal left forelimb by doing a general scrub with chlorhexidine scrub and water on sterile gauze. Apply a second clean scrub with chlorhexidine to the left hemithorax, starting from the intended incision site in the fourth intercostal space and radiating outward to and including the left axilla. Next, apply isopropyl alcohol to the entire surgical site, and allow it to dry fully.

3.3.2 Left Lateral Thoracotomy and Cell Implantation Procedure

1. Commence close monitoring of ECG, respiratory rate, $EtCO_2$, SpO_2, and blood pressure.
2. Place antimicrobial incise drape over the incision site on the left hemithorax.
3. Administer additional buprenorphine analgesia (7.5 μg/kg BW IV at this point).

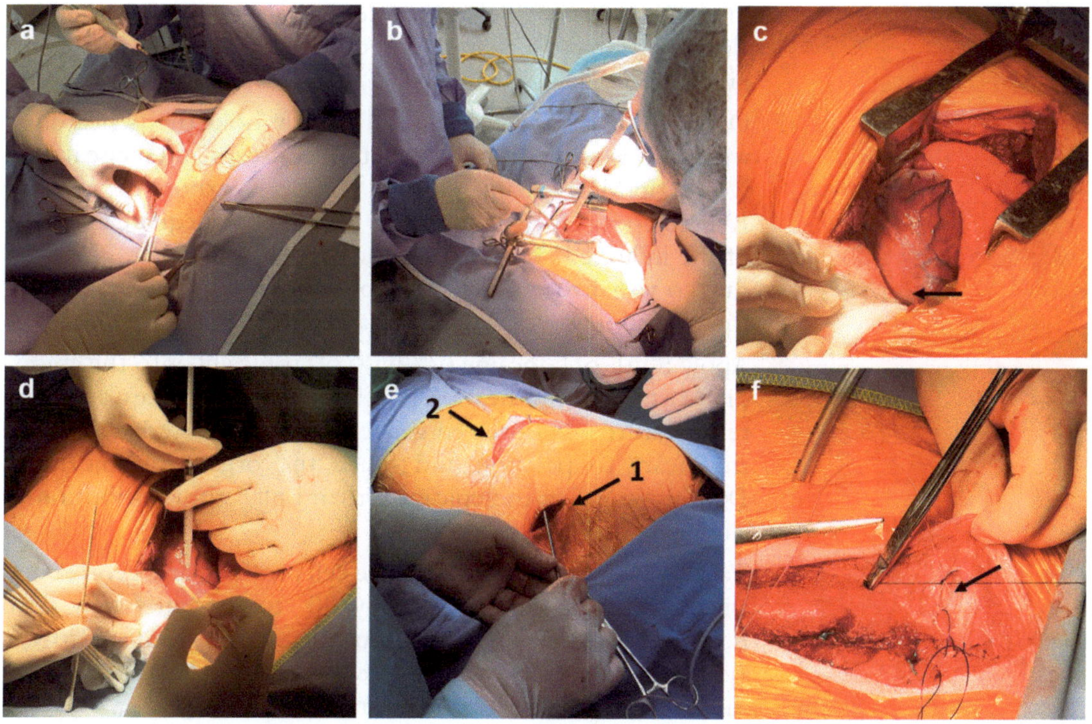

Fig. 3 Images of the surgical field during thoracotomy and cell transplantation procedure. (**a**) Incision in the fourth intercostal space for left lateral thoracotomy. (**b**) Subsequent view of thoracotomy showing surgical window created by deployment of the Weitlaner retractor. (**c**) Anterolateral view of the exposed heart with epicardially visible infarct scar indicated by arrow. Note that saline-wetted gauzes are being inserted under the posterior aspect of the heart, which improves exposures and helps brace the organ for cell injection. (**d**) After periocardiotomy, intramyocardial cell injections into the scar region. (**e**) Incision between the right and left scapular blades (arrow 1), where the telemetry device is being implanted. A trocar is used to tunnel the ECG leadwires medially through the muscular and soft tissues to the still-open thoracotomy site in the anterior chest (arrow 2). (**f**) Thoracotomy site with looped ECG lead wire (arrow) being secured to soft tissues

4. Make an approximately 8 cm incision along the fourth intercostal space, beginning approximately midway between the tip of the scapula and the sternum and extending medially to the inferomedial border of the sternum (Fig. 3a). Before entering the thoracic space, infiltrate the wound site with bupivacaine hydrochloride (0.25% wt/vol at 1–2 mg/kg BW) to provide local analgesia. Use an electrocautery to transect the subcutaneous and muscular layers and to control any bleeding during the dissection. Carefully ligate vessels if required to control bleeding.
5. Dip two cotton gauzes in warm saline and use these to gently displace and protect the left lung in the left thoracic space.
6. Insert a self-retaining Weitlaner retractor into the intercostal space. Gradually open the retractor to enlarge the fourth intercostal space, and fully expose the heart by transecting the

muscular layers with electrocautery and extending the incision along the intercostal space as required create an adequate surgical window (Fig. 3b).

7. Once the heart is fully exposed, gently hold the pericardium with an Adson or atraumatic Debakey forceps and incise it with round-tip Metzenbaum surgical scissors to expose the anterior LV and apex. Open as necessary to provide adequate exposure of the infarcted and border zones of the heart (Fig. 3c). Take care to visualize and avoid the phrenic neve which runs along the posterior aspect of the pericardium.
8. Apply 5 mL of lidocaine (2% wt/vol) topically onto the epicardial surface of the heart. Place two to three pieces of room-temperature saline-wetted gauze under the posterior aspect of the heart to position and brace the organ for injections. Directly visualize the epicardium and plan the distribution of sites to be injected with cells.
9. Take tuberculin syringes with the cell product, and gently bend the shaft of the 27G needle by 30° to prevent overly deep penetration of the myocardium. Insert the needle into the epicardium, and slowly inject the cell product into the myocardium (Fig. 3d). After withdrawing the needle, place a dry swab at the injection site and apply light pressure to minimize trans-epicardial cell leakage.
10. Repeat **step 9** for each injection. Each injection site should be at least 0.5–1.0 cm apart to maximize distribution of the cell product and minimize cell leakage. (*See* **Note 3** for interventions that may be required in the event of electrocardiographic or hemodynamic changes during the cell delivery step.)
11. Once delivery of the cell product is completed, gently remove the wet gauzes that were placed earlier to expose the heart and protect the lung. Carefully inspect the heart and thoracic cavity, and control bleeding, as necessary.
12. Make a new approximately 3 cm incision adjacent to the thoracotomy site in the lower chest and advance a 12F chest tube through the incision. Position the tip of the chest tube at the apex of the left lung, and the proximal end is clamped onto the surface of the pig.
13. Close the thoracotomy in layers, approximating the ribs with 5-0 Ethibond sutures and the muscular layers with 0 PDS*II sutures.
14. If implanting a subcutaneous telemetry device, pause after closing the muscular layers but prior to closure of the skin and subcutaneous tissues, and follow the procedure detailed in Subheading 3.3.3. If not, proceed directly to **step 15** for final wound closure.

15. Close the skin using a continuous suture with 2–0 Monocryl. (If a telemetry device has been implanted, do not use an electrocautery thereafter, as this may cause electrical damage of the device).
16. After closing the skin, remove the clamp on the chest tube and connect to a chest drainage system with negative pressure at approximately −20 cm water. Make a purse-string suture around the chest tube with a 2-0 Prolene suture and inflate the lungs. Gently remove the chest tube and simultaneously close the incision tightly with the purse-string suture.
17. Clean the incision sites with a wet gauze, and cover with a nonadherent, waterproof dressing pad.
18. Administer buprenorphine extended-release (0.18 mg/kg BW), meloxicam (0.3 mg/kg BW), and ceftiofur (5 mg/kg BW IM).
19. Allow the animal to slowly recover from anesthesia and extubate it after recovery of spontaneous jaw movements. Monitor closely thereafter, including by telemetric ECG if available.

3.3.3 Implantation of Subcutaneous Telemetry Device

1. Make an approximately 5 cm long incision in the dorsal neck between the right and left scapular blades and use blunt dissection to make a pocket in the trapezius muscle. Place the telemetry device to be implanted into the created pocket.
2. Use a 9G trocar to create a tunnel that spans medially through the muscular and soft tissues to the thoracotomy site in the anterior chest (Fig. 3e). Tunnel the lead wires of the implant device through this trocar, then remove the trocar.
3. Return to the thoracotomy site and position the ECG lead wires in a modified lead II configuration by placing the positive lead in the subcutaneous tissues overlying the apex of the heart and the negative lead over the base.
4. Once each lead has been pulled through and directed to its proper location, cut it to the appropriate length, then remove the silicone insulating covering to expose the wire end. Form a 1 cm-diameter loop of bare wire at the end of the lead and secure it with a 2-0 silk suture (Fig. 3f). Place a second suture around the lead at a point 4–5 mm from the end of the silicone covering to prevent fluid migration. Finally, secure the loop of wire to the underlying soft tissues using at least three simple interrupted 2-0 silk sutures.
5. Secure the pod of the telemetry device into the intramuscular pocket using 2-0 silk suture and close the skin and subcutaneous layers of the pocket using 2-0 *Vicryl sutures.*
6. Return to Subheading 3.3.2 and follow **steps 15–19** for closure of the skin and postoperative care. After recovery of the animal, verify the noninvasive recording of ECG signals using the telemetric ECG system.

4 Notes

1. The pig can rapidly go into ventricular tachycardia and/or ventricular fibrillation (VF) during the MI procedure, so the animal should be closely monitored in case rapid intervention is required. In our experience, most arrhythmias occur approximately 30 min after LAD occlusion or shortly following reperfusion. In the event of sustained ventricular tachycardia, lidocaine 2% wt/vol (at 2–4 mg/kg IV) and/or amiodarone (4–10 mg/kg IV, diluted in 10 mL of saline, and infused slowly over 10 minutes) should be promptly administered. If the animal goes into VF, defibrillation must be performed immediately with external paddles set at 360 J.
2. To ensure the survival of hPSC-CM xenografts in infarcted pigs as previously described [12], cell recipients are treated with an immunosuppression from 5 days pre-transplantation until euthanasia. Our immunosuppression regimen included abatacept (Orencia, given at 12.5 mg/kg on the day of cell transplantation and every 2 weeks thereafter); methylprednisolone (given as 250 mg PO on day of cell transplantation followed by a taper to 125 mg per day over 2 weeks, then daily maintenance thereafter); and cyclosporine A (given as a starting dose of 25 mg/kg PO twice per day and adjusted as necessary to achieve through concentrations of 250–350 μg/L, administered from 5 days prior to hESC-CM transplantation daily until sacrifice). Cyclosporine A levels should be measured 1–2 times per week, and blood can be sampled for this through the VAP. Never administer cyclosporine A through the VAP, as the drug can absorb to the line and lead to erroneous level measurements. While this immunosuppression regimen has proved effective with hPSC-CMs in infarcted swine, it will obviously need to be adjusted depending on the nature of the cell product being tested. Close clinical and microbiological monitoring is required to minimize the risk of opportunistic infections in immunosuppressed swine.
3. The pig should be closely monitored for arrhythmias and hemodynamic changes throughout the cell implantation procedure. Diluted epinephrine should be prepared prior to the procedure by mixing stock epinephrine (1 mg/mL) 1:10 (v/v) with normal saline in a 10 mL syringe. 0.5–1.0 mL of this diluted epinephrine should be infused intravenously if the heart rate drops below 50–60 bpm at any point. The cell injection step itself is obviously a particularly critical phase. A few premature ventricular contractions (PVCs) are commonly encountered during insertion of the injection needle into the myocardium, but these typically cease immediately after

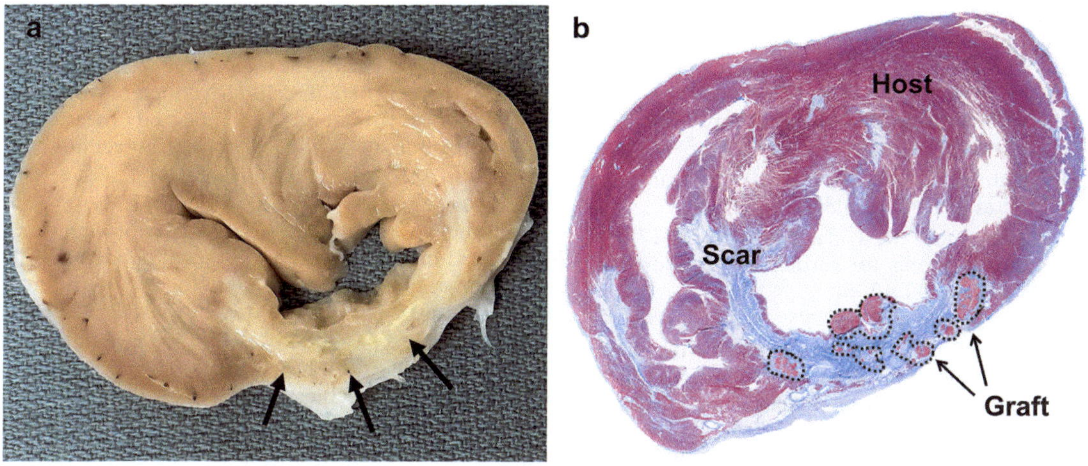

Fig. 4 Representative gross image of infarcted heart and histology from hPSC-CM engrafted heart. **(a)** Transverse slice with focally transmural infarct scar (arrow). **(b)** Whole-mount histological section of an engrafted heart stained for human-specific nuclear marker Ku80 (brown), sarcomeric myosin heavy chain (red), and infarct scar (blue). Graft tissue is circled by dotted lines. Scale bar = 5 mm

withdrawal of the needle. If more serious arrhythmias occur and/or the animal's blood pressure or oxygen saturation decline, the injection should be paused, gauze should be removed from the posterior aspect of the heart, and the heart allowed to recover for 1–2 min before proceeding. Increasing the titration of oxygen is sometimes also helpful. If arrhythmias persist despite these interventions, administer amiodarone (4 mg/kg bolus) and/or lidocaine (0.5 mg/kg bolus). If the animal goes into VF, administer 0.5–1 mL of diluted epinephrine, and defibrillate immediately. We have in multiple instances rescued animals that experienced major ECG disturbances (including VF) during the cell implantation procedure. These animals recovered fully and later exhibited a normal post-transplant course and excellent cell grafts (Fig. 4).

Acknowledgments

This work was supported by funding from an Ontario Institute for Regenerative Medicine Disease Team award, the McEwen Stem Cell Institute, the Peter Munk Cardiac Centre, the John R. Evans Leaders Fund/Canada Foundation for Innovation, and the University of Toronto's Medicine by Design initiative, which receives funding from the Canada First Research Excellence Fund. The authors would also like to thank Alyssa Goldstein and Karen Ho for their technical expertise and invaluable assistance with the preparation of this manuscript, and Christoph Haller for helpful comments.

References

1. Virani SS, Alonso A, Benjamin EJ et al (2020) Heart disease and stroke statistics—2020 update: a report from the American Heart Association. Circulation 141:139–596
2. Laflamme MA, Murry CE (2011) Heart regeneration. Nature 473:326–335
3. Romagnuolo R, Laflamme MA (2017) Programming cells for cardiac repair. Curr Opin Biotechnol 47:43–50
4. Laflamme MA, Chen KY, Naumova AV et al (2007) Cardiomyocytes derived from human embryonic stem cells in pro-survival factors enhance function of infarcted rat hearts. Nat Biotechnol 25:1015–1024
5. Caspi O, Huber I, Kehat I et al (2007) Transplantation of human embryonic stem cell-derived cardiomyocytes improves myocardial performance in infarcted rat hearts. J Am Coll Cardiol 50:1884–1893
6. Shiba Y, Fernandes S, Zhu W-Z et al (2012) Human ES-cell-derived cardiomyocytes electrically couple and suppress arrhythmias in injured hearts. Nature 489:322–325
7. Shiba Y, Filice D, Fernandes S et al (2014) Electrical integration of human embryonic stem cell-derived cardiomyocytes in a Guinea Pig chronic infarct model. J Cardiovasc Pharmacol Ther 19:368–381
8. van Laake LW, Passier R, Doevendans PA, Mummery CL (2008) Human embryonic stem cell–derived cardiomyocytes and cardiac repair in rodents. Circ Res 102:1008–1010
9. Chong JJH, Yang X, Don CW et al (2014) Human embryonic-stem-cell-derived cardiomyocytes regenerate non-human primate hearts. Nature 510:273–277
10. Shiba Y, Gomibuchi T, Seto T et al (2016) Allogeneic transplantation of iPS cell-derived cardiomyocytes regenerates primate hearts. Nature 538:388–391
11. Liu Y-W, Chen B, Yang X et al (2018) Human embryonic stem cell-derived cardiomyocytes restore function in infarcted hearts of non-human primates. Nat Biotechnol 36: 597–605
12. Romagnuolo R, Masoudpour H, Porta-Sánchez A et al (2019) Human embryonic stem cell-derived cardiomyocytes regenerate the infarcted pig heart but induce ventricular tachyarrhythmias. Stem Cell Rep 12:967–981
13. Wessels A, Sedmera D (2003) Developmental anatomy of the heart: a tale of mice and man. Physiol Genomics 15:165–176
14. Malinow MR, Hill JD, Ochsner AJ (1977) Heart rate in caged Macaca fascicularis. Effects of short-term physical exercise. J Med Primatol 6:69–75
15. Hughes HC (1986) Swine in cardiovascular research. Lab Anim Sci 36:348–350
16. Lelovas PP, Kostomitsopoulos NG, Xanthos TT (2014) A comparative anatomic and physiologic overview of the porcine heart. J Am Assoc Lab Anim Sci 53:432–438
17. Ghugre NR, Ramanan V, Pop M et al (2011) Quantitative tracking of edema, hemorrhage, and microvascular obstruction in subacute myocardial infarction in a porcine model by MRI: quantitative tracking of edema, hemorrhage, and MVO. Magn Reson Med 66: 1129–1141
18. Michael Swindle M, Smith AC (2008) Swine in biomedical research. In: Conn PM (ed) Sourcebook of models for biomedical research. Humana Press, Totowa, NJ, pp 233–239
19. Koudstaal S, Jansen MS, Lorkeers S, Gho JMIH et al (2014) Myocardial infarction and functional outcome assessment in pigs. J Vis Exp 86:e51269

Chapter 14

Defined Engineered Human Myocardium for Disease Modeling, Drug Screening, and Heart Repair

Malte Tiburcy, Tim Meyer, Pierre-Luc Satin, and Wolfram-Hubertus Zimmermann

Abstract

Different engineered heart muscle formats have been developed for applications in disease modeling, drug screening, and heart repair. The advantage of 3D engineered versus 2D monolayer and 3D aggregate cardiomyocyte cultures is a clearly advanced degree of maturation, which in many aspects resembles the postnatal rather than the embryonic or fetal heart, in the most advanced 3D culture formats. According to the desired in vitro (disease modeling or drug screening) and in vivo (heart repair) application, scale and geometry of tissue engineered heart muscle must be adapted. In this updated methods paper, we report a simple and scalable (up and down) collagen-based protocol for the construction of Engineered Human Myocardium (EHM) under defined, serum-free conditions.

Key words Tissue engineering, Heart, Pluripotent stem cells

1 Introduction

With the introduction of human embryonic (ESC) and induced (iPSC) pluripotent stem cells and methods for their directed differentiation, heart muscle engineering could be readily translated from rat [1, 2] to human models [3–6]. We and others have previously demonstrated several applications of ESC- and iPSC-based engineered human myocardium (EHM) models for in vitro studies of cardiomyocyte maturation and hypertrophy [7], disease modeling [8], and drug screening [9] as well as in vivo heart repair [10].

The most commonly applied methods to support cardiomyocyte self-assembly into 3D tissue can be categorized as scaffold-free and scaffold-supported models. Scaffold-free approaches include (1) cell-sheet engineering [11] and (2) aggregate cultures [12, 13]. Scaffold-supported approaches typically make use of either collagen or fibrin, with [3, 4, 6, 14, 15] or without [7, 16,

Kareen L. K. Coulombe and Lauren D. Black III (eds.), *Cardiac Tissue Engineering: Methods and Protocols*, Methods in Molecular Biology, vol. 2485, https://doi.org/10.1007/978-1-0716-2261-2_14,

17] additional supplementation with Matrigel™. Synthetic scaffold materials may be applied as alternative, but resultant tissue must be compared to state-of-the-art engineered heart muscle models, which all apply biological scaffold materials, to support their use. Similar as in rodent models [18], an essential role for non-myocytes and here in particular for fibroblasts or stromal cells with fibroblast-activity was identified [3, 7, 15, 17, 19].

Mechanical, electrical, or combined electromechanical stimuli are commonly applied to support organotypic development in heart muscle constructs [7, 17]. While mechanical loading is an absolute requirement for the engineering of heart muscle with advanced maturation, electrical stimulation may provide complementary or additional support depending on the applied culture conditions. Numerous parameters have been suggested to benchmark maturation [20, 21], which include structural (sarcomere structure, t-tubules, cardiomyocyte dimension), molecular (myosin and troponin isoform switches), metabolic (glucose and fatty acid utilization), and functional (force-frequency-response, drug responsiveness) measures. Finally, we posit that functional parameters and here in particular physiological responses to (1) increasing electrical stimulation frequencies (positive force-frequency response) and (2) pharmacological stimuli such as isoproterenol (positive inotropic, lusitropic, and chronotropic catecholamine-response) are the most predictive and least ambiguous parameters in the characterization of engineered heart muscle maturation.

Defining culture conditions, by replacement of undefined serum- or Matrigel™ supplements, and scalability (up and down) is key to ensure robust and versatile applications of tissue engineering technologies in vitro (disease modeling and drug screening) and in vivo (heart repair). In this updated methods chapter, we report on a readily scalable collagen type I-based protocol [7] for the construction of engineered human myocardium (EHM) from pluripotent stem cells under defined conditions for applications in disease modeling, drug screening, and heart repair (Fig. 1).

2 Materials

2.1 Cell Culture Components

1. Iscove's MEM with GlutaMax (Thermo Scientific, Waltham, USA).
2. B27 supplement without insulin (Thermo Scientific, Waltham, USA).
3. 10,000 U/mL Penicillin—10,000 μg/mL Streptomycin (P/S; Thermo Scientific, Waltham, USA)—optional.
4. 100× MEM non-essential amino acids (Thermo Scientific, Waltham, USA).

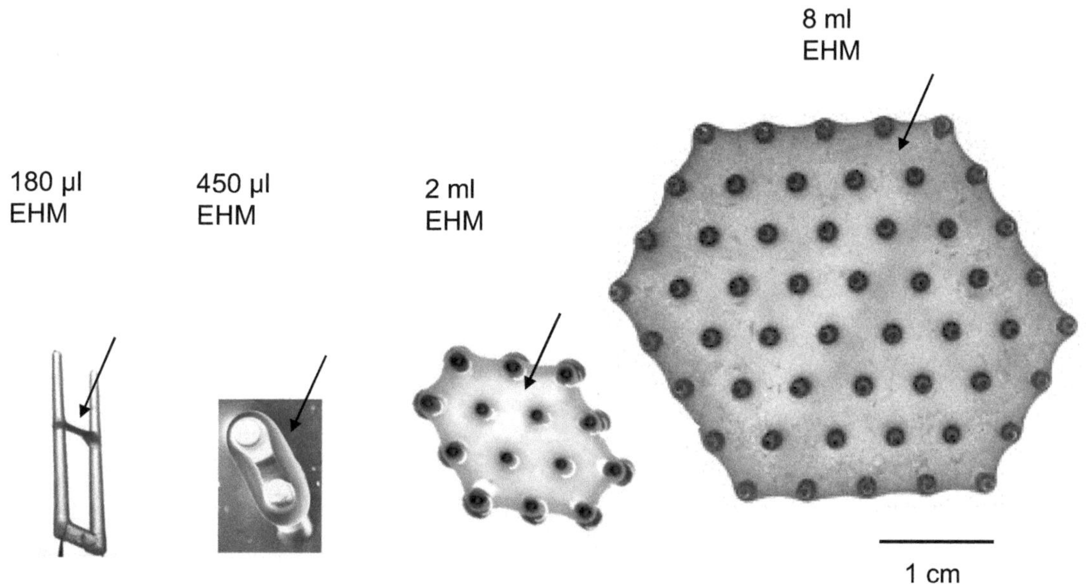

Fig. 1 Scalability of EHM according to desired applications. Photographs depict typical EHM scales used by us for disease modeling and drug screening (scale down) and heart repair (scale up). The indicated volumes represent the initial volumes of the respectively used EHM reconstitution mixtures. Mechanical loading is realized by the integration of flexible poles into the respective tissue formulations (black/translucent poles). Geometries and scale can further be adapted as needed. After completion of the self-organization process, EHM thickness (*z*-dimension) is typically 0.5–1 mm in the displayed formulations (x/y-dimensions can be freely scaled by adaptation of the casting molds). Scale bar: 1 cm

5. L-ascorbic acid 2-phosphate sesquimagnesium salt hydrate (Sigma-Aldrich, St. Louis, USA).
6. Recombinant human IGF-1 (Peprotech, New Jersey, USA).
7. Recombinant human FGF-2 (Peprotech, New Jersey, USA).
8. Recombinant human $VEGF_{165}$ (Peprotech, New Jersey, USA).
9. Recombinant human TGF-β1 (Peprotech, New Jersey, USA).

2.2 EHM Generation Components

1. Collagen type I (typically 3–7 mg/mL; *see* **Note 1**):
2. 2× Concentrated medium (2× RPMI).
3. NaOH 0.1 N.
4. 50 mL centrifuge tubes.
5. 2 and 5 mL serological pipettes.
6. (a) Multiple use silicone molds with flexible holders (*see* Fig. 2a; **Note 2**).

   Silicone molds can be autoclaved and reused.

   (b) Single use multi-well plate (*see* Fig. 2b; **Note 2**).

   48-well plate with removable flexible holders and defined pole stiffness (myrPlate, myriamed GmbH, Göttingen, Germany).
7. EHM culture medium.

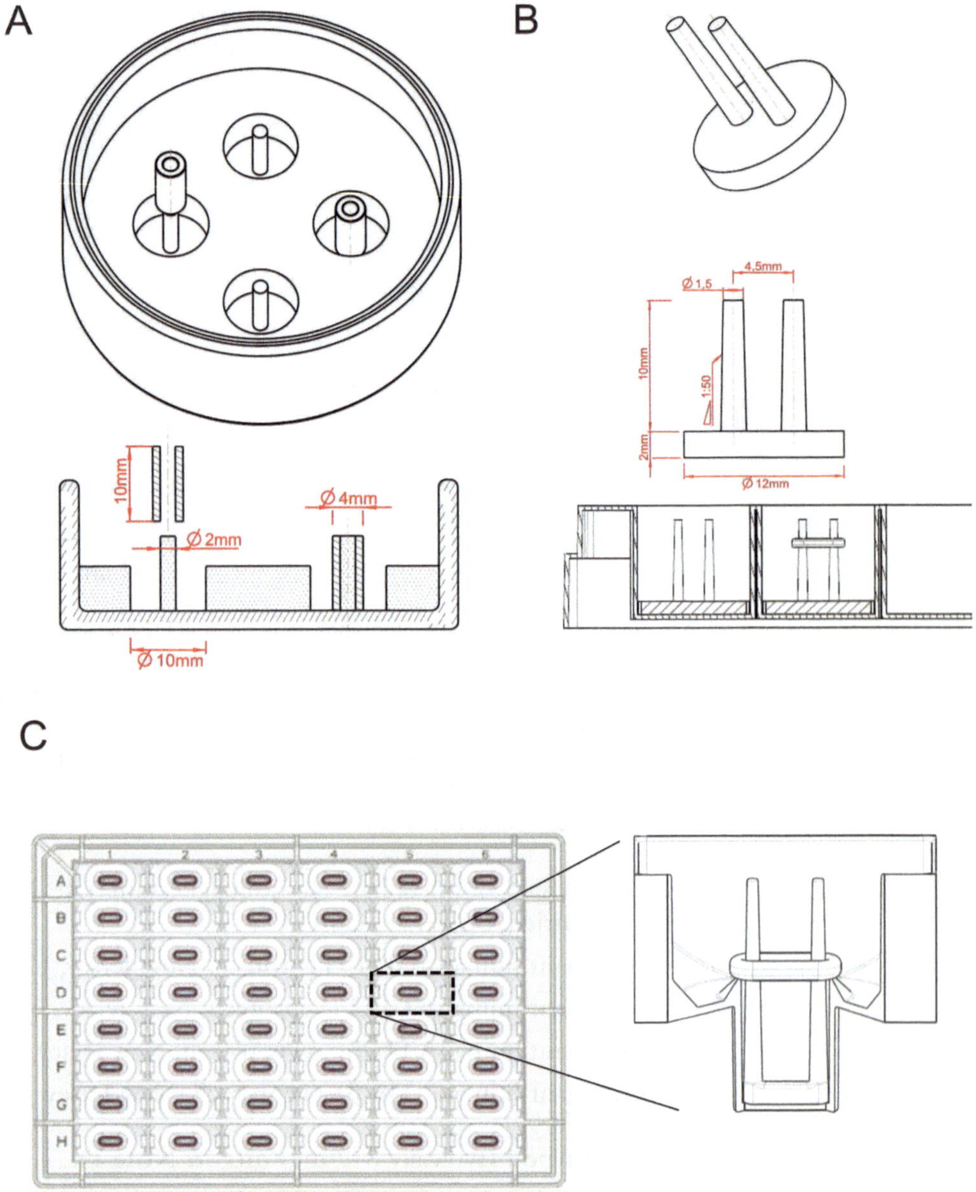

Fig. 2 EHM casting molds with stretchers. (**a**) Custom-made mold with four circular recesses to support the formation of circular EHM (450 μL casting volume per EHM); we prepare molds from polydimethylsiloxane (PDMS, also known as silicone) and removable PDMS or polytetrafluoroethylene (Teflon®) tubes in glass dishes (diameter 60 mm); removable tubes allow for easy retrieval of circular EHM from the casting molds. (**b**) Custom-made stretchers (flexible poles fixed on a base plate) prepared by 3D prototype printing or cast-molding to support auxotonic contractions of EHM (right); loading can be adapted by the choice of material with desired biophysical properties and the pole design (thickness, geometry, length). (**c**) Multi-well plate format for automated casting, parallel culture, and video-optic analysis of 48 EHM with close-up view of a single well showing the cavity for casting and the flexible poles for maturation under defined mechanical loading (refer to [22] for additional details)

3 Methods

3.1 Preparation of EHM Reconstitution and Culture Media

1. 10× culture medium (10× RPMI): solubilize 1.04 g RPMI powder (Thermo Scientific, Waltham, USA) in 10 mL deionized water. Filter through a 0.22 μm syringe filter.
2. 2× medium (2× RPMI): Mix 2 mL 10× RPMI, 0.8 mL B27 minus insulin (Thermo Scientific, Waltham, USA), 0.2 mL of P/S (optional) and 7 mL deionized water. Filter through a 0.22 μm syringe filter.
3. EHM culture medium: add 20 mL B27 minus insulin, 5.3 mL P/S (optional), 5.3 mL 100× MEM non-essential amino acids, 100 ng/mL IGF-1, 10 ng/mL FGF-2, 5 ng/mL $VEGF_{165}$, and 54 mg L-ascorbic acid 2-phosphate sesquimagnesium salt hydrate to 500 mL Iscove's medium (*see* **Note 3**). For EHM casting and during the first 3 days of culture also add 5 ng/mL TGF-β1 (*see* **Note 4**).

3.2 Preparation of Cells for EHM Generation

1. Any source of human cardiomyocytes from pluripotent stem cells can be applied (fresh or frozen). Prepare a single cell suspension of cardiomyocytes by enzymatic digestion. We typically apply a mix of Accutase and TrypLE or Versene to detach cardiomyocytes grown as monolayers [22].
2. Any source of human fibroblasts or mesenchymal stromal cells (primary or stem cell derived) can be applied (fresh or frozen). Prepare a single cell suspension of fibroblasts by enzymatic digestion using TrypLE [7].
3. Cells need to be mixed at a desired ratio in EHM medium with TGF-β1 and kept on ice. We typically use a cardiomyocyte: fibroblast ratio of 70:30 with a total cell concentration of 5.3×10^6/mL. We recommend to use cell pools of high purity, i.e., >90% by flow cytometry for ACTN2 and TNNT2 as well as THY1 and VIM in the cardiomyocyte and fibroblast populations, respectively. For different EHM geometries and non-myocyte sources (including one or multiple non-myocyte cell types) optimal cardiomyocyte:non-myocyte ratios must be determined experimentally.
   (a) Cryopreserved cells should be thawed using standard protocols, washed in EHM culture medium, pelleted by centrifugation (200 × *g*, 7 min, 20–23 °C), and resuspended in EHM culture medium at the desired cell density.
   (b) We recommend the use of cells freshly retrieved and enzymatically dispersed from monolayer or suspension bioreactor cultures.
4. We recommend to test for viability using, for example, trypan blue exclusion in a Neubauer chamber, nucleocounting

Table 1
Example for an EHM master mix composition for applications in disease modeling and drug screening

Master mix components[a]	Volume (μL)
Collagen (6.84 mg/mL)[b]	2112
2× RPMI	2112
0.1 N NaOH[c]	288
Total cell number (26.4×10^6)[d]	4992
Total volume of master mix[e]	9504
Number of EHM (450 μL EHM)	20
Number of EHM (180 μL EHM)	48

EHM master mix volumes can be adapted as desired for further up- and down-scaling. Cell composition and concentration as well as collagen content have to be adapted for optimal results

[a]Master mix can be scaled up and down as needed. This example is for applications in disease modeling and drug screening

[b]We typically use a final collagen concentration of 0.7–1.8 mg/mL

[c]We recommend a dropwise addition and adjustment of the volume according to the phenol red indicator color change

[d]Resuspend cells in EHM medium with TGFβ1. Adjust volume to reach a target cell concentration of 1.25×10^6 or 0.5×10^6 cells in 450 μL (silicon molds) or 180 μL (multi-well plate) EHM, respectively

[e]We recommend to prepare an excess amount of master mix (+5–10%) to compensate for the loss of the viscous master mix during pipetting

(Nucleocounter®), or current exclusion measurements (CASY-Counter®). Note, that these different assays will provide different viability levels. For optimal use in tissue engineering viability of >70% is recommended.

5. Determine the cell number to calculate the volume of master mix needed (Table 1). For drug screening and disease modeling applications, we prepare EHM with an individual volume of 180 μL or 450 μL in either self-made or commercially available multi-well plates (Fig. 2). Further down- [9] and up-scaling [7] is possible, but may require adaptations of the cell composition and concentration as well as the collagen concentration for optimal results.

3.3 EHM Generation

All materials should be prechilled at 4 °C and the following steps should be performed on ice. The amount of master mix needed depends on the required number and geometry of EHM. We typically prepare master mixes for up to 48 EHM (in case of loop-shaped EHM for drug screening or disease modeling) with a collagen concentration of 0.7–1.8 mg/mL and a cell density of $2.5–5 \times 10^6$ cells per mL.

1. With a serological pipette, transfer required volume of collagen solution into a prechilled 50 mL centrifuge tube (Table 1). The master mix volumes are adjusted to account for volume lost (5–10%) during pipetting of the viscous master mix.

2. Add the same volume of 2× RPMI and mix well by shaking the tube. Do not use the pipette to mix as there will be too much hydrogel lost in the pipette.
3. Neutralize the pH by adding 0.1 N NaOH. NaOH should be added dropwise while shaking the tube until the phenol red indicator turns from yellow to red/pink.
4. Resuspend the cells with a serological pipette and add to the master mix. Mix well to ensure homogenous distribution of cells.
5. Pipette the master mix manually or with an automated pipetting device into the desired casting molds (Fig. 3a).
6. Allow the hydrogel to consolidate for 1 h at 37 °C in a humidified incubator with 5% CO_2 (*see* Fig. 3b, **Note 5**).
7. After consolidation, carefully add pre-warmed EHM culture medium without disrupting the still sensitive hydrogel-cell mixture.
8. Change medium the next day and then every other day.
9. (a) Silicone molds for casting (Fig. 2a) and flexible poles for mechanical loading of EHM (Fig. 2b):

   Within 3–5 days in culture EHMs compact around the central pole of the casting mold and start to contract spontaneously and rhythmically (*see* **Note 6**). When fully condensed EHMs can be transferred to flexible poles to support diastolic loading and auxotonic contractions for optimal EHM maturation (*see* **Note 7**).

   (b) 48-well tissue plate (Fig. 2c):

   The EHM master mix can be pipetted into specially designed 48-well culture plates with integrated flexible poles [22]. In this model, manual transfer from the casting mold onto a distinct stretch device is not required as the EHM compact around the flexible poles at a spatially defined position (Fig. 3c, d).
10. We perform studies of EHM contractility typically from culture day 10 onwards using either isometric force measurements in organ baths or video-optic analyses of contractions.

3.4 Analysis (Quality Control) by Video-Optic Recordings

Video-optic recordings of pole bending is a valid first approach to screen for contractile parameters (Table 2) in spontaneously, electrically, or optogenetically paced EHM, but should be supplemented by isometric force measurements for deep functional phenotyping.

1. EHM should be contracting visibly and synchronously (i.e., the whole tissue beats in unison) at latest by day 10 in culture. Force of contraction increases with time in culture with a steady state reached after 4–6 weeks (may vary depending on the cell

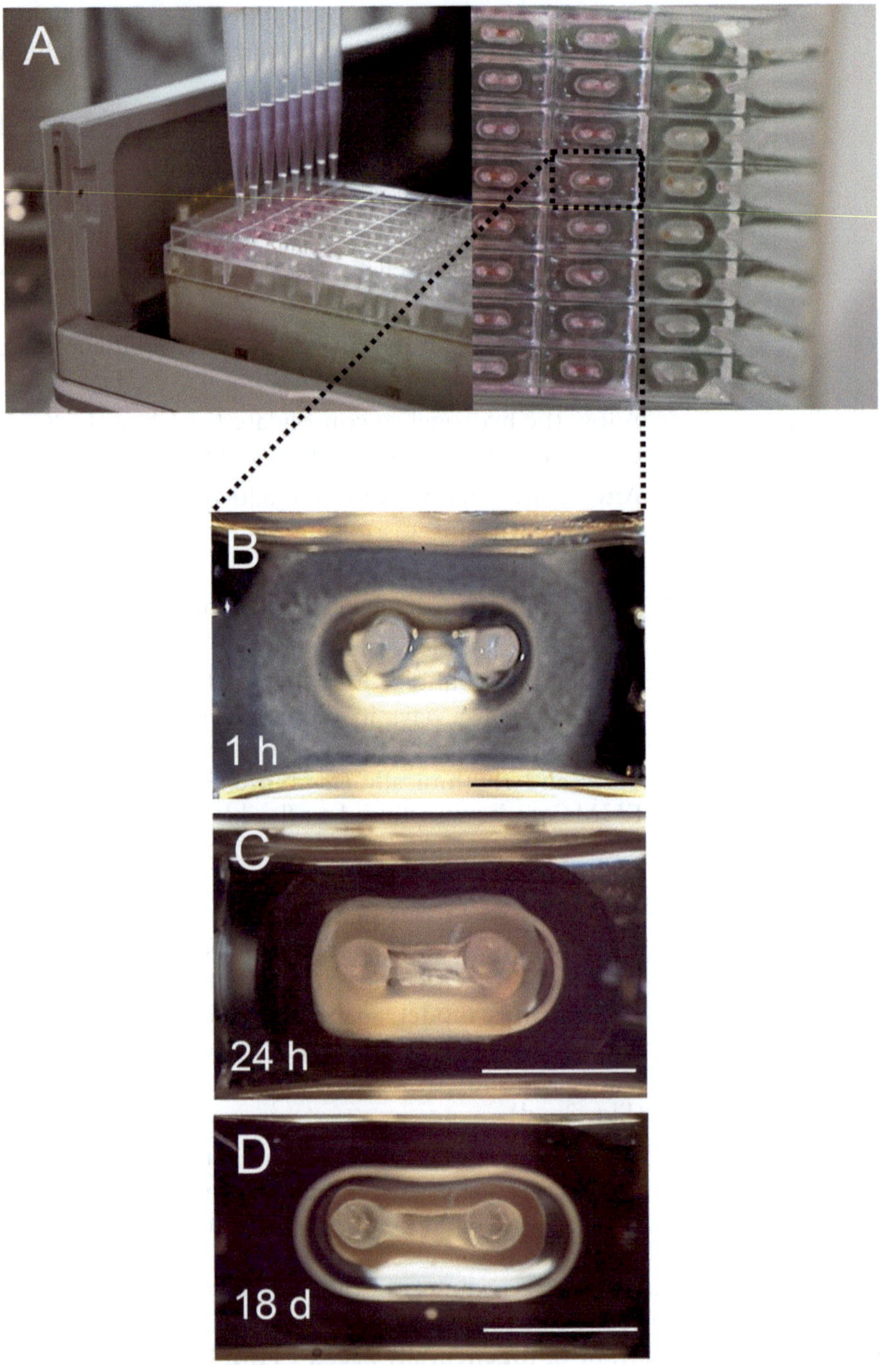

Fig. 3 EHM formation. (**a**) Automated casting of EHM in a multi-well plate. (**b**) 1 h after casting, EHM condense (collagen gelation results in an opaque appearance of the EHM reconstitution mixture). (**c**) Notable EHM compaction and suspension on flexible poles for mechanical loading 24 h after casting. (**d**) EHM after completion of compaction on culture day 18 with EHM fully suspended on the flexible poles to impose mechanical loading (afterload adaptable by pole design) and to support auxotonic contractions under preloaded conditions (adaptable by pole design). Scale bars: 5 mm

Table 2
Technical parameters and experimental endpoints assessed by video-optic (primary assay) and isometric (secondary assay for deep phenotyping) contractility assays

		Video-optical analysis	Organ bath
Technical parameters	Mode of contraction	Auxotonic	Isometric
	Data sampling frequency	30–100 Hz	100–1000 Hz
	Force measurement	Indirect (pole-bending)	Direct (force transducer)
	Readout	Pixel shift (%)[a]	Force (mN)
	Range (typical)	0.5–15%[b]	0.1–3 mN
	Parallel analysis	Yes	Limited
	Automation ready	Yes	Limited
	Continuous analysis (sterile)	Yes	Limited
Experimental Endpoints	Force-preload studies	Limited	Yes
	Force-frequency studies	Limited	Yes
	Force of contraction (systolic; mN)	Yes	Yes
	Resting (diastolic) force (mN)	Yes	Yes
	Contraction time (ms)	Yes	Yes
	Relaxation time (ms)	Yes	Yes
	Contraction velocity (+dF/dt)	Yes	Yes
	Relaxation velocity (−dF/dt)	Yes	Yes
	Beating frequency (bpm)	Yes	Yes
	Beating regularity (R-R interval)	Yes	Yes

[a]Distance of flexible poles in systole related to pixel distance in diastole in %

[b]Pole bending depends on tissue force development and biophysical properties of the applied poles/holders. Flexible poles with appropriate biophysical properties as to stiffness and bending kinetics have to be identified experimentally to ensure that pole-bending measurements are within a linear range

model). EHM can be kept in culture for more than 1 year without loss of function (*see* **Note 8**).

2. Video-optic recordings should be performed with a high speed (>50 fps), high resolution (pixel size <50 μm), and large field-of-view (4 K) camera with telecentric optics as well as hard- and software developed for fast and parallel image processing to allow for continuous image recordings and real-time analyses.
3. Contactless video-optical measurements provide the advantage of continuous analyses of EHM maturation and/or response to interventions.

3.5 Analysis (Quality Control) by Isometric Force Measurements

Isometric force measurements at defined preloading and electrical stimulation is the gold standard for deep contractile phenotyping. Refer for an overview of the typically recorded parameters to Table 2.

1. EHM are transferred individually from the silicon mold or multi-well plate onto hooks connected to a force transducer in thermostatted (*see* **Note 9**) organ baths filled with oxygenated Tyrode's solution [1].
2. After 15 min equilibration, EHM are subjected to electrical field stimulation at 1–1.5 Hz, i.e., well above the endogenous EHM beating rate to not confound assessments of inotropy by chronotropic effects (force-frequency responses have to be considered), and preloaded to a tissue length (L_{max}) with maximal (active) force of contraction (contraction amplitude) according to the Frank-Starling law. At Lmax and 1.8 mmol/L calcium, isometric force should be >1 mN and the active (i.e., systolic minus diastolic force) to passive (diastolic) force ratio should be >1 (*see* **Note 10**).

After completion of the contractility assessments EHM can be subjected to histological, biophysical, and molecular follow-up studies.

4 Notes

1. Make sure to use acid-solubilized collagen I. We are typically using commercially available bovine collagen type I, but have used collagen from other animal sources and expect any high-quality collagen type I source with undisturbed gelling capacity to work as well. We recommend to test newly acquired batches of collagen as to their gelling capacity prior use in EHM manufacturing. Collagen must be kept at 4 °C at all times to decelerate the naturally occurring collagen degradation process.
2. Silicone molds of desired geometries can be readily and cheaply prepared without specialized equipment. For a one-step casting, culture, and analysis procedure with minimal inter-EHM variability as well as process automation we recommend the use of multi-well plate formats designed according to ANSI/SLAS standards (Fig. 2b; [22]).
3. We prefer serum-free culture of human pluripotent stem cell-derived EHM, but note that serum-containing culture medium may also be applied. In serum-containing medium, we commonly observed functional features of heart failure, such as a blunted force-frequency response and therefore do not recommend serum-containing media if EHM with a non-failing heart muscle phenotype are to be investigated.
4. The addition of TGF-β1 is instrumental for the support of stroma cell-mediated collagen compaction in EHM. Exposure to effective concentrations of TGF-β1 may be adapted as needed to support the cell-dependent tissue compaction process [23] in different EHM geometries.

5. Collagen condensation (i.e., cell-independent collagen gelation; [23]) is an essential prerequisite for homogenous cell entrapment in the forming EHM. If the collagen is not condensed after 1 h at 37 °C (Fig. 3b), it is either of low quality (degraded), not properly neutralized (pH), or too diluted. We typically use 0.7–1.8 mg/mL final concentration of collagen in the EHM master mix. Note: acid-solubilized collagen has a limited shelf-life with notable degradation starting at 6 months even if kept constantly at 4 °C.
6. Cell-dependent compaction of the collagen-hydrogel during the first 3–5 culture days is essential to obtain functional EHM (Fig. 3a–c). Compaction depends on the non-myocyte (fibroblast) cell content, type, and function [7, 23]. Depending on the fibroblast content, type, and function maximal EHM compaction can be reached between 6 h and 5 days after casting.
7. The transfer to stretchers (Fig. 2) has been described before [5]. We prefer flexible mechanical stretchers (Fig. 2b) over static stretcher to facilitate auxotonic contractions [24].
8. Electrical stimulation may be required to assess contractility in long-term (several months) EHM cultures. This applies particularly to EHM generated from the most commonly used directed differentiation protocols, which provide cardiomyocytes with primarily ventricular properties (reviewed in [25, 26]). Alternative protocols may be used to provide atrial/nodal cell populations, which will result in EHM with a higher and sustained endogenous beating rate [27].
9. EHMs are very temperature sensitive and stop beating quickly if temperature drops below 37 °C. Even subtle hypothermia may obscure contractile parameters (slowing of beating rate, alterations in force of contraction, prolongation of contraction kinetics).
10. General consideration when performing and reporting contractility studies: we recommend to always report absolute twitch force values in isometric contraction experiments in EHM with highly standardized cross-sectional area [7]. While normalization to cross-sectional area is redundant in contractility assessments under acute interventions, we do recommend normalization to tissue cross-sectional area and/or output cardiomyocyte content in case of chronic interventions or in disease models, which may exhibit differences in EHM compaction or cell composition (selection/survival bias). In this context, it should also be noted that force of contraction and engineered tissue diameter do not scale linearly, resulting in an apparent higher cross-sectional force in smaller vs. larger EHMs, despite similar cardiomyocytes content and absolute force values.

Acknowledgments

M.T. is supported by the German Research Foundation (DFG TI 956/1-1, SFB 1002 C04). T.M. is supported by the DZHK (German Center for Cardiovascular Research; Shared Expertise Funding Scheme). P.L.S. is a member of DFG IRTG 1816. W.H.Z. is supported by the DZHK, the Leducq Foundation, the German Ministry for Science and Education (BMBF - IndiHeart), and the German Research Foundation (DFG SFB 1002 C04 and S01, IRTG 1816, MBExC - EXC 2067/1).

Conflict of Interest

M.T., T.M., and W.H.Z. are listed as inventors of several granted and pending patent applications in the field of stem cell differentiation, tissue engineering, and tool development. The University Medical Center Göttingen has licensed related IP to myriamed GmbH and Repairon GmbH. W.H.Z. is founder and advisor of myriamed GmbH (active in drug screening) and Repairon GmbH (active in the clinical translation of tissue engineered heart repair). M.T. and T.M. are advisors of myriamed GmbH and Repairon GmbH.

References

1. Zimmermann WH, Fink C, Kralisch D, Remmers U, Weil J, Eschenhagen T (2000) Three-dimensional engineered heart tissue from neonatal rat cardiac myocytes. Biotechnol Bioeng 68:106–114
2. Zimmermann WH, Schneiderbanger K, Schubert P, Didie M, Munzel F, Heubach JF, Kostin S, Neuhuber WL, Eschenhagen T (2002) Tissue engineering of a differentiated cardiac muscle construct. Circ Res 90:223–230
3. Kensah G, Roa Lara A, Dahlmann J, Zweigerdt R, Schwanke K, Hegermann J, Skvorc D, Gawol A, Azizian A, Wagner S et al (2013) Murine and human pluripotent stem cell-derived cardiac bodies form contractile myocardial tissue in vitro. Eur Heart J 34: 1134–1146
4. Schaaf S, Shibamiya A, Mewe M, Eder A, Stohr A, Hirt MN, Rau T, Zimmermann WH, Conradi L, Eschenhagen T et al (2011) Human engineered heart tissue as a versatile tool in basic research and preclinical toxicology. PLoS One 6:e26397
5. Soong PL, Tiburcy M, Zimmermann WH (2012) Cardiac differentiation of human embryonic stem cells and their assembly into engineered heart muscle. Curr Protoc Cell Biol Chapter 23(Unit23):28
6. Tulloch NL, Muskheli V, Razumova MV, Korte FS, Regnier M, Hauch KD, Pabon L, Reinecke H, Murry CE (2011) Growth of engineered human myocardium with mechanical loading and vascular coculture. Circ Res 109:47–59
7. Tiburcy M, Hudson JE, Balfanz P, Schlick S, Meyer T, Chang Liao ML, Levent E, Raad F, Zeidler S, Wingender E et al (2017) Defined engineered human myocardium with advanced maturation for applications in heart failure modeling and repair. Circulation 135:1832–1847
8. Hanses U, Kleinsorge M, Roos L, Yigit G, Li Y, Barbarics B, El-Battrawy I, Lan H, Tiburcy M, Hindmarsh R et al (2020) Intronic CRISPR repair in a preclinical model of noonan syndrome-associated cardiomyopathy. Circulation 142(11):1059–1076
9. Mills RJ, Parker BL, Quaife-Ryan GA, Voges HK, Needham EJ, Bornot A, Ding M, Andersson H, Polla M, Elliott DA et al (2019) Drug screening in human PSC-cardiac organoids identifies pro-proliferative

compounds acting via the mevalonate pathway. Cell Stem Cell 24(895–907):e896

10. Riegler J, Tiburcy M, Ebert A, Tzatzalos E, Raaz U, Abilez OJ, Shen Q, Kooreman NG, Neofytou E, Chen VC et al (2015) Human engineered heart muscles engraft and survive long term in a rodent myocardial infarction model. Circ Res 117:720–730
11. Sasaki D, Matsuura K, Seta H, Haraguchi Y, Okano T, Shimizu T (2018) Contractile force measurement of human induced pluripotent stem cell-derived cardiac cell sheet-tissue. PLoS One 13:e0198026
12. Giacomelli E, Meraviglia V, Campostrini G, Cochrane A, Cao X, van Helden RWJ, Krotenberg Garcia A, Mircea M, Kostidis S, Davis RP et al (2020) Human-iPSC-derived cardiac stromal cells enhance maturation in 3D cardiac microtissues and reveal non-cardiomyocyte contributions to heart disease. Cell Stem Cell 26(862–879):e811
13. Pointon A, Pilling J, Dorval T, Wang Y, Archer C, Pollard C (2017) From the cover: high-throughput imaging of cardiac microtissues for the assessment of cardiac contraction during drug discovery. Toxicol Sci 155:444–457
14. Nunes SS, Miklas JW, Liu J, Aschar-Sobbi R, Xiao Y, Zhang B, Jiang J, Masse S, Gagliardi M, Hsieh A et al (2013) Biowire: a platform for maturation of human pluripotent stem cell-derived cardiomyocytes. Nat Methods 10: 781–787
15. Zhang D, Shadrin IY, Lam J, Xian HQ, Snodgrass HR, Bursac N (2013) Tissue-engineered cardiac patch for advanced functional maturation of human ESC-derived cardiomyocytes. Biomaterials 34:5813–5820
16. Munarin F, Kaiser NJ, Kim TY, Choi BR, Coulombe KLK (2017) Laser-etched designs for molding hydrogel-based engineered tissues. Tissue Eng Part C Methods 23:311–321
17. Ronaldson-Bouchard K, Ma SP, Yeager K, Chen T, Song L, Sirabella D, Morikawa K, Teles D, Yazawa M, Vunjak-Novakovic G (2018) Advanced maturation of human cardiac tissue grown from pluripotent stem cells. Nature 556:239–243
18. Naito H, Melnychenko I, Didie M, Schneiderbanger K, Schubert P, Rosenkranz S, Eschenhagen T, Zimmermann WH (2006) Optimizing engineered heart tissue for therapeutic applications as surrogate heart muscle. Circulation 114:I72–I78
19. Kreutziger KL, Muskheli V, Johnson P, Braun K, Wight TN, Murry CE (2011) Developing vasculature and stroma in engineered human myocardium. Tissue Eng A 17:1219–1228
20. Karbassi E, Fenix A, Marchiano S, Muraoka N, Nakamura K, Yang X, Murry CE (2020) Cardiomyocyte maturation: advances in knowledge and implications for regenerative medicine. Nat Rev Cardiol 17:341–359
21. Liaw NY, Zimmermann WH (2016) Mechanical stimulation in the engineering of heart muscle. Adv Drug Deliv Rev 96:156–160
22. Tiburcy M, Meyer T, Liaw NY, Zimmermann WH (2020) Generation of engineered human myocardium in a multi-well format. STAR Protoc 1:100032
23. Schlick SF, Spreckelsen F, Tiburcy M, Iyer LM, Meyer T, Zelarayan LC, Luther S, Parlitz U, Zimmermann WH, Rehfeldt F (2019) Agonistic and antagonistic roles of fibroblasts and cardiomyocytes on viscoelastic stiffening of engineered human myocardium. Prog Biophys Mol Biol 144:51–60
24. Zimmermann WH, Melnychenko I, Wasmeier G, Didie M, Naito H, Nixdorff U, Hess A, Budinsky L, Brune K, Michaelis B et al (2006) Engineered heart tissue grafts improve systolic and diastolic function in infarcted rat hearts. Nat Med 12:452–458
25. Burridge PW, Keller G, Gold JD, Wu JC (2012) Production of de novo cardiomyocytes: human pluripotent stem cell differentiation and direct reprogramming. Cell Stem Cell 10: 16–28
26. Zimmermann WH (2021) Engineered heart muscle Models in phenotypic drug screens. Handb Exp Pharmacol 265:143–156. https://doi.org/10.1007/164_2020_385. PMID: 33136187
27. Cyganek L, Tiburcy M, Sekeres K, Gerstenberg K, Bohnenberger H, Lenz C, Henze S, Stauske M, Salinas G, Zimmermann WH, Hasenfuss G, Guan K (2018) Deep phenotyping of human induced pluripotent stem cell-derived atrial and ventricular cardiomyocytes. JCI Insight 3(12):e99941. https://doi.org/10.1172/jci.insight.99941. PMID: 29925689; PMCID: PMC6124434

Chapter 15

Tubular Cardiac Tissue Bioengineered from Multi-Layered Cell Sheets for Use in the Treatment of Heart Failure

Hidekazu Sekine and Teruo Okano

Abstract

This chapter describes a method for creating tubular cardiac tissue in vitro. Thick cardiac tissue in a tubular configuration is prepared by stacking cell sheets stepwise on the inner wall of a segment of small intestine, which functions as a blood vessel bed. The capillaries of the small intestinal segment are fed by an artery and drained by a vein. Therefore, perfusion culture of the cardiac tissue is achieved by continuously infusing culture medium into the arterial vessel that supplies the segment of small intestine. The aim of this technique is to fabricate tubular cardiac tissue that can function as a pump by sequentially implanting and culturing cardiac cell sheets on the inner wall of a perfused segment of small intestine.

Key words Regenerative medicine, Cell sheet engineering, Temperature-responsive culture dish, Cardiac tissue regeneration, Extracellular matrix, 3D, Vascular network, Transplantation, Cardiomyocyte, Cardiac function, Endothelial cell, Neovascularization, Ischemic heart, Myocardial infarction, Animal model, Cardiac assist device

1 Introduction

We have been producing myocardial tissue using cell sheet engineering. This technique involves the preparation, manipulation, and transplantation of cell sheets from single cells using a temperature-responsive culture dish followed by the stacking of multiple cell sheets to construct a three-dimensional (3D) tissue. The temperature-responsive culture dish has a surface modified with poly(N-isopropylacrylamide) (PIPAAm) [1, 2], which is a temperature-sensitive polymer with a lower critical solution temperature of 32 °C. Therefore, the surface of the temperature-responsive culture dish is relatively hydrophobic at 37 °C, which allows cells to adhere to it during culture. However, the surface of the culture dish becomes hydrophilic when the temperature is reduced to 32 °C or lower, which results in cell detachment. Conventional methods to harvest single (disaggregated) cells

Kareen L. K. Coulombe and Lauren D. Black III (eds.), *Cardiac Tissue Engineering: Methods and Protocols*,
Methods in Molecular Biology, vol. 2485, https://doi.org/10.1007/978-1-0716-2261-2_15,

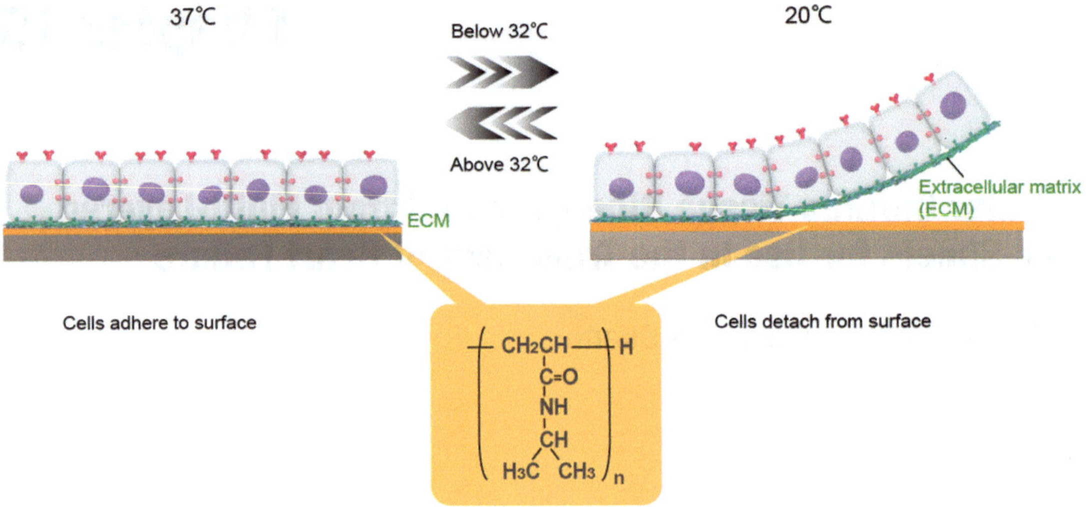

Fig. 1 Temperature-dependent changes in the hydration state of poly(N-isopropylacrylamide) (PIPAAm) can be used to regulate the attachment and detachment of a cell sheet in a culture dish

PIPAAm is a polymer that exhibits temperature-dependent phase transition behavior. The polymer side chains of PIPAAm undergo reversible hydration and dehydration as the temperature is varied. As a result, PIPAAm dissolves in aqueous solution at low temperatures due to hydration of its side chains, but it becomes insoluble in water at temperatures higher than 32 °C. A temperature-responsive culture dish is prepared by modifying a polystyrene dish with PIPAAm using electron beam polymerization. Confluent cells can be cultivated on a temperature-responsive culture dish at 37 °C because the surface of the dish is hydrophobic at this temperature, allowing adhesion of the cells. However, the surface of the dish becomes hydrophilic when the temperature is reduced below 32 °C, permitting detachment of the cells in the form of a sheet while maintaining cell–cell junctions

from a culture dish involve the use of enzyme such as trypsin or dispase to disrupt the cell–cell junctions and the cell culture plate connections. By contrast, when the temperature-responsive culture dish is used, it is possible to separate only the connections between the cells and the culture dish, thereby maintaining the cell–cell junctions and allowing the cells to be harvested and collected in the form of a sheet (see Fig. 1).

Cell sheet engineering [3, 4] enables the construction of a tissue with a high cell density because the construct can be generated using cells alone. Therefore, cell sheet engineering is ideally suited to the production of tissues with scaffold free high cell densities such as the heart, liver, and kidney. Cardiac cell sheet transplantation in a rat model of heart failure resulted in the transplanted cell sheets and host myocardial cells connecting with each other to form gap junctions [5]. Furthermore, the rate of engraftment was found to be substantially higher for transplanted cell sheets than for cells administered by injection [6]. In addition, co-culture techniques can be used to generate cell sheets that also contain endothelial cells, which not only provide a sustained release of angiogenic factors after cell sheet transplantation but also directly contribute to neovascularization and acceleration of the recovery of ischemic heart function [7]. Clinical and preclinical studies have investigated the potential use of cell sheet engineering for organs and tissues such as the heart, esophagus, cornea, cartilage, lungs, and ears, and treatment methods based on cell sheet engineering are beginning to be applied to various tissues [8–14].

Cell sheet engineering can generate planar cardiac tissue constructs simply by stacking multiple cell sheets. However, this technique can also be used to create tubular cardiac tissue by tubularizing the cell sheets. It is thought that the production of tubular cardiac tissue that functions as a pump will make it possible to develop a new therapy to replace heart transplantation for patients with refractory heart failure, including those with ischemic heart disease or congenital heart disease (see Fig. 2). We have created tubular cardiac tissue by winding a cardiac cell sheet made of newborn rat cardiomyocytes around a fibrin gel tube. The prepared tubular cardiac tissue pulsated spontaneously and generated detectable changes in its intraluminal pressure [15]. In a separate study, a tubularized construct was made in vitro using six cell sheets derived from newborn rat cardiomyocytes, and the tubular construct was then transplanted so as to replace part of the abdominal aorta of a rat. When both ends of the transplanted tissue were clamped and the pressure inside the tubular cardiac tissue was measured, pressure changes due to pulsation of the tubular cardiac tissue were detected [16]. The above findings are proof-of-concept that further stacking of cell sheets could be used to generate tubular cardiac tissue that produces sufficiently large increases in internal pressure for clinical use as a ventricular assist device.

One disadvantage of cell sheet stacking is the limitation in the thickness of the tissue that can be fabricated. When cell sheets are layered to construct a tissue with a thickness of 80 μm or more, necrosis occurs within the central region of the tissue due to a deficiency of nutrients [17]. Therefore, the production of thick myocardial tissue by the stacking of cardiac cell sheets requires the induction of a vascular network throughout the tissue to provide an adequate supply of oxygen and nutrients. We have devised a method to achieve this that involves the sequential transplantation of multiple cardiac cell sheets at time intervals that allow sufficient angiogenesis to occur from the host tissue (see Fig. 3a). This technique makes it possible to fabricate thicker (about 1 mm) cardiac tissue that generates greater contractile force in vivo (see Fig. 3b). Notably, we successfully used repeated grafting of cardiac cell sheets onto existing blood vessels (superficial caudal epigastric artery and vein) to construct thick cardiac tissue with accompanying blood vessels that was suitable for ectopic anastomotic transplantation into another site (see Fig. 3c) [18].

Following on from the above in vivo study, we have devised an in vitro method of creating thick cardiac tissue that is based on the

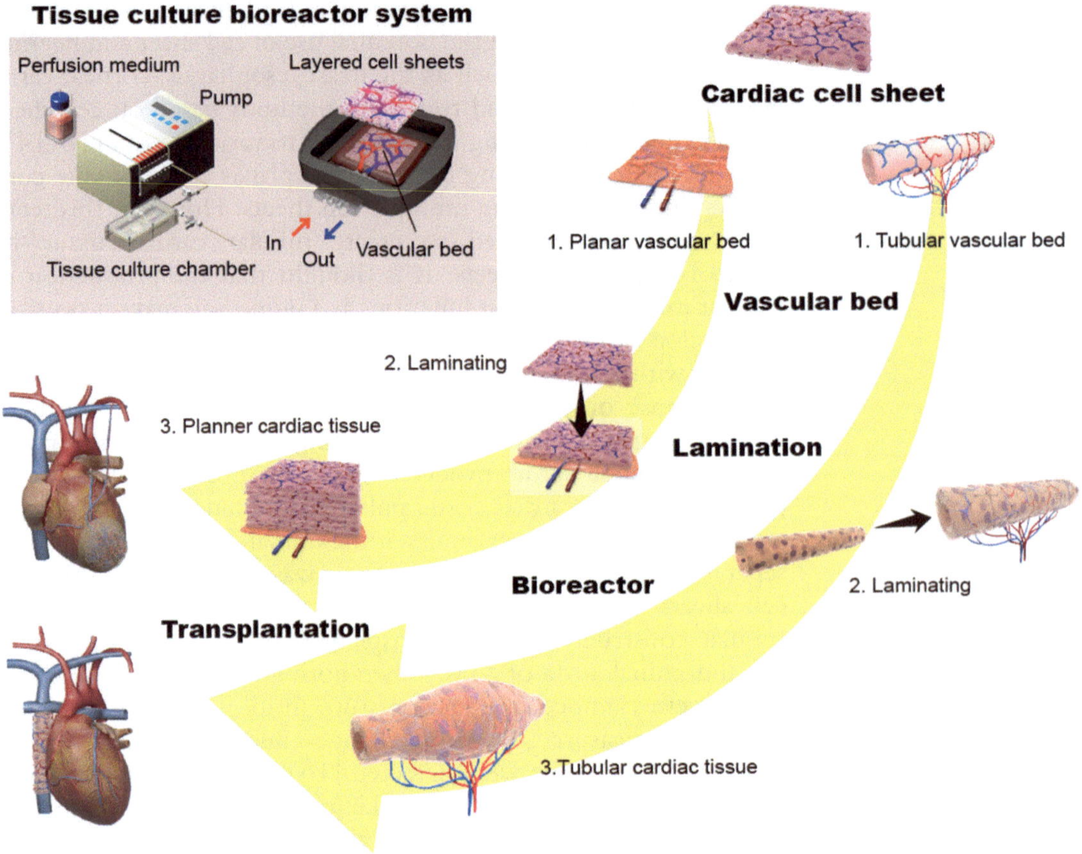

Fig. 2 Development of a clinically applicable technology based on cell sheet engineering that allows the manufacture of three-dimensional organs containing blood vessels
To create planar or tubular cardiac tissue that contains blood vessels and functions as a pump, human cardiac cell sheets are laminated stepwise on a planar or tubular vascular bed, and perfusion culture is performed in vitro using a bioreactor that mimics the biological environment. One of our current research aims is to develop a method of stacking myocardial sheets to form a tubular cardiac structure. A bioreactor system for long-term, stable perfusion culture will be constructed by incorporating methods for monitoring electrical, physical, (pulsation and pressure) and metabolic function. We anticipate that future technological developments will allow us to engineer tubular or planar bioartificial organs that could be used as auxiliary pumps to assist the function of the failing heart in patients with ischemic heart disease or dilated cardiomyopathy

stepwise stacking of cell sheets on a perfusable vascular bed. The first step involves the placement of a triple layer of cardiac cell sheets onto a perfusable vascular bed: the vascular network in the cardiac tissue construct connects with the blood vessels in the vascular bed during perfusion culture, allowing oxygen and nutrients to be supplied throughout the cardiac tissue. In each subsequent step, three-layered cardiac cell sheets are added on top of the existing tissue, with sufficient time given between steps to allow the blood vessels within the existing tissue to connect with the vascular network in the newly added cardiac cell sheets. Using this technique, it

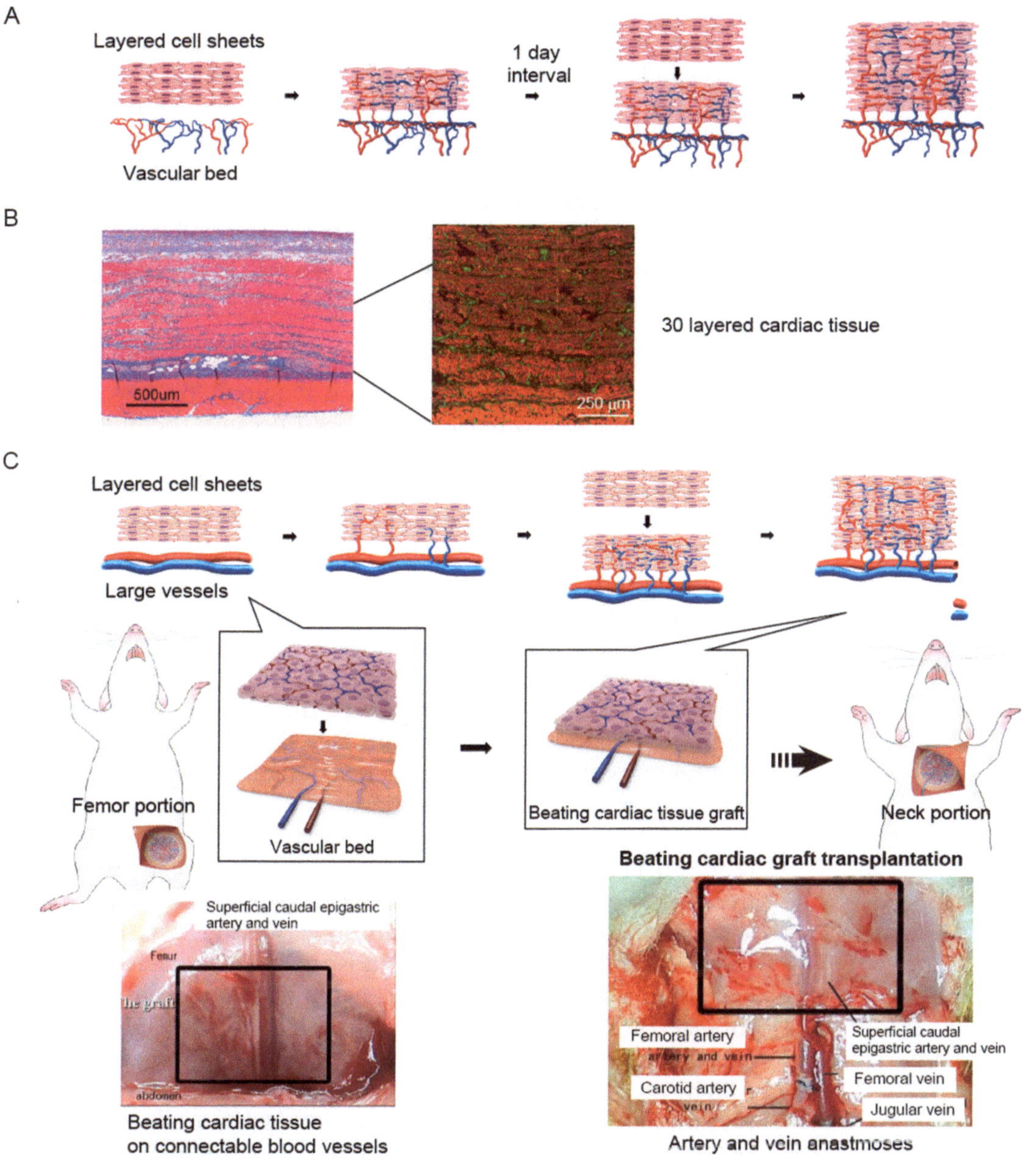

Fig. 3 Fabrication of a thick cardiac tissue graft containing blood vessels by multi-step transplantation of laminated cardiac cell sheets. (**a**) A laminated (three-layered) cardiac cell sheet is transplanted onto a vascular bed. One day after transplantation, during which time a capillary network develops in the cardiac tissue graft, a new laminated cardiac cell sheet is transplanted onto the original cardiac cell sheet. By repeating this process numerous times, thick myocardial tissue with a capillary network can be constructed. (**b**) Multi-step transplantation of up to 10 laminated cardiac cell sheets at intervals of 1 day can generate cardiac tissue that is about 1 mm in thickness (0.84 $\pm$ 0.16 mm) and contains well-formed capillaries. Adapted from [15] with permission. (**c**) Creation of cardiac tissue suitable for ectopic transplantation. Three-layered cardiac cell sheets were repeatedly grafted onto existing blood vessels (superficial caudal epigastric artery and vein, which are branches of the femoral artery and vein) in vivo to fabricate beating cardiac tissue supplied by the femoral artery. The cardiac tissue graft and femoral blood vessels were then resected and transplanted into

is possible to fabricate myocardial tissue with a thickness of 100 μm or more, which exceeds the lamination limit. To date, we have successfully applied this method to fabricate thick cardiac tissue using the rat femoral muscle as a vascular bed *ex vivo* (culture medium was perfused into the artery supplying the femoral muscle and drained via the vein) [19]. In addition, we have succeeded in generating thick tissue de novo by constructing a perfused collagen-derived vascular bed. Thus, it is possible to fabricate thick cardiac tissue that contains a vascular network by stacking cardiac cell sheets in a stepwise manner on a perfusable vascular bed [20].

In order to create thick tubular cardiac tissue using cell sheet engineering, a tubular vascular bed is required. We have used the small intestine as a tubular vascular bed because the inner wall of the small intestine has a rich network of capillaries that are capable of supplying oxygen and nutrients to cell sheets that are stacked on it. Furthermore, the small intestine can be easily isolated along with its feeding artery and draining vein, allowing it to be perfused during culture [21]. The above features make the small intestine well suited for use as a tubular vascular bed during perfusion culture of stacked cell sheets.

When a vascular bed derived from living tissue is used, it is considered necessary to decellularize the tissue in order to avoid immune rejection by the recipient. Decellularization is a technique that removes the cellular components of a tissue to leave only the extracellular matrix. Decellularization is most often carried out using a chemical method (e.g., a surfactant), biological method (e.g., an enzyme such as DNAse), or a physical method (e.g., freeze-thawing). By sequentially stacking cell sheets on the inner wall of the decellularized small intestine (which is used as a tubular vascular bed), it is possible to construct thick cardiac tissue that exceeds the limits of simple lamination. Thus, functional human tubular cardiac tissue can be fabricated by decellularizing the isolated small intestine, laminating immune pluripotent stem cell (iPSC)-derived cardiac cell sheets onto its inner wall and performing perfusion culture. However, such a production process requires a technique for tubularizing a cell sheet and laminating it onto the inner wall of a tubular vascular bed. In this chapter, we describe a method for producing human tubular cardiac tissue that has the potential to be developed into a technique for creating bioartificial organs for therapeutic use as an alternative to heart transplantation.

Fig. 3 (continued) the neck by anastomosis of the femoral artery and vein with host blood vessels (carotid artery and jugular vein). Pulsation of the graft restarted immediately after anastomosis of the blood vessels, and the graft survived for 2 weeks after ectopic transplantation. Adapted from [18] with permission. Copyright 2006 Federation of American Societies for Experimental Biology

2 Materials

The following animals, cells, and reagents were used in the present example.

2.1 Cell Culture

1. Cardiomyocytes: The human iPSC line 201B7, which is transfected with the puromycin-resistance gene under the control of the mouse alpha-myosin heavy chain (α-MHC) promoter and the neomycin-resistance gene under the control of the rex-1 promoter [22, 23] (*see* **Note 3**).
2. 200 ng/mL puromycin for purification of the cardiomyocytes.
3. Human umbilical vein endothelial cells (HUVECs).
4. Normal human dermal fibroblasts (NHDFs).
5. 35-mm temperature-responsive culture dish (UpCell®).
6. 10% fetal bovine serum (FBS).
7. Dulbecco's Modified Eagle Medium (DMEM).
8. Endothelial Cell Growth Medium-2 (EGM-2, BulletKit).
9. Penicillin-streptomycin solution.
10. 1/500 dilution of rabbit anti-cardiac troponin T primary antibody to detect cardiomyocytes by immunohistochemistry.
11. 1/200 dilution of Alexa Fluor 568-conjugated goat anti-rabbit IgG secondary antibody.
12. 1/500 dilution of phycoerythrin-conjugated anti-human CD31 primary antibody to detect endothelial cells.

2.2 Vascular bed

1. 2.5–4.0% isoflurane and a regulated nebulizer to anesthetize the rat.
2. Sprague-Dawley (SD) male rat for resection of a 5-cm length of small intestine with a branch of the superior mesenteric artery and a branch of the superior mesenteric vein.
3. Scissors and forceps.
4. Thermal knife.
5. 100 units heparin sodium.
6. Hanks' Balanced Salt Solution (HBSS).
7. Ultrapure water (e.g., MilliQ) for demucosalization.

2.3 Tissue Perfusion Bioreactor

1. Bioreactor system: a one-pass system consisting of a microprocessor-controlled delivery pump, temperature transmitter, a flow transmitter, and pressure transmitter.
2. Tissue culture chamber custom-made by Tokai Hit.
3. Bioreactor tubing: Micro-Renathane (inner diameter, 0.3 mm and 0.6 mm) to be used as connecting tubing and Pharmed Ismaprene (inner diameter, 1.3 mm) to be used as pump tubing.

2.4 Tubularization of Cardiac Tissue

1. Metal coil with an external diameter of 4 mm and a length of 20 mm.
2. Eppendorf 1000-μL pipette tip.
3. HBSS.

2.5 Perfusion Medium

1. Leibovitz's L-15 with L-glutamine, phenol red, and sodium pyruvate cell culture medium.
2. 10% FBS.
3. 20 ng/mL basic fibroblast growth factor (bFGF).
4. Penicillin-streptomycin solution.

2.6 Measurement of Electric Potential and Internal Pressure

1. Millar 1.4-F Mikro-Tip catheter transducer.
2. Millar 1.1-F electrophysiology catheter, 4.5 cm in length.
3. Millar pressure signal conditioner.
4. PowerLab data acquisition and analysis systems.

2.7 Histological Analysis

1. 4% paraformaldehyde fixative.
2. Hematoxylin and eosin.
3. 1/100 dilution of mouse anti-cardiac troponin T monoclonal primary antibody.
4. 1/10 dilution of rabbit anti-CD31 polyclonal primary antibody.
5. 1/200 dilution of Alexa Fluor 568-conjugated goat anti-mouse IgG.
6. 1/200 Alexa Fluor 488-conjugated goat anti-rabbit IgG.
7. Confocal laser-scanning microscope (e.g., Olympus FV1200/IX83).

3 Methods

All animal studies must be approved by the institutional animal experiment ethics committee and performed in accordance with institutional animal use guidelines.

3.1 Bioreactor Set-Up

1. A tissue perfusion culture chamber is constructed for stable perfusion culture of the vascular bed and tubular cardiac tissue (see Fig. 4a).
2. Three rotary pumps are used for the bioreactor, and the Leibovitz's L-15 medium is independently fed into the vascular bed, intestinal tract, and chamber.

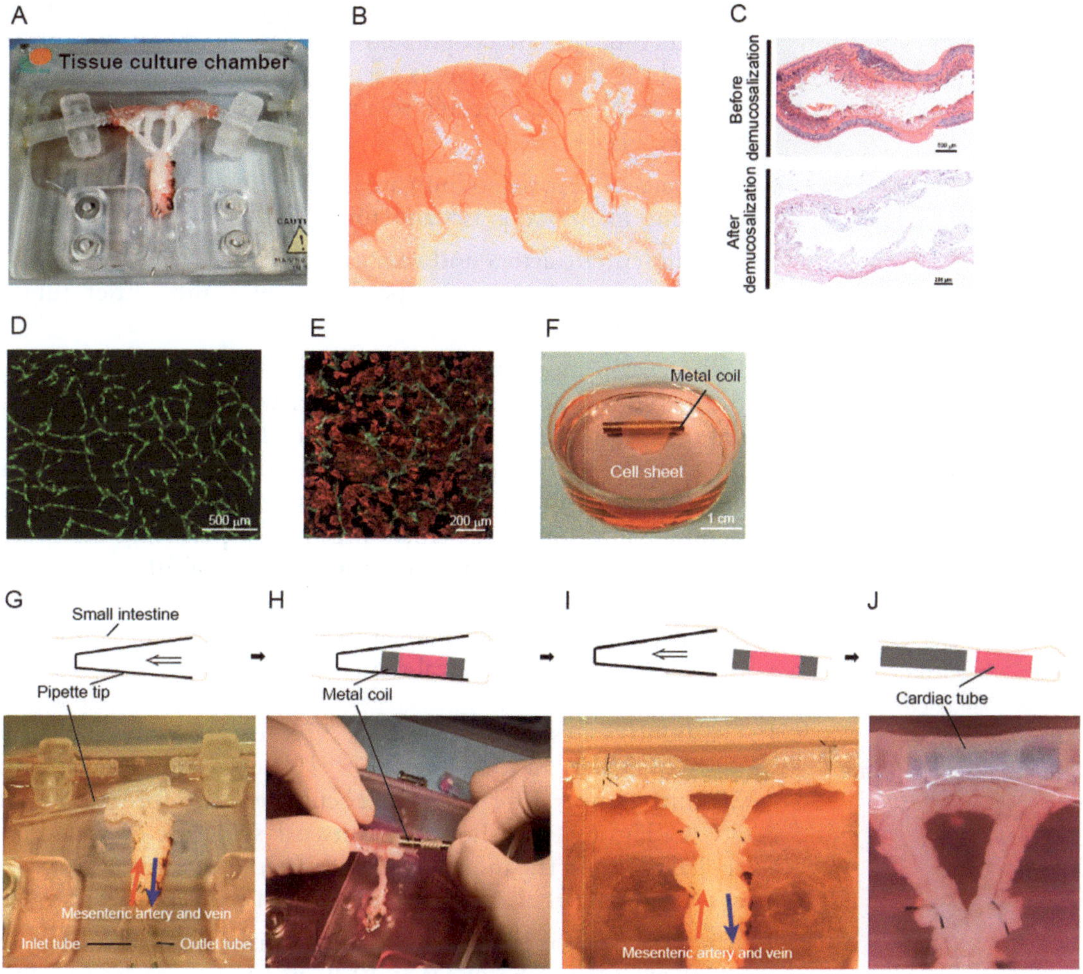

Fig. 4 Fabrication of tubular cardiac tissue by cell sheet engineering using the small intestine as a vascular bed. (**a**) A segment of small intestine in the chamber of the bioreactor. (**b**) The capillary network of the small intestine. (**c**) Hematoxylin-eosin-stained sections of the small intestine obtained before or after demucosalization with ultrapure water for 24 h. (**d**) Network structure of the vascular endothelial cells co-cultured with human iPSC-derived cardiomyocytes for 5 days. (**e**) Immunostaining of endothelial cells (CD31: green) and cardiomyocytes (cardiac troponin T: red). (**f**) Wrapping of a three-layered cardiac cell sheet around a metal coil. (**g**) Insertion of a 1000-μL pipette tip into the lumen of the small intestine from one side. (**h**) Insertion of the metal coil wrapped with cell sheets into the lumen of the pipette tip. (**i**) Extraction of the pipette tube in the opposite direction to its insertion, resulting in the attachment of the cell sheet to the surface of the demucosalized intestinal lumen. (**j**) The metal coil is removed 24 h later, after engraftment of the tubular cardiac tissue to the lumen of the small intestine

3.2 Fabrication of the Vascular Bed

1. The small intestine is resected from a rat for use as a vascular bed. Anesthesia is induced with 4.0% isoflurane and the dissection is performed under continuous anesthesia with 2.5% isoflurane.
2. The excised tissue is the ileum of an 8-week-old rat.

3. Blood vessels other than those supplying the required segment of small intestine are coagulated with a thermal knife (*see* **Note 1**), and the intestinal tract is excised to a length of about 5 cm (*see* **Note 2**) and connected to a tissue perfusion culture chamber (see Fig. 4a, b).
4. 100 units of heparin sodium are injected into the dorsal vein of the penis for blood anticoagulation.
5. The mesenteric artery and vein associated with the small intestine are connected to the perfusate inlet and outlet tubes, respectively (see Fig. 4g).
6. Demucosalization of the small intestine is performed to allow the cell sheet to engraft onto the inner wall of the small intestinal tract. Specifically, the small intestinal mucosa is exposed to ultrapure water (a hypotonic solution) so that the mucosal cells absorb water, swell, and die. In this example, we performed demucosalization of the small intestine by perfusing ultrapure water into the intestinal tract at a flow rate of 50 μL/min for 24 h. Decellularization of the mucosa is evident in small intestine perfused with ultrapure water but not in small intestine perfused with physiological saline (see Fig. 4c).

3.3 Fabrication of Tubular Cardiac Tissue

1. Cardiomyocytes, HUVECs and NHDFs are combined at a ratio of 5:1:5. A total of 1.0×10^6 cells are seeded on a 35-mm temperature-responsive culture dish for 5 days at 37 °C. Figure 4d, e shows the network structure of the endothelial cells and the morphology of the co-cultured cardiomyocytes and endothelial cells after 5 days of culture, respectively.
2. After culture for 5 days, the culture dishes are incubated in another CO_2 incubator set at 20 °C in order to release confluent cells as intact sheets (*see* **Note 3**).
3. One of three collected cell sheets is placed on a cell culture dish and allowed to stand for 1 h in an incubator at 37 °C so that the cell sheet adheres to the surface of the culture dish (*see* **Note 4**).
4. Next, a second cell sheet is placed on top of the first cell sheet and allowed to stand for 30 min in an incubator at 37 °C after aspiration of the culture medium.
5. Then, a third cell sheet is added in the same manner to fabricate a triple-layered construct.
6. The triple-layered cell sheet is wrapped around the outer circumference of the metal coil (*see* **Note 5**). Figure 4f shows the wrapping process.
7. A 1000-μL pipette tip is inserted into the luminal surface of the small intestine from the right side of the small intestinal stump (*see* Fig. 4g and **Note 6**).

8. The metal coil wrapped with cardiac cell sheets is inserted into the larger lumen of the pipette tip (*see* Fig. 4h and **Note 6**).
9. The pipette tip is withdrawn from the left side to leave the cell sheet attached to the luminal surface of the small intestine (see Fig. 4i).
10. The metal coil is withdrawn 24 h later (the coil can be pulled out once the cell sheet has attached to the small intestine) (see Fig. 4j).

3.4 Perfusion Culture of Tubular Cardiac Tissue

1. Triple-layered cardiac cell sheets are placed onto the vascular bed, and the resulting tissue construct is subjected to perfusion culture in the bioreactor system.
2. Medium is perfused via the inlet tube that is connected to the small intestinal artery (*see* **Note** 7), and the volume of medium leaving the outlet is measured by a digital weighing scale (see Fig. 5a).
3. Perfusion medium is flowed at a rate of 50–70 μL/min (*see* **Note 8**) so that the arterial pressure increases from 5 mmHg to 30 mmHg (see Fig. 5b).
4. Perfusion medium is also flowed through the intestinal lumen (*see* **Note 8**) so as to maintain the shape of the tubular cardiac tissue (see Fig. 5c).
5. The temperature, arterial pressure, and volume of the perfused medium are recorded during perfusion culture.

3.5 Measurements of Electric Potential and Internal Pressure

1. A 1.1-F electrophysiology catheter and 1.4-F Mikro-Tip catheter transducer are inserted through the channel into the lumen of the tubular cardiac tissue.
2. The pressure catheter is connected to the PowerLab system, which acts as an amplifier and A/D converter to measure the internal pressure changes caused by pulsation of the cardiac tissue.
3. After 1 week of perfusion culture, the pulsation of the tubular cardiac tissue is confirmed visually (see Fig. 5c, d), and the corresponding changes in electric potential and internal pressure are detected (see Fig. 5i).

3.6 Histological Analysis

1. After perfusion culture, the fabricated tissues (rat small intestinal vascular bed and implanted cardiac cell sheets) are fixed with 4% paraformaldehyde.
2. Tissue sections are stained with hematoxylin-eosin using conventional methods.
3. To detect cardiomyocytes and endothelial cells, deparaffinized sections are incubated overnight at 4 °C with a 1/100 dilution of mouse anti-cardiac troponin T monoclonal primary

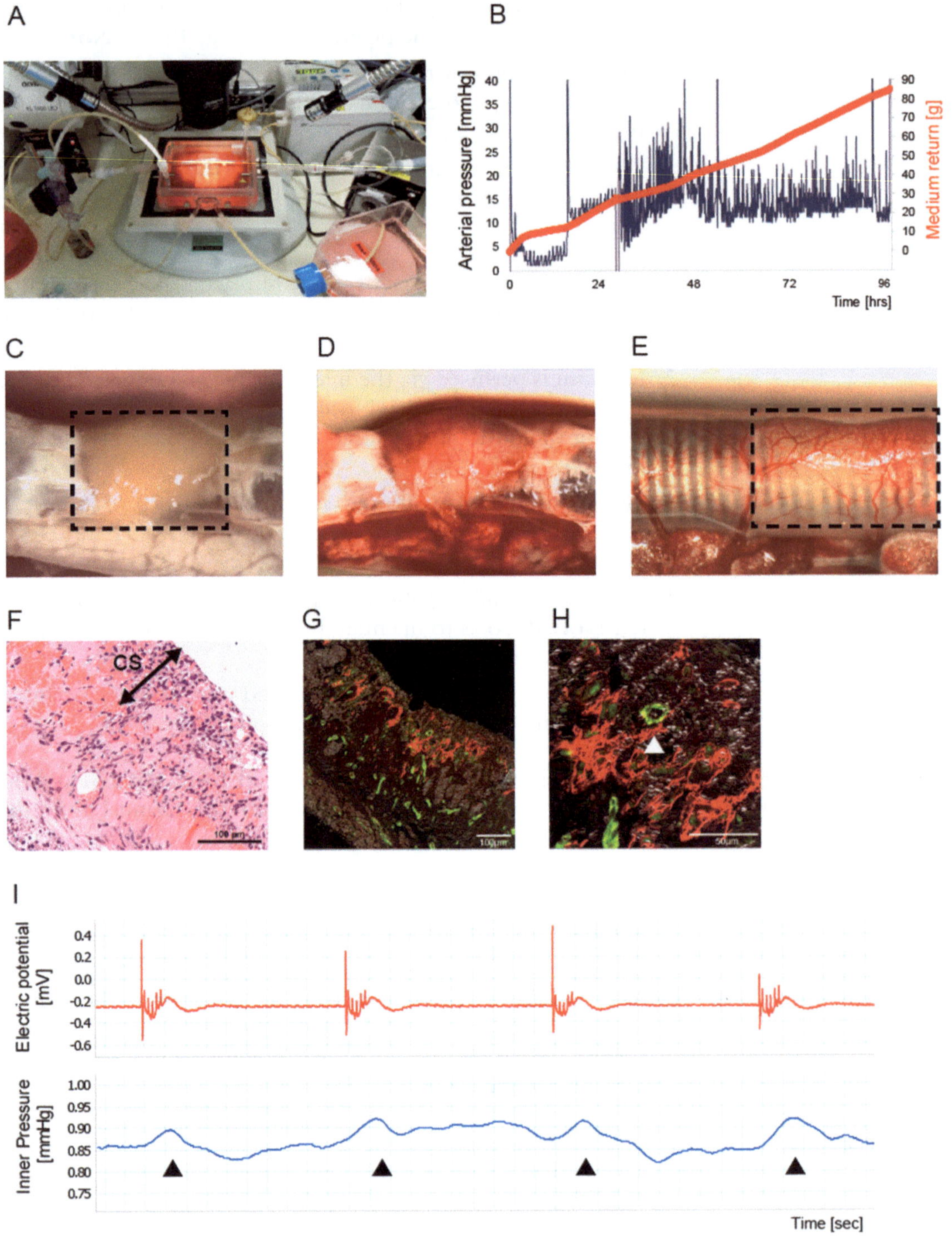

Fig. 5 Perfusion culture and functional evaluation of tubular myocardial tissue. (**a**) Photograph showing the overall view of the bioreactor. (**b**) A representative example showing the arterial pressure changes and amount of perfusion fluid returned from the vein during the perfusion of tubular cardiac tissue. (**c**) Macroscopic image showing tubular cardiac tissue (dashed rectangle) 1 week after perfusion culture. (**d**) Macroscopic image of tubular cardiac tissue after reperfusion with blood. (**e**) New blood vessels within the laminated cardiac cell sheet were visualized after perfusion of the tubular cardiac tissue with blood. The dashed rectangle indicates

antibody and a 1/10 dilution of rabbit anti-CD31 polyclonal primary antibody.

4. The specimens are subsequently treated with 1/200 dilutions of Alexa Fluor 568-conjugated goat anti-mouse IgG and Alexa Fluor 488-conjugated goat anti-rabbit IgG secondary antibodies for 1 h at room temperature.
5. The sections are observed using a confocal laser-scanning microscope. Figure 5e, g and h show that new blood vessels have been induced within the cardiac tissue and that these new blood vessels are connected with the vascular network of the rat small intestine.

4 Notes

1. During harvesting of the small intestine from the rat for use as a vascular bed, blood vessels that are not needed can be cauterized by applying pressure with a thermal scalpel so as to seal the tissue. The tissue can then be dissected. This technique achieves safe and rapid hemostasis.
2. The small intestine is easily damaged when handled directly, but it can be safely manipulated with a cotton swab.
3. Since large number of cardiomyocytes are used to produce cell sheets, differentiation of iPS cells into cardiomyocytes by bioreactors and mass culture is required. Protocols for cardiac differentiation of hiPS cells, purification of cardiomyocytes, and generation of cardiac cell sheets have been shown previously [22, 23]. Briefly, human iPS cells 201B7 (RIKEN) were grown in Primate ES Cell Medium supplemented with 5 ng/mL of bFGF on mitomycin C-treated mouse embryonic fibroblasts maintain under a humidified atmosphere of 37 °C and 5% CO_2.

   To purify cardiomyocytes after cardiac differentiation, lentiviral vectors (α-MHC-pure rex-1-neo) containing puromycin-resistance genes under the control of the mouse α-MHC promoter and neomycin-resistance genes under the control of the rex-1 promoter were used and introduced into

Fig. 5 (continued) the position of the laminated cell sheet. (**f**) Hematoxylin-eosin-stained section of tubular cardiac tissue 1 week after perfusion culture. The double-headed arrow indicates part of the cell sheet. (**g**) The presence of blood vessels (CD31: green) and myocardium (cardiac troponin T: red) in the tubular cardiac tissue. (**h**) High-magnification images showing immunostained sections of tubular cardiac tissue. The arrowhead shows a red blood cell that has reached a capillary within the myocardial tissue. This result indicates that functional blood vessels are constructed within the cardiac tissue during perfusion culture. (**i**) Measurements of the electric potentials and internal pressure of the constructed tubular cardiac tissue. The arrowheads indicate increases in internal pressure generated by pulsation of the tubular cardiac tissue

human iPS cells [24]. The differentiated cardiomyocytes are then purified by puromycin treatment. The myocardial differentiation protocol in the bioreactor system (ABLE) is as follows: embrionic bodies are cultured in the bioreactor system with mTeSR1 and 2 days after starting culture in the bioreactor system with 50 μg/mL ascorbic acid, 2 mM L-glutamine, and cultured in StemPro34 medium containing 400 μM 1-thioglycerol. Cells were then treated with 0.5 ng/mL bone morphogenetic protein-4 (days 0–1), 10 ng/mL bone morphogenetic protein-4 (days 1–4), 5 ng/mL bFGF (days 1–4), 3 ng/mL Activin A (days 1–4), 4 μM IWR-1 (days 4–6), 5 ng/mL vascular endothelial growth factor (days 6–16), and 10 ng/mL bFGF (days 6–16).

4. When a cardiac cell sheet harvested from a temperature-responsive dish is placed onto the surface of a new dish or another cell sheet, it is important that the culture medium is completely removed from the culture dish. If a large amount of culture medium is present, adhesion of the new cell sheet to the dish surface or preexisting cell sheet (extracellular matrix) will be slowed, and anchorage-dependent cell death may occur.
5. The entire procedure should be performed on a clean bench in order to avoid bacterial contamination of the tubular cardiac tissue preparation.
6. When inserting a pipette tip or cell sheet into the lumen of the small intestine, a magnifying glass or stereomicroscope should be used to facilitate insertion of the pipette tip or cell sheet into the center of the lumen as far as possible.
7. An air trap should be provided on the arterial tube because air bubbles generated in the perfusate due to temperature changes can potentially cause embolization of peripheral blood vessels and perfusion failure.
8. Since this is a one-pass system, the cost of consumables such as growth factors and culture medium is high. A future aim is to construct a system that can reuse perfusate and other consumables.
9. A major issue regarding the clinical use of these constructs in human patients is that transplanted xenogeneic tissue can elicit an immune response and be rejected unless it is first decellularized. Furthermore, it will be necessary to establish a method for reseeding endothelial cells into the vascular scaffold of a decellularized tissue.

Acknowledgments

This research was supported by the "Development of innovative manufacturing technology for three-dimensional tissues and organs based on cell sheet engineering" from the Japan Agency for Medical Research and Development (AMED, https://www.amed.go.jp/en/index.html; grant no. JP17he0702249) and JSPS KAKENHI grant no. 19H04453. The authors thank OXMED COMMS (www.oxmedcomms.com) for writing assistance. We also acknowledge the assistance of Prof. Tatsuya Shimizu, Prof. Jun Homma, Dr. Kazunori Sano, Mr. Shogo Shimizu, Prof. Katsuhisa Matsuura, and Prof. Eiji Kobayashi.

References

1. Yamada N et al (1990) Thermo-responsive polymeric surfaces: control of attachment and detachment of cultured cells. Makromol Chem Rapid Commun 11:571–576. Hüthig & Wepf Verlag
2. Okano T et al (1993) A novel recovery system for cultured cells using plasma-treated polystyrene dishes grafted with poly (N-isopropylacrylamide). J Biomed Mater Res 27(10):1243–1251
3. Yang J et al (2005) Cell sheet engineering: recreating tissues without biodegradable scaffolds. Biomaterials 26(33):6415–6422
4. Kobayashi J et al (2019) Cell sheet tissue engineering: cell sheet preparation, harvesting/manipulation, and transplantation. J Biomed Mater Res A 107(5):955–967
5. Sekine H et al (2006) Cardiomyocyte bridging between hearts and bioengineered myocardial tissues with mesenchymal transition of mesothelial cells. J Heart Lung Transplant 25(3):324–332
6. Sekine H et al (2011) Cardiac cell sheet transplantation improves damaged heart function via superior cell survival in comparison with dissociated cell injection. Tissue Eng Part A 17(23–24):2973–2980
7. Sekine H et al (2008) Endothelial cell coculture within tissue-engineered cardiomyocyte sheets enhances neovascularization and improves cardiac function of ischemic hearts. Circulation 118(Suppl. 14):S145–S152
8. Yang J et al (2006) Cell delivery in regenerative medicine: the cell sheet engineering approach. J Control Release 116(2):193–203
9. Sawa Y et al (2012) Tissue engineered myoblast sheets improved cardiac function sufficiently to discontinue LVAS in a patient with DCM: report of a case. Surg Today 42(2):181–184
10. Nishida K et al (2004) Corneal reconstruction with tissue-engineered cell sheets composed of autologous oral mucosal epithelium. N Engl J Med 351(12):1187–1196
11. Sato M et al (2019) Combined surgery and chondrocyte cell-sheet transplantation improves clinical and structural outcomes in knee osteoarthritis. NPJ Regen Med 4:4
12. Kanzaki M et al (2008) Functional closure of visceral pleural defects by autologous tissue engineered cell sheets. Eur J Cardiothorac Surg 34(4):864–869
13. Yamamoto K et al (2017) Middle ear mucosal regeneration by tissue-engineered cell sheet transplantation. NPJ Regen Med 2:6
14. Iwata T et al (2018) Periodontal regeneration with autologous periodontal ligament-derived cell sheets—a safety and efficacy study in ten patients. Regen Ther 9:38–44
15. Kubo H et al (2007) Creation of myocardial tubes using cardiomyocyte sheets and an in vitro cell sheet-wrapping device. Biomaterials 28(24):3508–3516
16. Sekine H et al (2006) Pulsatile myocardial tubes fabricated with cell sheet engineering. Circulation 114(Suppl. 1):I87–I93
17. Shimizu T et al (2002) Fabrication of pulsatile cardiac tissue grafts using a novel 3-dimensional cell sheet manipulation technique and temperature-responsive cell culture surfaces. Circ Res 90(3):e40
18. Shimizu T et al (2006) Polysurgery of cell sheet grafts overcomes diffusion limits to produce thick, vascularized myocardial tissues. FASEB J 20(6):708–710
19. Sekine H et al (2013) In vitro fabrication of functional three-dimensional tissues with perfusable blood vessels. Nat Commun 4:1399

20. Sakaguchi K et al (2013) In vitro engineering of vascularized tissue surrogates. Sci Rep 3: 1316
21. Inui A et al (2019) Generation of a large-scale vascular bed for the in vitro creation of three-dimensional cardiac tissue. Regen Ther 11: 316–323
22. Matsuura K et al (2016) TRPV-1-mediated elimination of residual iPS cells in bioengineered cardiac cell sheet tissues. Sci Rep 6: 21747
23. Seta H et al (2017) Tubular cardiac tissues derived from human induced pluripotent stem cells generate pulse pressure in vivo. Sci Rep 7: 45499
24. Kita-Matsuo H et al (2009) Lentiviral vectors and protocols for creation of stable hESC lines for fluorescent tracking and drug resistance selection of cardiomyocytes. PLoS One 4(4): e5046

Chapter 16

Quantifying Cardiomyocyte Proliferation and Nucleation to Assess Mammalian Cardiac Regeneration

Emma B. Brandt and Ahmed I. Mahmoud

Abstract

Neonatal mice display a remarkable ability to regenerate their heart following an injury during the first week of life. A key facet of successful cardiac regeneration is the proliferation of cardiomyocytes to replace the lost cells. Stimulating cardiomyocyte proliferation in the adult heart is a very promising approach to restore cardiac structure and function following injury. Here, we outline our approach to assess cardiomyocyte proliferation following a myocardial injury via the cell cycle markers phospho-histone H3 and Aurora B. We additionally discuss how we assess successful regeneration using wheat germ agglutinin to measure cardiomyocyte size, nuclear staining to quantify cardiomyocyte nucleation, and Trichrome staining to identify myocardial regeneration and scarring in the myocardium.

Key words Cardiac regeneration, Cardiomyocyte proliferation, Immunohistochemistry, Cardiomyocyte nucleation

1 Introduction

Cardiomyocyte proliferation following a cardiac injury is readily observed in lower level species, such as newts and zebrafish, which results in complete regeneration of the damaged heart [1–3]. In contrast, cardiac injuries in the adult mammalian heart result in significant myocardial damage and formation of a fibrotic scar [4]. However, neonatal mice have the ability to regenerate their hearts following an injury during the first week of life [5]. Common injury models used in neonatal mice include ligation of the left anterior descending artery (LAD), apical resection, and cryoinjury [6]. Cardiac regeneration requires that new cardiomyocytes be generated to replace those lost following injury, resulting in minimal to no scarring in the myocardium. During endogenous heart regeneration, the newly formed myocardium arises from the division of pre-existing cardiomyocytes, which was further confirmed by dual lineage tracing to track contributions of cardiomyocytes

Kareen L. K. Coulombe and Lauren D. Black III (eds.), *Cardiac Tissue Engineering: Methods and Protocols*, Methods in Molecular Biology, vol. 2485, https://doi.org/10.1007/978-1-0716-2261-2_16,

and non-cardiomyocytes to the regenerated myocardium [5, 7]. Therefore, detection and quantification of cardiomyocyte proliferation following an injury is a critical component of assessing a cardiac regenerative response.

However, in vivo detection of dividing cardiomyocytes in mammals is problematic due to the ability of cardiomyocytes to undergo multinucleation and polyploidy as the cells mature [8]. Thus, all common measurements of cardiomyocyte division in tissue have their pitfalls, and these are summarized in a comprehensive review by Derks et al. [9]. Flow cytometry, 3D imaging, disassociated cardiomyocytes, FISH, karyotyping are all examples of commonly used techniques to investigate ploidy and nuclear content. Assessment of cell cycle typically includes incorporation of nucleotide analogs such as EDU/BRDU for S-phase, phospho-histone H3 (pH 3) for mitosis, and Aurora B or anillin for cytokinesis. The continued gold standard to demonstrate cardiomyocyte division is time-lapse imaging of cultured cardiomyocytes [9, 10]. For in vivo detection, the use of the FUCCI (Fluorescence Ubiquitination-based Cell Cycle Indicators) system in transgenic mice has been a promising approach for in vivo cell cycle imaging to capture cardiomyocytes at all stages of the cell cycle during the time of harvest [11]. Other investigators have combined the FUCCI system with time-lapse imaging of beating zebrafish hearts to track these cells in real time [12, 13]. However, currently, this live time-lapse imaging is not technically feasible for use in mammals. Thus, we acknowledge our current technologies can only capture snapshots of proliferation in vivo and illustrate the need to utilize multiple methodologies to prove enhanced cardiomyocyte proliferation between experimental groups. Here, we outline our approach for assessing mammalian cardiomyocyte proliferation in heart tissue sections using the mitotic marker pH 3 in conjunction with the cytokinesis marker Aurora B, as a measure of cardiomyocyte cell cycle activity in vivo. Additionally, quantification of cardiomyocyte size and nucleation following injury provides additional evidence of myocyte cell cycle activity since proliferating cardiomyocytes tend to be smaller, mononuclear, and diploid [14]. Cardiomyocyte cell cycle withdrawal is characterized by failure to undergo cytokinesis, and indeed, polyploidy in cardiomyocytes is shown to inhibit cardiac regeneration [15, 16].

The protocol for myocardial infarction (MI) surgeries have already been published in detail for neonatal mice [6] and adults [17–19], so we will outline how to assess cardiomyocyte proliferation in heart tissue following injury as an important hallmark of cardiac regeneration. This protocol uses different markers of cell division in heart tissue sections to detect cardiomyocytes positive for mitosis (phospho-histone H3 or pH 3) and cytokinesis (Aurora B) following cardiac injury. Neonatal cardiomyocyte proliferation peaks at 7 days post injury (DPI) during regeneration, thus we use

7 DPI as a cut off for collection and assessment of proliferation markers following injury. Furthermore, Masson's Trichrome Stain allows for scar size quantification to demonstrate myocardial regeneration and reduction of fibrosis. It takes approximately 3 weeks for neonatal hearts to fully regenerate following an injury at P1, so we harvest hearts at 21 DPI to assess scar size via Trichrome staining. For adults, useful timepoints for assessing myocardial regeneration and fibrosis would range from 14 days post-injury to several weeks. Additionally, measuring cardiomyocyte cell size via wheat germ agglutinin (WGA) staining provides additional information about hypertrophy, a common hallmark of many cardiac pathologies, in non-regenerating hearts. Finally, we demonstrate how to isolate cardiomyocytes from fixed heart tissue to assess cardiomyocyte nuclear content as an additional indicator of newly formed cardiomyocytes.

2 Materials

2.1 Heart Isolation

1. Forceps, dissecting scissors, hemostat, curved forceps, 5 mL syringe, 27 G $\times 1/2''$ needle, PBS, 4% paraformaldehyde (PFA), 70% ethanol, dissecting microscope.
2. Paraffin wax embed according to standard protocols.

2.2 Heart Sectioning

1. Microtome, slides, forceps, brushes, flotation water bath, slide heater.

2.3 Immunohistochemistry

1. Xylene, 100% ethanol, 95% ethanol, 70% ethanol, ddH_2O, PBS, microwave or boiling water bath, slide staining jars/containers, slide box, antigen retrieval solution: (10 mM sodium citrate solution, pH 6.0), liquid blocking pen, 10% goat serum, primary antibodies: phospho-histone H3 (rabbit), cTnT (mouse), Aurora B (rabbit), secondary antibodies: AF488 (goat anti-rabbit) and AF555(goat anti-mouse), AF488-conjugated wheat germ agglutinin (WGA), DAPI, mounting medium, cover slips, fluorescent microscope.

2.4 Masson's Aniline Blue Trichrome Staining

1. Xylene, 100% ethanol, 95% ethanol, distilled water, Trichrome Aniline Blue staining kit: Bouin Fluid, Ferric Chloride acidified, Hematoxylin 1%, Biebrich Scarlet-Acid Fuchsin Stain Aqueous, Phosphomolybdic-Phosphotungstic Acid, Aniline Blue Stain Aqueous, Acetic acid 0.5% Aqueous, permanent resin-based mounting medium, cover slips, Coplin jars.
2. We used Newcomer Supply Trichrome, Masson, Aniline Blue Staining Kit for our procedures.

2.5 Cardiomyocyte Isolation from Fixed Hearts

1. 4% PFA, Collagenase D (2.4 mg/mL), Collagenase B (1.8 mg/mL), 37 °C incubator, centrifuge, Connexin 43 (rabbit), desired secondary antibody to Connexin 43, DAPI or Hoechst, hemocytometer, fluorescent microscope.
2. Enzymes should be reconstituted according to manufacturer.
3. Blocking buffer: 4% BSA, 0.2% Triton X-100, 1 mM EDTA, 0.02% sodium azide.

3 Methods

3.1 Heart isolation from Day 7 or Day 21 Post-Injury of Both Neonatal and Adult Hearts

1. Prepare a 5 mL syringe filled with PBS for perfusion, and a petri dish filled with PBS for washing.
2. Euthanize mice according your institutional guidelines for neonates and adults.
3. Immobilize mouse on dissection surface with tape and spray the abdomen with 70% ethanol.
4. Lift the skin just below the ribs with forceps and using small dissecting scissors create an incision below the ribs under the diaphragm.
5. Puncture and resect the diaphragm carefully to avoid damaging the heart, then cut up each side of the ribs towards the neck.
6. Once the heart is easily visualized, fold back the ribs and hold in place using a hemostat to keep the chest cavity open for the remainder of the procedure (*see* **Note 1**).
7. Identify the left and right atria, then make small cuts in each with the dissecting scissors.
8. Use curved forceps to grasp the base of the heart, insert needle with syringe full of PBS into the base of the left ventricle, then gently perfuse 2.5 mL of solution. Repeat this procedure on the right ventricle (*see* **Note 2**).
9. After perfusion, gently lift the heart with the curved forceps and use small dissecting scissors to carefully cut the heart from the chest working from the posterior to the anterior just above the atria.
10. Place the heart in a petri dish full of PBS to wash, then use a dissecting microscope if desired to trim off atria, major blood vessels, and any remaining tissue stuck to the scar site.
11. Once heart is perfused, it can be placed into a 4% PFA solution overnight at 4 °C for fixation. (See Subheading 3.7 for cardiomyocyte isolation for fixation times.)
12. After fixation, hearts are washed several times in PBS then transferred into a 70% ethanol solution for storage at 4 °C.

13. Hearts may then be embedded for a two-chamber view (ventricle openings on block face) in paraffin wax according to standard procedures (*see* **Note 3**).

3.2 Heart Sectioning

1. Prior to sectioning the hearts, place the paraffin blocks at 4 °C to chill and facilitate easier cutting.
2. Place the chilled blocks into the microtome and trim the block face until the suture can be observed in the section (*see* **Note 4**).
3. Starting at the suture site and continuing down to the apex of the heart, cut 5 μm sections every 50–100 μm through the heart.
4. Aim to get 2–3 sections on each slide.
5. Allow slides to dry on heater for several hours to promote tissue adherence on the slide.

3.3 Immunohistochemistry of 7 DPI Neonatal or Adult Heart Sections to Assess Cardiomyocyte Proliferation

1. For each heart sample, select multiple paraffin sections that have little to no section folding and are similar in size, suggested 3–6 replicate sections.
2. Deparaffinize the slides.
   (a) Place slides for 3–5 min in the following order of solutions: 3 changes of Xylene, 100% ethanol, 95% ethanol, 70% ethanol, distilled H_2O.
3. Immerse slides in PBS.
4. Prepare antigen retrieval solution.
5. Microwave slides in 10% antigen retrieval solution for 10 min in a slide staining box with a cover or in a boiling water bath for 20 min.
6. Bring slides to room temperature by replacing any liquid lost in the heating process with PBS.
7. Prepare a humidifying chamber for the slides (*see* **Note 5**).
8. Carefully dry off slides with compressed air, wipes, flicking the slide, or any combination of these.
9. Outline the area around the heart tissue samples on the slide with a liquid blocking pen.
10. Add ~500 μL of 10% goat serum for blocking and incubate slides at room temperature for 20 min in the humid chamber.
11. Primary antibodies are diluted in PBS as follows: pH 3 (1:100 anti-rabbit) or Aurora B (1:50 anti-rabbit), and cTnT (1:200 anti-mouse). Use ~200 μL of primary antibody solution per slide.
12. Incubate slides in humidified chamber with primary antibodies overnight at 4 °C.

13. The next day, remove antibody and wash slides three times for 5 min in PBS.
14. Secondary antibodies are diluted in PBS as follows: goat anti-rabbit AF448 (1:400), goat anti-mouse AF555 (1:400), and if using, WGA AF488-conjugated (50 μg/mL).
15. Incubate slides with ~200 μL of secondary antibody solution in humid chamber for 1 h at room temperature in the dark.
16. Protect the slides from light for the rest of the following steps.
17. Wash slides in PBS four times for 5 min in the dark.
18. Add DAPI (1:10,000 dilution) to the heart sections for 2–5 min.
19. Wash slides with an additional PBS wash for 5 min.
20. Mount slides with desired mounting medium and cover slip. Store slides at 4 °C in the dark until ready to image.

3.4 Imaging and Quantifying Proliferating Cardiomyocytes in 7 DPI Neonatal or Adult Heart Sections

1. Stained slides for pH 3 and cTnT or Aurora B and cTnT can be imaged on a fluorescence microscope or confocal microscope. Images for quantification can be taken at 10× or 20×. For each heart sample, image the entire heart section for quantification.
2. Quantify the number of pH 3+/cTnT+ or Aurora B+/cTnT+ cells present. These cells tend to have condensed troponin structures indicative of sarcomere disarray, a hallmark of cardiomyocyte proliferation (Fig. 1a) (*see* **Note 6**).
3. Average the number of pH 3+/cTnT+ or Aurora B+/cTnT+ per section across the replicates. This quantification of proliferating cardiomyocytes can be reported as the average number of positive cardiomyocytes per visual field or the entire heart section.

3.5 Staining and Imaging Wheat Germ Agglutinin (WGA) at 21 DPI of Neonatal or Adult Hearts to Quantify Cardiomyocyte Size

1. Stained slides with cTnT, WGA, and DAPI can be imaged on a fluorescence microscope. Images for counting can be taken at 10× or 20×, representative images to illustrate size differences can be obtained at higher magnifications. Capture images of cardiomyocytes in cross-section around the heart section, ensuring to capture both border zone and remote zone areas in hearts that underwent MI.
2. Using ImageJ to quantify the cell size based on the scale bar, measure a minimum of 100 cells per heart section.
3. Cardiomyocytes undergoing pathological hypertrophy will tend to have a larger cross-sectional area (Fig. 1b).

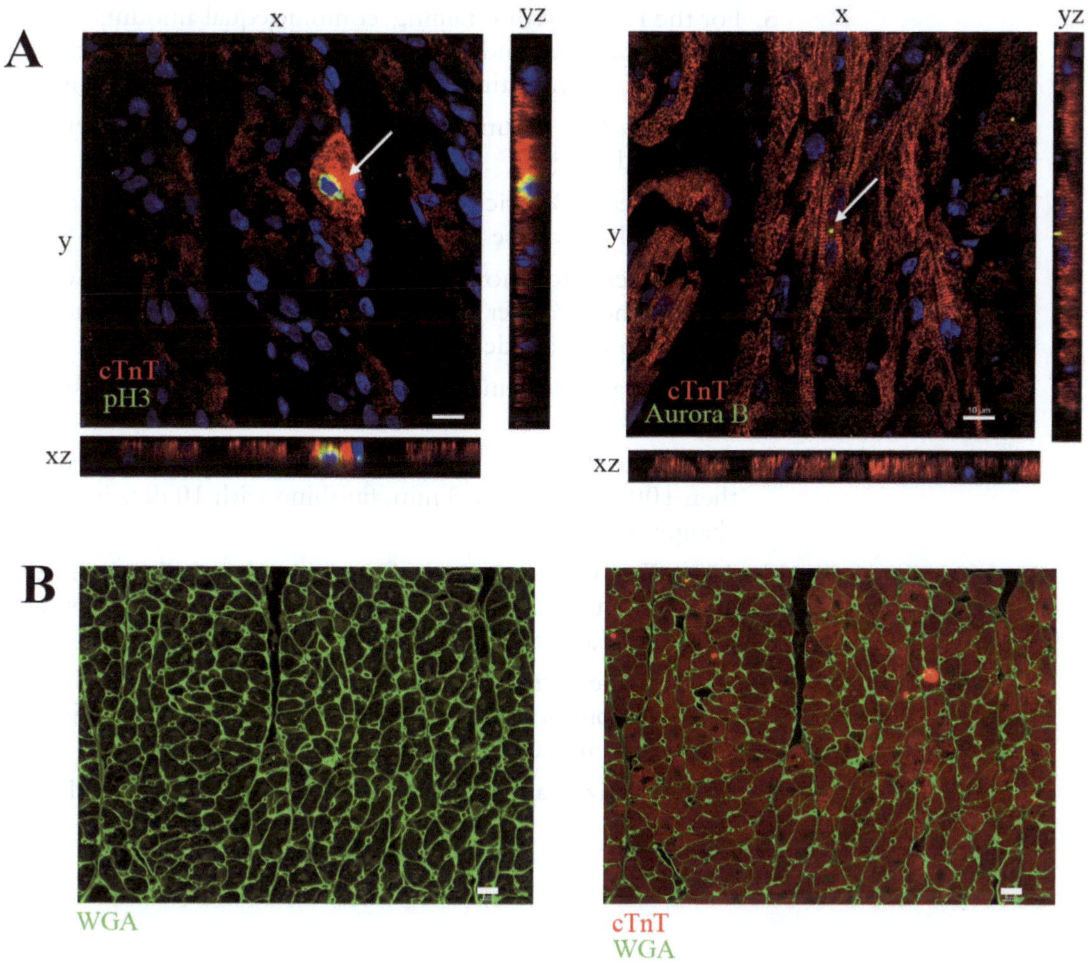

Fig. 1 Examples of pH 3+ or Aurora B+ cardiomyocytes and WGA staining for cardiomyocyte size. (**a**) Representative z-stack confocal images of pH 3+ cardiomyocytes and Aurora B+ cardiomyocytes. The white arrows indicate positive cells and the z-stacks are illustrated in orthogonal views showing the XZ plane on the bottom and the YZ plane on the right. Scale bar = 10 μM. (**b**) Representative images of WGA staining alone and with cTnT. Scale bar = 10 μM

3.6 Assessing Scar Formation in 21 DPI Neonatal or Adult Hearts Using Trichrome Staining

1. Preheat Bouin fluid to 60 °C in a water bath.
2. Deparaffinize slides through three changes of Xylene for 3 min each, then through 3 min of 100% ethanol, then 95% ethanol. Wash well in distilled water.
3. Place slides in the heated Bouin fluid and keep at 60 °C for 1 h in a water bath. Alternatively, slides can be placed in room temperature Bouin fluid and allowed to sit overnight.
4. Cool at room temperature if heated, wash well in running tap water and rinse in distilled water.

5. For the Hematoxylin staining, combine equal amounts of ferric chloride, acidified and Hematoxylin 1% and mix well. Make this solution fresh each time. Allow slides to stain for 10 min.
6. Wash the slides in running tap water for 10 min and then rinse in distilled water.
7. Place slides in Biebrich Scarlet-Acid Fuchsin Stain Aqueous for 2 min and then rinse well in distilled water.
8. Place slides into Phosphomolybdic-Phosphotungstic Acid for 10 min, then transfer slides directly into the Aniline Blue Stain for 5 min. Rinse slides well in distilled water.
9. Finally, place slides into acetic acid 0.5% for 3–5 min. Rinse in distilled water.
10. Dehydrate the slides by immersing in 95% ethanol for 3 min, then 100% ethanol for 3 min, finishing with 10 dips into three changes of Xylene.
11. Allow Xylene to evaporate off in a fume hood, then still working in the fume hood, cover sections in appropriate mounting medium and cover slip.
12. Once imaged with a light microscope, scar tissue within the heart sections is readily discerned by blue staining within the red-stained myocardium.
13. This scar size can be quantified using scale bars and ImageJ (Fig. 2).

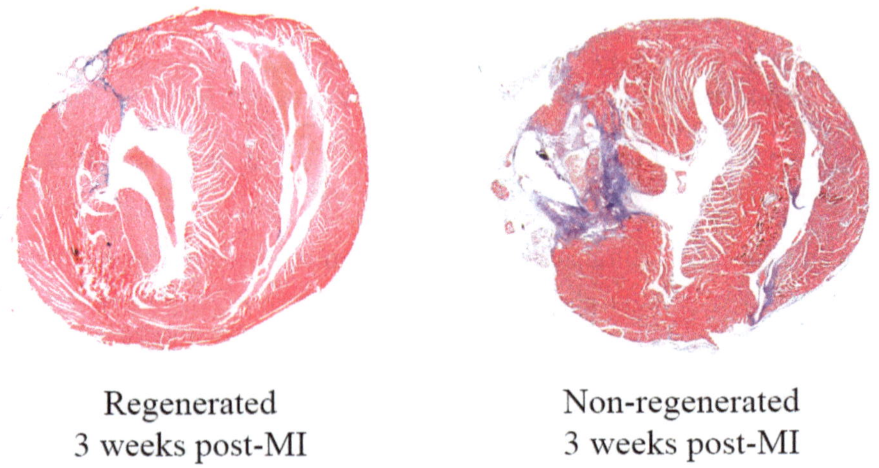

Fig. 2 Examples of Trichrome staining in regenerating and non-regenerating hearts. The left heart was harvested 21 DPI and illustrates successful regeneration with little scarring. The right heart was harvested 21 DPI and illustrates unsuccessful regeneration with residual scarring

3.7 Cardiomyocyte Isolation from Fixed Hearts for Nucleation Analysis

1. Fresh hearts are harvested and placed in 4% PFA for 2–3 h at room temperature.
2. Hearts are washed in three changes of PBS.
3. Hearts are then incubated with Collagenase D at 2.4 mg/mL and Collagenase B at 1.8 mg/mL for 12 h at 37 °C.
4. The resulting supernatant is collected and spun down at 500 rpm for 2 min to collect the isolated cardiomyocytes. These isolated cardiomyocytes can then be stored at 4 °C.
5. The remaining heart tissue is minced into smaller pieces and the enzyme incubation is repeated until no more cardiomyocytes are released from the tissue.
6. For blocking, incubate the cardiomyocytes with blocking buffer at room temperature with agitation for 5 min.
7. The cardiomyocytes are stained with a Connexin 43 antibody (1:100, anti-mouse) in PBS for 1 h at room temperature with agitation.
8. Wash the cells twice with blocking buffer by spinning down at 500 rpm for 2 min and resuspending.
9. For fluorescent detection of the cardiomyocytes, the Connexin 43 antibody can be probed with any anti-rabbit desired secondary. Secondary antibody incubation plus Hoechst or DAPI in PBS is done at room temperature with agitation for 1 h.
10. Cells are washed twice again with blocking buffer and resuspended in PBS.
11. To count the cardiomyocytes, use a wide-pore pipette tip to put 10 μL of the cardiomyocyte suspension into a hemocytometer and image under a fluorescent light microscope.
12. For counting mononuclear versus multinucleated cardiomyocytes, ensure at least three different counts are done on each sample and at least three individual samples per group, aiming to get ~1000 cardiomyocytes counted for each sample (*see* **Note** 7).

4 Notes

1. In the mice that underwent MI, scar tissue from the infarct site results in attachment of the heart to the ribs. Carefully trim this scar tissue with scissors and use blunt dissection with forceps to free the heart to avoid losing the heart suture or any heart issue.
2. Successful perfusion will result in clearing of all blood in the heart chambers, blanching and expansion of the lungs, and blanching of the liver.

3. Hearts should be oriented with the apex facing the bottom of the block and the suture site closest to the block face. Several 7 DPI hearts from neonatal mice can be placed in one block due to their smaller size. In this scenario, it is important when staining and imaging to diligently keep track of each heart and their corresponding replicate sections.
4. For sectioning the hearts, it's helpful to use sutures that have an easily recognizable color such as blue to make identifying the suture site easier.
5. A slide box lined with wet paper towels can be used as a humidified chamber.
6. Aurora B may need higher magnification to identify positive cells. Z-stacks can provide confirmation that these proteins reside within the cardiomyocytes.
7. The total number of cardiomyocytes present in a 10 μL aliquot is ~150–200 cells.

References

1. Jopling C, Sleep E, Raya M, Marti M, Raya A, Izpisua Belmonte JC (2010) Zebrafish heart regeneration occurs by cardiomyocyte dedifferentiation and proliferation. Nature 464(7288): 606–609. https://doi.org/10.1038/nature08899
2. Laube F, Heister M, Scholz C, Borchardt T, Braun T (2006) Re-programming of newt cardiomyocytes is induced by tissue regeneration. J Cell Sci 119(22):4719–4729. https://doi.org/10.1242/jcs.03252
3. Becker RO, Chapin S, Sherry R (1974) Regeneration of the ventricular myocardium in amphibians. Nature 248(5444):145–147. https://doi.org/10.1038/248145a0
4. Lutgens E, Daemen MJAP, de Muinck ED, Debets J, Leenders P, Smits JFM (1999) Chronic myocardial infarction in the mouse: cardiac structural and functional change1. Cardiovasc Res 41(3):586–593. https://doi.org/10.1016/s0008-6363(98)00216-8
5. Porrello ER, Mahmoud AI, Simpson E, Hill JA, Richardson JA, Olson EN, Sadek HA (2011) Transient regenerative potential of the neonatal mouse heart. Science 331(6020): 1078–1080. https://doi.org/10.1126/science.1200708
6. Mahmoud AI, Porrello ER, Kimura W, Olson EN, Sadek HA (2014) Surgical models for cardiac regeneration in neonatal mice. Nat Protoc 9(2):305–311. https://doi.org/10.1038/nprot.2014.021
7. Li Y, He L, Huang X, Bhaloo SI, Zhao H, Zhang S, Pu W, Tian X, Li Y, Liu Q, Yu W, Zhang L, Liu X, Liu K, Tang J, Zhang H, Cai D, Ralf AH, Xu Q, Lui KO, Zhou B (2018) Genetic lineage tracing of nonmyocyte population by dual recombinases. Circulation 138(8):793–805. https://doi.org/10.1161/CIRCULATIONAHA.118.034250
8. Alkass K, Panula J, Westman M, Wu T-D, Guerquin-Kern J-L, Bergmann O (2015) No evidence for cardiomyocyte number expansion in preadolescent mice. Cell 163(4): 1026–1036. https://doi.org/10.1016/j.cell.2015.10.035
9. Derks W, Bergmann O (2020) Polyploidy in cardiomyocytes. Circ Res 126(4):552–565. https://doi.org/10.1161/CIRCRESAHA.119.315408
10. Mohamed TMA, Ang Y-S, Radzinsky E, Zhou P, Huang Y, Elfenbein A, Foley A, Magnitsky S, Srivastava D (2018) Regulation of cell cycle to stimulate adult cardiomyocyte proliferation and cardiac regeneration. Cell 173(1):104–116. e112. https://doi.org/10.1016/j.cell.2018.02.014
11. Alvarez R Jr, Wang BJ, Quijada PJ, Avitabile D, Ho T, Shaitrit M, Chavarria M, Firouzi F, Ebeid D, Monsanto MM, Navarrete N, Moshref M, Siddiqi S, Broughton KM, Bailey BA, Gude NA, Sussman MA (2019) Cardiomyocyte cell cycle dynamics and proliferation revealed through cardiac-specific transgenesis of fluorescent ubiquitinated cell cycle indicator

(FUCCI). J Mol Cell Cardiol 127:154–164. https://doi.org/10.1016/j.yjmcc.2018.12.007

12. Uribe V, Ramadass R, Dogra D, Rasouli SJ, Gunawan F, Nakajima H, Chiba A, Reischauer S, Mochizuki N, Stainier DYR (2018) In vivo analysis of cardiomyocyte proliferation during trabeculation. Development 145(14):dev164194. https://doi.org/10.1242/dev.164194
13. Choi W-Y, Gemberling M, Wang J, Holdway JE, Shen M-C, Karlstrom RO, Poss KD (2013) In vivo monitoring of cardiomyocyte proliferation to identify chemical modifiers of heart regeneration. Development 140(3):660–666. https://doi.org/10.1242/dev.088526
14. Patterson M, Barske L, Van Handel B, Rau CD, Gan P, Sharma A, Parikh S, Denholtz M, Huang Y, Yamaguchi Y, Shen H, Allayee H, Crump JG, Force TI, Lien CL, Makita T, Lusis AJ, Kumar SR, Sucov HM (2017) Frequency of mononuclear diploid cardiomyocytes underlies natural variation in heart regeneration. Nat Genet 49(9):1346–1353. https://doi.org/10.1038/ng.3929
15. Gonzalez-Rosa JM, Sharpe M, Field D, Soonpaa MH, Field LJ, Burns CE, Burns CG (2018) Myocardial polyploidization creates a barrier to heart regeneration in Zebrafish. Dev Cell 44(4):433–446. e437. https://doi.org/10.1016/j.devcel.2018.01.021
16. Han L, Choudhury S, Mich-Basso JD, Ammanamanchi N, Ganapathy B, Suresh S, Khaladkar M, Singh J, Maehr R, Zuppo DA, Kim J, Eberwine JH, Wyman SK, Wu YL, Kuhn B (2020) Lamin B2 levels regulate polyploidization of cardiomyocyte nuclei and myocardial regeneration. Dev Cell 53(1):42–59.e11. https://doi.org/10.1016/j.devcel.2020.01.030
17. Kolk MV, Meyberg D, Deuse T, Tang-Quan KR, Robbins RC, Reichenspurner H, Schrepfer S (2009) LAD-ligation: a murine model of myocardial infarction. J Vis Exp (32). https://doi.org/10.3791/1438
18. Reichert K, Colantuono B, McCormack I, Rodrigues F, Pavlov V, Abid MR (2017) Murine Left Anterior Descending (LAD) coronary artery ligation: an improved and simplified model for myocardial infarction. J Vis Exp 122:55353. https://doi.org/10.3791/55353
19. Lugrin J, Parapanov R, Krueger T, Liaudet L (2019) Murine myocardial infarction model using permanent ligation of left anterior descending coronary artery. J Vis Exp (150). https://doi.org/10.3791/59591

Chapter 17

Injectable ECM Scaffolds for Cardiac Repair

Jervaughn D. Hunter, Todd D. Johnson, Rebecca L. Braden, and Karen L. Christman

Abstract

Injectable biomaterials have been developed as potential minimally invasive therapies for treating myocardial infarction (MI) and heart failure. Christman et al. first showed that the injection of a biomaterial alone into rat myocardium can improve cardiac function after MI. More recently, hydrogel forms of decellularized extracellular matrix (ECM) materials have shown substantial promise. Here, we present the methods for fabricating an injectable cardiac-specific ECM biomaterial with demonstrated positive outcomes in small and large animal models for cardiac repair as well as initial safety in a Phase I clinical trial. This chapter also covers the methods for the injection of a biomaterial into rat myocardium using a surgical approach through the diaphragm. Although the methods shown here are for injection of an acellular biomaterial, cells or other therapeutics could also be added to the injection for testing other regenerative medicine strategies.

Key words Injectable, Extracellular matrix, Hydrogel, Cardiac repair, Rat, Decellularization

1 Introduction

A minimally invasive approach for cardiac repair has numerous positive benefits including decreasing local tissue trauma, surgery times, risk due to surgery, hospital stays, and recovery times. These positive attributes have led to the investigation of injectable therapies for treating myocardial infarction (MI). Christman et al. in 2004 showed that injection of a biomaterial alone directly into the myocardium could lead to beneficial outcomes for cardiac repair post-MI [1, 2]. Since this initial study, numerous naturally derived biomaterials including alginate, collagen, chitosan, decellularized tissues, fibrin, hyaluronic acid, keratin, and Matrigel, as well as several synthetic biomaterials composed of polyethylene glycol or poly (N-isoproylacrylaminde) have been investigated [3, 4]. Ideally, these injectable biomaterials would be injected utilizing current catheter technology for quicker translation to the clinic. However, this mode of delivery provides unique challenges and design

Kareen L. K. Coulombe and Lauren D. Black III (eds.), *Cardiac Tissue Engineering: Methods and Protocols*, Methods in Molecular Biology, vol. 2485, https://doi.org/10.1007/978-1-0716-2261-2_17,

parameters for the biomaterial such as incorporating the ability to pass through a 27G needle and the appropriate kinetics to not gel at body temperature for up to an hour due to the duration of these procedures [3]. In 2009, Singelyn et al. developed a decellularized biomaterial derived from porcine myocardial tissue, which provides tissue-specific cues for cardiac repair [5]. In brief, fresh porcine myocardium is decellularized by spinning the chopped tissue in detergents, and then the decellularized tissue is lyophilized and milled into a fine powder. This powder is then partially digested in acidic conditions by pepsin into a liquid form that once brought to physiological conditions (salt, pH, and temperature), gels with the appropriate kinetics for catheter delivery [6, 7]. This myocardial matrix hydrogel was initially tested by injection into rat myocardium post-MI and was shown to maintain cardiac function, increase the size of cardiomyocyte islands within the infarcted region, decrease negative left ventricle remodeling, and improve hemodynamics [6, 8]. The matrix was also shown to be deliverable through numerous transendocardial injections via catheter delivery in a porcine model [6] and led to increasing cardiac function, decreased infarct fibrosis, and increased cardiac muscle at the endocardium [7]. Furthermore, this myocardial matrix hydrogel was successfully translated in a Phase I clinical trial in subacute and chronic MI patients [9]. In this chapter, the methods for decellularization, material digestion and processing of the matrix into an injectable liquid form are presented. Also, detailed instructions for injecting a biomaterial into rat myocardium with a surgical approach through the diaphragm are included. Here, the injection is occurring into a healthy rat heart but several methods for modeling myocardial infarction could be applied before the injection with either total coronary occlusion, coronary occlusion followed by reperfusion, or cryo-injury. Although the specific approach is for a biomaterial alone, both growth factors, cells, and/or other therapeutics could be included in this procedure for further study options.

2 Materials

Use ultrapure water for all solutions and rinsing steps. All materials and supplies for the material processing should be sterile or as clean as possible to prevent contamination. Any surgical supplies or tools that will be in contact with the animal during surgery should be autoclaved.

2.1 Decellularization Materials

1. Sharp knife and cutting board.
2. Decellularization Solution: 1% Sodium dodecyl sulfate (SDS), 1× Phosphate-buffered saline (PBS). Dissolve 80 g of SDS powder (*see* **Note 1**) in 800 mL of water to make a 10% stock solution of SDS. In an autoclaved 4 L beaker combine

3400 mL of water, 400 mL of the 10% SDS stock, and 200 mL of a 20× PBS stock solution. Stir until dissolved.

3. Plastic cryomolds and Optimal cutting temperature (OCT) compound.
4. Autoclaved 1 L beakers with 3/8″ × 2 1/2″ stir bars.
5. Stir plate that can be set to 125 rpm and can hold a 1 L beaker.
6. Penicillin/Streptomycin or PenStrep (PS): 10,000 Units/mL Penicillin and 10,000 μg/mL Streptomycin.
7. Autoclaved fine mesh metal strainer.
8. Autoclaved 1 L bottles.
9. Sterile 50 mL plastic conical tubes.

2.2 Digestion and Injection Preparation Materials

1. Lyophilizer and Wiley® Mini-Mill.
2. Digestion Solution: 0.1 M HCl, 1 mg/mL pepsin from porcine gastric mucosa (2500 units/mg protein). Fully dissolve the pepsin in acid by shaking the solution in a 50 mL conical by hand or on a mechanical shaker on a low setting (no more than 100 rpm). Then sterile filter through a 0.22 μm pore size filter.
3. Autoclaved 20 mL scintillation vial and 5/16″ × 1/2″ stir bar.
4. Sterile filtered solutions for pH and salt adjustments: 0.1 M NaOH, 1 M NaOH, 0.1 M HCl, 1 M HCl, 1× PBS, and 10× PBS.

2.3 Cardiac Surgical Injection Materials

1. Isoflurane anesthesia system.
2. Ventilator with temperature probe.
3. Heating pump connected to warmed surgical table.
4. Electric razor and vacuum.
5. Sterile syringes: 1 mL, 3 mL, and 10 mL.
6. Sterile needles: 25G, 27G, and 30G.
7. IV Catheter 14G × 2″.
8. Aspiration tube constructed from a 20G Intramedic tip with 10 cm of PE 100 tubing attached.
9. Autoclaved surgical tools: scalpel, scissors, fine forceps, standard forceps, needle holder, and towel clamp.
10. Topical: Betadine, Isopropyl Alcohol, Artificial Tears, Surgilube, and Triple Antibiotic Ointment.
11. Injectable: Lactated Ringers, 1% Lidocaine, and Buprenex.
12. Suture: Vicryl 4-0 FS-1, Vicryl 5-0 FS-2, and Prolene 5-0 RB-1.
13. Lab tape or masking tape.

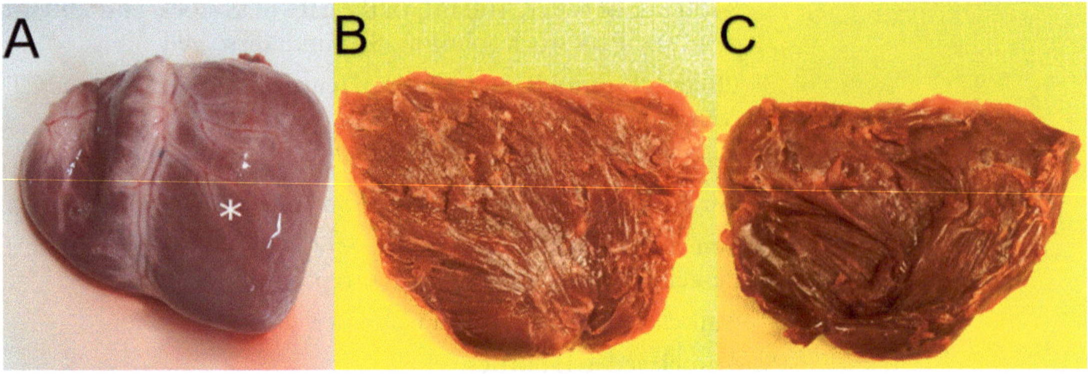

Fig. 1 (**a**) A fresh harvested porcine heart before processing. Labeled with a (*) is the left ventricular myocardium used in the protocol. The fresh porcine heart is trimmed down such that only the myocardial tissue of the left ventricle remains. (**b**) Epicardial view, showing that the epicardium, large vessels, and superficial fat have been removed. (**c**) Endocardial view showing the endocardium, mitral valve, chordae tendineae, and papillary muscles are also removed leaving behind the red myocardial tissue

3 Methods

3.1 Tissue Processing and Decellularization

1. Starting with a fresh unfrozen (*see* **Note 2**) porcine heart (Fig. 1a), isolate only the left ventricle by removing the thinner right ventricular free wall tissue, the septal wall, both atria, and the valves.
2. Remove any superficial fat, fascia, chordae tendineae, and papillary muscles, leaving behind only the red myocardial tissue (Fig. 1b).
3. Cut the remaining tissue into small regular cubed pieces about $2 \times 2 \times 2$ mm (*see* **Note 3**) (Fig. 2a).
4. Divide cut tissue into 1 L autoclaved beakers with 20–35 g of tissue per beaker (*see* **Note 4**).
5. Add 800 mL of water and a 3/8″ × 2 1/2″ stir bar to each beaker.
6. Stir beakers at 125 rpm for 30–45 min. Cover beaker with parafilm (*see* **Note 5**).
7. Strain tissue through the autoclaved fine mesh metal strainer and rinse with water.
8. Place tissue back into beaker with stir bar. Add 800 mL of decellularization solution and 4 mL of PS.
9. Stir beakers at 125 rpm for 2 h. Cover beaker with parafilm.
10. Repeat **steps** 7 and **8**. Then stir beakers at 125 rpm 24 h.
11. Repeat **step 10** for 2–4 times (for a total of 3–5 days in the decellularization solution) until the tissue has become completely decellularized or turned fully white in color (*see* **Note 6**) (Fig. 2b).

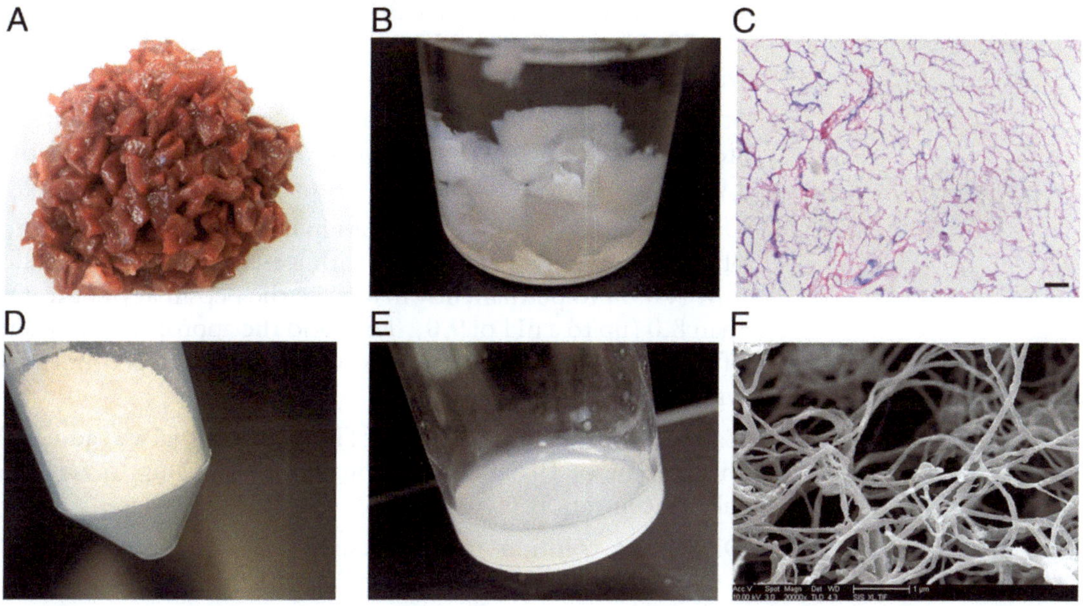

Fig. 2 (**a**) Fresh myocardial tissue chopped into fine pieces. (**b**) Fully decellularized, white myocardial tissue. (**c**) An H&E stained section of fully decellularized myocardial tissue showing no remaining nuclei. (**d**) Lyophilized and milled porcine myocardial matrix. (**e**) Once partially digested in acid by pepsin, the material is now in a liquid form. (**f**) When brought to physiological conditions including salt, temperature, and pH, the material forms a nanofibrous gel as viewed by scanning electron microscopy. Modified and reprinted from [3] with permission from Elsevier

12. Repeat **step** 7 and then place tissue back into beaker with stir bar and 800 mL of water. Stir at 125 rpm for 24 h.
13. Repeat **step 12** but stir in water three times for 15 min, 30 min, and 1 h, respectively.
14. Repeat **step** 7 and then transfer decellularized tissue into 1 L autoclaved bottles with 800 mL of water. Vigorously shake for 1 min.
15. Repeat **step 14** and then check for bubble formation after shaking. Lack of bubbles indicates removal of SDS solution. If bubbles remain then keep rinsing and shaking in the 1 L bottle with water to remove SDS from tissue.
16. Strain tissue with fine mesh metal strainer and transfer each into a 50 mL conical as seen in Fig. 2c and freeze at −80 °C (*see* **Note** 7).

3.2 Digestion and Injection Preparation

1. Using a lyophilizer, freeze dry the decellularized tissue.
2. Once fully dry, with a Wiley® Mini Mill process the material through a #40 mesh filter (pore size of 0.422 mm) into an autoclaved 20 mL vial (Fig. 2d). Before using the mill, make sure to fully clean all surfaces of the mill and mesh with 70% ethanol to minimize contamination of the decellularized tissue.

3. In a sterile environment (such as a tissue culture hood), mass out 30–50 mg of milled ECM into an autoclaved 20 mL vial with a 5/16″ × 1/2″ stir bar.
4. Still in a sterile environment, add an appropriate amount of sterile filtered digestion solution to the 20 mL vial until the ECM is at 10 mg/mL. Stir 125 rpm for 48–56 h (*see* **Note 8**).
5. Once digested (Fig. 2e), the ECM in liquid form is brought to physiological conditions on ice with ice-chilled solutions. Add 1 M NaOH to permanently inactivate the pepsin at pH greater than 8.0 (up to a pH of 9.0). Then add the appropriate amount of 1 M HCl and thoroughly mix until a pH of 7.4 is confirmed with a pH strip.
6. Calculate the new total volume of liquid in the vial and add 1/9 this volume of 10× PBS to bring the solutions up to 1× PBS salt concentration.
7. Dilute the solution to the desired ECM concentration of 6 mg/mL for the injection with 1× PBS.
8. The liquid form of the ECM can either be kept on ice for immediate injection preparation or frozen and lyophilized for long-term storage at −80 °C with desiccant. If stored in a lyophilized state, the material can simply be resuspended with the appropriate amount of sterile water back to 6 mg/mL ECM (*see* **Note 9**).
9. Liquid form of the ECM is kept on ice and then loaded into the sterile syringe approximately 1 h before it is injected into the myocardium. Upon injection into the animal the material will gel or self-assemble into a nanofibrous structure (Fig. 2f).

3.3 Characterization of Decellularized Hydrogel

Before using the material for cardiac injection, the macromolecular composition and mechanical properties should be characterized. In 2015, Ungerleider et al. described the process in which properly decellularized hydrogels can be characterized [10]. Qualitative confirmation of decellularized material can be achieved by observing the absence of nuclei in hematoxylin and eosin images and nuclear staining (DAPI) of decellularized tissue. If intact nuclei are observed in decellularized samples, the decellularization is insufficient. To assess double-stranded DNA (dsDNA), sulfated glycosaminoglycan (sGAG), and protein content a standard pico green, dimethylmethylene blue (DMMB), and SDS PAGE assay can be used, respectively [10]. As reported by Reing et al., acceptable dsDNA content for decellularized hydrogels is suggested to be below 50 ng dsDNA/mg ECM [11] although many clinically used decellularized ECM patches have significantly higher concentrations [12]. The methods described above typically generate a concentration of 0.1–5 ng dsDNA/mg ECM for the myocardial matrix hydrogel. Additionally, sGAG content was seen in the range of

5–15 μg sGAGS/mg ECM [10]. Mechanical properties of the resuspended and gelled ECM should also be evaluated. For ECM gels, the storage and loss moduli at 1 rad/s have been reported 5–10 Pa and 1–5 Pa, respectively [10]. Furthermore, for syringe and catheter delivery, the material should be shear thinning at frequencies 0.1–100 Hz with complex viscosities of 1–0.002 Pa s [10].

3.4 Cardiac Surgical Injection

All animal work should be done under an approved animal protocol as governed by your institution. For our studies, all experiments were performed in accordance with the guidelines established by the Committee on Animal Research at the University of California, San Diego, and the American Association for Accreditation of Laboratory Animal Care.

1. Anesthetize a female Sprague Dawley rat (225–250 g) for 3–5 min with 5% Isoflurane.
2. Intubate with a 14G × 2″ IV catheter. Secure the rat in a supine position on the surgical table and connect to the ventilator where anesthesia is maintained with 1–3% Isoflurane.
3. Apply ophthalmic ointment (Artificial Tears) to the eyes.
4. Insert the temperature probe coated in Surgilube into the rectum of the rat.
5. Administer two 1.5 mL subcutaneous injections of Lactated Ringers through a 25G needle away from the incision region.
6. With the electric razor, shave and vacuum the abdomen and chest region free of hair. Disinfect the incision region with three successive rounds of Betadine followed by 70% Isopropyl alcohol swabs.
7. Use a 25G needle to inject 3–4 beads of 1% Lidocaine subcutaneously along the length of the initial incision region, below but parallel to the left rib cage beginning along the midline.
8. Use the scalpel to make a 3–4 cm cutaneous incision beginning inferior to the xiphoid process and continuing lateral left approximately 1 cm caudal to the ribcage (Fig. 3).
9. Carefully cut through the muscle to expose the xiphoid process without damaging the liver. Once the xiphoid process is exposed, cut through the muscle along the length of the cutaneous incision.
10. Expose the diaphragm by lifting and anchoring the xiphoid process. This can be accomplished by running half the length of an appropriate suture (Vicryl 4-0 FS-1) through the xiphoid process and taping the free ends of the suture to a nearby high point (*see* **Note 10**).

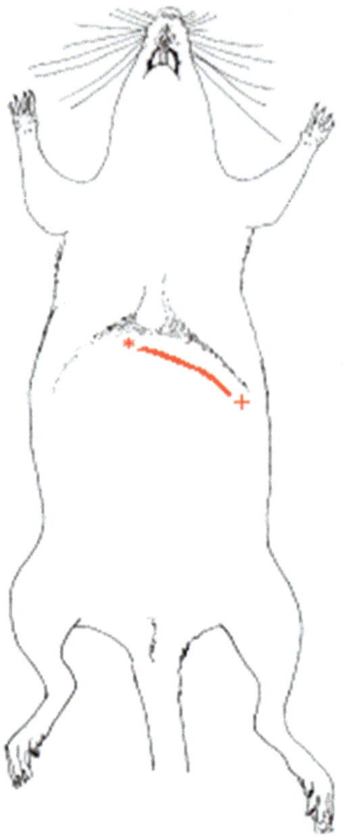

Fig. 3 The initial cutaneous incision beginning inferior to the xiphoid process (*) and continuing lateral left approximately 1 cm caudal to the ribcage (+). Modified and reprinted from Yang et al. [13] with permission from Elsevier

11. Use fine forceps to hold the diaphragm dorsal to the heart apex and slightly pull the diaphragm caudal in order to prevent damaging lung or heart tissue. Cut a small opening in the diaphragm at a point dorsal to the heart apex. From this opening make a ventral incision that extends 1–1.5 cm being careful not to cut the lungs (Fig. 4).
12. Tear the pericardium at the apex without damaging the lungs to expose the left ventricle free wall (*see* **Note 11**).
13. Hold the heart lightly enough to provide stability but not interfere with its regular beating. Inject up to 75 μL of the biomaterial through a 27G or 30G needle at a steady rate with the bevel or bore hole oriented toward the left ventricle lumen. The needle tip should enter the left ventricle-free wall almost parallel to the epicardial surface to ensure the injection does not enter the lumen (*see* **Note 12**) (Fig. 5).
14. Close the diaphragm with a taper needle suture (Prolene 5-0 RB-1). Anchor the suture using a double knot at the dorsal edge of the incision and close three fourth of the incision with a running stitch (*see* **Note 13**).

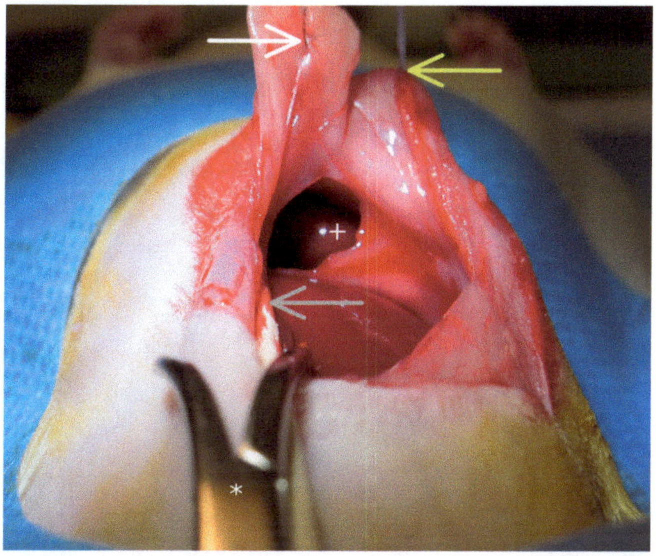

Fig. 4 The diaphragm is exposed by lifting xiphoid process by running half the length of an appropriate suture (Vicryl 4-0 FS-1) through the xiphoid process (white arrow) and taping the free ends of the suture to a nearby high point. Diaphragm visibility can be increased by repeating this process with another suture placed approximately 1 cm lateral left from the xiphoid process (yellow arrow). The heart apex is visible through the diaphragm (+). The diaphragm incision is made beginning dorsal to the heart apex and extends ventral 1–1.5 cm. A sterile cotton swab (grey arrow, cotton head of swab not visible) anchored with a towel clamp (*) is used to keep the right lung from obscuring the heart apex

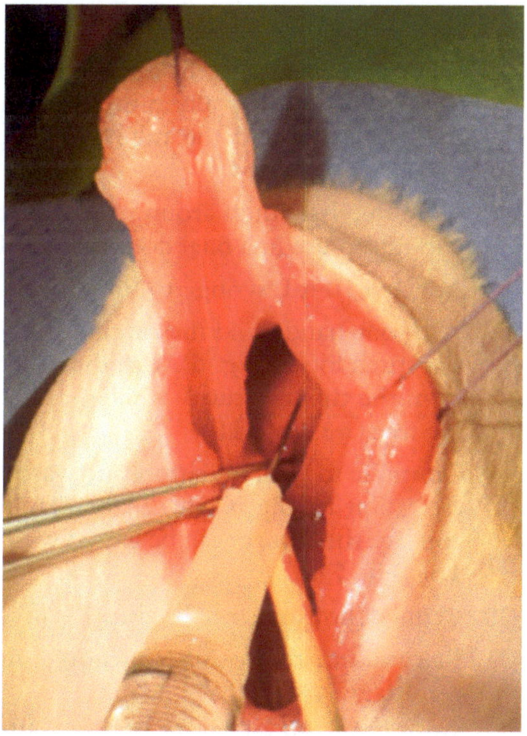

Fig. 5 The material is then injected into the left ventricle of the heart as shown. A 27G needle is used to deliver the material into the wall of the heart. If too much force is used, the needle can fully puncture into the lumen of the ventricle, thus leading to a failed delivery

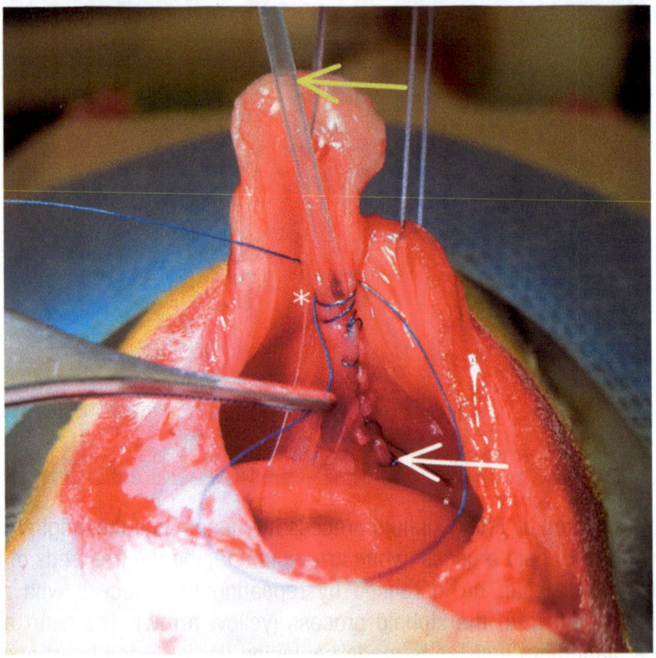

Fig. 6 The diaphragm is sutured closed beginning at the base of the incision using a running stitch (white arrow). The running stitch extends around the aspiration tubing (yellow arrow) and is held taut while leaving the last stitch loose (*) for the closing knot once the aspiration tube is removed

15. When three fourth of the opening is closed, insert an aspiration tube into the cavity along the incision and continue the running stitch around the tube until the incision is completely closed (*see* **Note 14**).
16. Place one last stitch around the tube but leave the loop large enough to use in a final double knot (Fig. 6).
17. Attach a 10 mL syringe to the aspiration tube.
18. With one hand, hold the running stitch tight while leaving the last stitch loose (Fig. 6). Use the other hand to suction the air from the thoracic cavity while simultaneously pulling the tubing free (*see* **Note 15**).
19. Continue to hold the suture and watch the diaphragm once the aspiration tubing is free. If the diaphragm remains tight and concave the sutures can be closed with a double knot (*see* **Note 16**) (Fig. 7).
20. Reduce the Isoflurane level to 1% after closing the diaphragm.
21. Release and remove the anchored suture used to make the diaphragm visible.
22. Suture the muscle layer closed with an appropriate suture (Vicryl 4-0 FS-1) using close-spaced intermittent stitches.

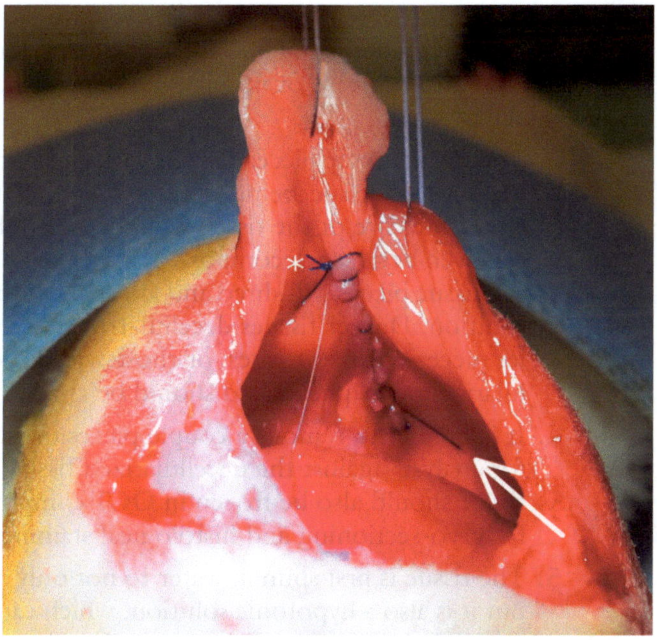

Fig. 7 The airtight running stitch closing the diaphragm (*) produces a concave diaphragm (white arrow) after aspirating air from the thoracic cavity

23. Close the skin with an appropriate suture (Vicryl 5-0 FS-2) using intermittent stitches and seal with tissue adhesive (Vetbond).
24. Apply triple antibiotic ointment over the wound.
25. Reduce the Isoflurane level to 0% and monitor breathing until the rat can breathe independent of the ventilator.
26. Administer an appropriate dose of postoperative analgesic such as Buprenex. Check with your animal care facility to determine which analgesic, administrative route, and dosage is recommended for your facility (*see* **Note 17**).

4 Notes

1. The quality of SDS powder can vary greatly between brands and can even have variation from the same brand. The SDS powder should be white and very fine. When dissolved, the SDS solution should turn completely clear. Our lab has previously received batches of SDS that had gathered into large aggregates due to exposure to water before storage and a batch that dissolved into a yellow solution. Double check the quality of SDS since this could have significant impact on the decellularization of the tissue. We have had consistent results with ordering Fisher Scientific, #S529–500.

2. After the heart is harvested, it should be kept on ice to minimize degradation of the ECM, which is an issue due to the immediate release of matrix metalloproteinases (MMPs) from cell death.
3. The size of tissue pieces for this process has been optimized for the type of tissue being decellularized. If the pieces are too large, the core of the tissue will not decellularize properly and can even begin to decompose. If the tissue pieces are too small, the tissue pieces will shred and fall apart in the decellularization solution. Denser tissues decellularize slower and require smaller pieces. Weaker tissues, or those with little ECM, can decellularize more quickly and should be cut into larger pieces.
4. It is helpful to remove a couple pieces of fresh tissue to freeze for analysis against the decellularized tissue. Some of the fresh tissue should also be frozen in OCT compound in a cryomold for cryo-sectioning and histological staining.
5. The tissue is first spun in water to not only rinse off the blood, but it is also a hypotonic solution, which can actually begin the decellularization process by rupturing the exposed cells.
6. The length of time needed to decellularize the tissue varies and is dependent on the tissue type and size of the cut tissue pieces. Larger pieces take longer than smaller ones and thus, after overnight in the decellularization solutions (18 h $\pm$ 6) the larger pieces can be cut in half if needed. While spinning in the decellularization solution the tissue will release the cellular content, as seen by a cloudy haze that accumulates over time in the beaker. Eventually, the tissue will turn completely white, leaving behind only the ECM scaffold of the tissue.
7. Like the fresh tissue, some of the decellularized tissue should be kept for analysis and frozen in OCT compound in a cryomold for cryo-sectioning and histological staining to confirm complete decellularization of the tissue (Fig. 2c).
8. The digest should be checked regularly throughout 48–56 h. The ECM material can creep up the walls of the glass vial and out of the digestion solution due to the motion of the stir bar. If this happens, the vial should be centrifuged for 30 s at 500 rpm and the material should be scraped down back into the solution using a small autoclaved metal spatula. It is best to centrifuge and scrape down at least 30 min after beginning digestion and then twice a day to ensure optimal digestion. Once digested, the material should have an increased viscosity and appear homogeneous, but it will not be transparent.
9. To properly resuspend the lyophilized aliquot of the material, place the aliquot on ice. With a pipette, add the appropriate amount of sterile water to achieve 6 mg/mL and allow to sit on ice for 5–10 min. Now, place the pipette in the aliquot and

alternate stirring with the tip and pipetting up and down. Continue pipetting up and down until homogenous and no small particulates can be identified. Centrifuging in a minicentrifuge for less than 5 s also assists with this process. To ensure homogeneity, use a 1 mL syringe with a 25G needle to shear the material by drawing it up and dispensing it 2–3 times or until there is less resistance when drawing into the syringe. Be careful to avoid introducing air bubbles.

10. To increase visibility, a second suture can be used in the same manner but placed along the ribcage edge approximately 1 cm lateral left from the xiphoid process. With the diaphragm exposed, the location of the heart apex will be visible through the diaphragm between the lungs.
11. Once the heart apex is free from the pericardium, visibility can be aided by anchoring a sterile cotton swab with a towel clamp along the midline keep the incision open and the right lung from blocking visibility.
12. Blanching of the left ventricular-free wall during the injection is a visible indication of injection success but may not always be visible.
13. The stitch placement should be close set to make an airtight closure to allow the lungs to properly inflate. If air is present in the thoracic cavity, the animal will not be able to breathe. Mouth breathing or gasping post surgery may be an indication of air in the thoracic cavity.
14. The aspiration tubing can be taped to a surgical lamp in order to keep the aspiration tube positioned next to the sutures. The placement of the tube along the diaphragm closure is necessary to limit the chances of aspirating lung tissue or pericardium.
15. Between 1 and 3 mL of total volume (liquid/air) may be removed from the cavity. If resistance is felt, the suction may be blocked by pericardium or lung tissue. If this occurs, stop suctioning and adjust the aspiration tubing. Keep the aspiration tubing next to the suture line to reduce the chances of aspirating any lung tissue.
16. If there is a leak in the sutures, the diaphragm will not remain tight and will balloon. The sutures can be loosened to reinsert the aspiration tube and repeat the process.
17. Buprenorphine is a controlled substance commonly sold under the brand name Buprenex and may require special licensing to purchase. When a 0.05 mg/kg dose is subcutaneously administered using a 27G needle, the animal will be mobile, sternal, and alert during their immediate recovery. Within an hour the animal's behavior and coat appearance should be normal. There should be no signs of distress such as hunched back, ruffled fur, or mouth breathing.

References

1. Christman KL, Fok HH, Sievers RE et al (2004) Fibrin glue alone and skeletal myoblasts in a fibrin scaffold preserve cardiac function after myocardial infarction. Tissue Eng 10(3–4):403–409
2. Christman K, Vardanian A, Fang Q et al (2004) Injectable fibrin scaffold improves cell transplant survival, reduces infarct expansion, and induces neovasculature formation in ischemic myocardium. J Am Coll Cardiol 44(3): 654–660
3. Johnson TD, Christman KL (2012) Injectable hydrogel therapies and their delivery strategies for treating myocardial infarction. Expert Opin Drug Deliv 10(1):59–72
4. Rane AA, Christman KL (2011) Biomaterials for the treatment of myocardial infarction: a 5-year update. J Am Coll Cardiol 58(25): 2615–2629
5. Singelyn JM, DeQuach JA, Seif-Naraghi SB et al (2009) Naturally derived myocardial matrix as an injectable scaffold for cardiac tissue engineering. Biomaterials 30(29):5409–5416. Elsevier Ltd
6. Singelyn JM, Sundaramurthy P, Johnson TD et al (2012) Catheter-deliverable hydrogel derived from decellularized ventricular extracellular matrix increases endogenous cardiomyocytes and preserves cardiac function post-myocardial infarction. J Am Coll Cardiol 59(8):751–763
7. Seif-Naraghi SB, Singelyn JM, Salvatore MA et al (2013) Safety and efficacy of an injectable extracellular matrix hydrogel for treating myocardial infarction. Sci Transl Med 5(173): 173ra25
8. Wassenaar JW, Gaetani R, Garcia JJ et al (2016) Evidence for mechanisms underlying the functional benefits of a myocardial matrix hydrogel for post-MI treatment. J Am Coll Cardiol 67(9):1074–1086
9. Traverse JH, Henry TD, Dib N et al (2019) First-in-man study of a cardiac extracellular matrix Hydrogel in early and late myocardial infarction patients. JACC: Basic Transl Sci 4(6):659–669
10. Ungerleider JL, Johnson TD, Rao N et al (2015) Fabrication and characterization of injectable hydrogels derived from decellularized skeletal and cardiac muscle. Methods 84: 53–59
11. Reing JE, Brown BN, Daly KA et al (2010) The effects of processing methods upon mechanical and biologic properties of porcine dermal extracellular matrix scaffolds. Biomaterials 31(33):8626–8633
12. Huleihel L, Hussey GS, Naranjo JD et al (2016) Matrix-bound nanovesicles within ECM bioscaffolds. Sci Adv 2(6):e1600502
13. Yang CH, Lee BB, Jung HS et al (2002) Effect of electroacupuncture on response to immobilization stress. Pharmacol Biochem Behav 72(4):847–855

Chapter 18

Encapsulation of Pediatric Cardiac-Derived C-Kit⁺ Cells in Cardiac Extracellular Matrix Hydrogel for Echocardiography-Directed Intramyocardial Injection in Rodents

Preety Shakya, Milton E. Brown, and Michael E. Davis

Abstract

Pediatric cardiac-derived c-kit$^+$ cell therapies represent an innovative approach for cardiac tissue repair that have demonstrated promising improvements in recent studies and offer multiple benefits, such as easy isolation and autologous transplant. However, concerns about failure of engraftment and transient paracrine effects have thus far limited their use. To overcome these issues, an appropriate cell delivery vehicle such as a cardiac extracellular matrix (cECM) hydrogel can be utilized. This naturally derived biomaterial can support embedded cells, allowing for local diffusion of paracrine factors, and provide a healthy microenvironment for optimal cellular function. This protocol focuses on combining cardiac-derived c-kit$^+$ cells and a cECM hydrogel to prepare a minimally invasive, dual therapeutic for in vivo delivery. We also outline a detailed method for ultrasound-guided intramyocardial injection of cell-laden hydrogels in a rodent model. Additional steps for labeling cells with a fluorescent dye for in vivo cell tracking are provided.

Key words Cardiac c-kit$^+$ cells, Cardiac extracellular matrix, Cell-biomaterial therapy, Ultrasound-guided injection, Cardiac repair

1 Introduction

Cell therapies have emerged as a promising strategy for tissue regeneration and restoration of impaired cardiac function in pediatric and adult patients. Among other cell types, cardiac-derived c-kit$^+$ cells are one of the most studied cell populations for cardiac repair [1–4]. Clinical and preclinical studies have demonstrated safety, feasibility, and modest improvements in human trials, and age-dependent therapeutic effects in juvenile rodent models after treatment with cardiac c-kit$^+$ cells [2]. These therapeutic benefits are primarily thought to be attributable to paracrine effects through release of reparative growth factors and exosomes [3, 4]. However,

Kareen L. K. Coulombe and Lauren D. Black III (eds.), *Cardiac Tissue Engineering: Methods and Protocols*, Methods in Molecular Biology, vol. 2485, https://doi.org/10.1007/978-1-0716-2261-2_18,

chronic improvement is limited in cell therapies due to poor retention as cells get washed out by the circulation system over time [5, 6]. In addition, these cells are also exposed to a diseased and hostile microenvironment which may not provide them with reparative signaling cues [6].

A naturally derived biomaterial such as decellularized cardiac extracellular matrix (cECM) hydrogel presents as an attractive option as cell delivery vehicle. This acellular, injectable hydrogel is capable of self-assembling into a fibrous and porous scaffold in vivo and has shown great promise in improving cardiac function in cardiomyopathy animal models [7]. A recently completed phase I clinical trial showed improvements in left ventricular remodeling in a subset of patients following myocardial infarction [8]. Furthermore, studies have shown that this biomaterial not only promotes endogenous repair, but also offers additional therapeutic benefits by degrading into chemoattractant and angiogenic products [9, 10]. In vitro studies have shown the promise of a novel combination therapy including cECM hydrogel and c-kit$^+$ cells that may improve cell engraftment and function by providing biochemical cues of a native microenvironment [4, 11–13].

This protocol outlines an in vivo application of the cell-biomaterial therapeutic in a minimally invasive manner. Briefly, cardiac c-kit$^+$ cells are isolated from human atrial appendage tissue by magnetic cell sorting, cloned and expanded as needed [14]. cECM hydrogel is prepared from porcine ventricular tissue through a series of decellularization, lyophilization, and digestion steps, and then freeze-dried for long-term storage [15]. Lyophilized cECM is rehydrated with sterile water at desired concentration (Subheading 3.2) and then used for encapsulation of c-kit$^+$ cells (Subheading 3.4). A precise dose of the cell-laden hydrogel is safely and accurately delivered to a desired region of rat myocardium using ultrasound-guided injection (Subheading 3.5). The workflow summary is illustrated in Fig. 1. As an optional step, the cells are also labeled with fluorescent dye such as DiR (1,1′-dioctadecyl-3,3,3′,3′-tetromethylindotricarbocyanine iodide) prior to encapsulation (Subheading 3.3) which enables tracking of cell retention using in vivo imaging. Echocardiographic assessments post injection can aid in evaluating functional recovery of the myocardium and establish the efficacy of c-kit$^+$ cells delivered in cECM hydrogel for cardiac repair.

2 Materials

2.1 Cell Culture

1. Cryopreserved c-kit$^+$ cells.
2. T-75 flasks.
3. Laminar flow hood.

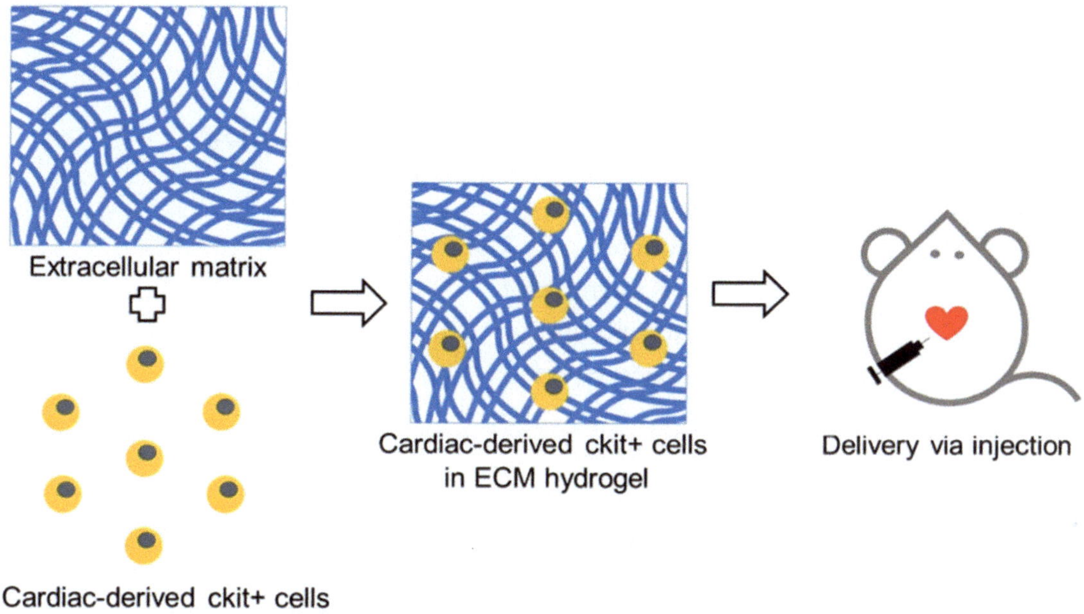

Fig. 1 Workflow of cellular encapsulation of c-kit^{+} cells in cardiac extracellular matrix hydrogel followed by in vivo delivery via intramyocardial injection

4. Cell culture media: Ham's F-12, 1× Penicillin-Streptomycin-Glutamine, 10% fetal bovine serum, 10 ng/mL basic fibroblast growth factor.
5. Sterile filter.
6. Cell culture incubator at 37 °C and 5% CO_2.
7. Serological pipettes.
8. Pipettes and tips.

2.2 Hydrogel Resuspension

1. Beaker.
2. Ice.
3. Lyophilized cECM hydrogel.
4. Sterile water.
5. Pipette and tips.
6. 1 mL syringe and 25G needle.
7. Minicentrifuge.

2.3 Cell Labeling

1. DiR dye.
2. Dimethyl sulfoxide (DMSO).
3. Trypsin.
4. Cell counter.
5. Universal centrifuge.

6. Pipettes and tips.
7. Serological pipettes.
8. Falcon tube.
9. Cell culture incubator at 37 °C and 5% CO_2.

2.4 Cell Encapsulation

1. Trypsin.
2. Cell counter.
3. Universal centrifuge.
4. Pipettes and tips.
5. Serological pipettes.
6. Microcentrifuge tube.
7. Centrifuge tube.
8. Resuspended cECM hydrogel.
9. C-kit$^+$ cells.

2.5 Intramyocardial Injection

1. Vevo2100 ultrasound imaging system.
2. Echo imaging platform.
3. Transducer MS-250.
4. Anesthesia induction chamber.
5. 2–3% isoflurane.
6. Oxygen tank.
7. Weighing scale.
8. Plastic beaker.
9. Syringe rail, holder, clamp, and controls.
10. 1 mL tuberculin syringe and 29G needle.
11. Ultrasound echo gel.
12. Eye ointment.
13. Hair remover.
14. Tape.

3 Methods

3.1 Cardiac C-kit$^+$ Cell Culture

Pediatric cardiac c-kit$^+$ cells are isolated as previously described [14]. The following steps must be performed under sterile conditions.

1. Prepare cell culture media by passing all constituents through a sterile filter.
2. Warm culture media for around 30 min in water bath.

3. Thaw previously cryopreserved vial containing 1–2 million cells per vial.
4. Label a T-75 flask with cell line number, passage number, and date.
5. Add 10 mL of warm media into the flask.
6. Pipette the cells from vial into the flask.
7. Place the flask in 37 °C incubator set at 5% CO_2.
8. Change media after 24 h and every 2 days subsequently.
9. Allow cells to proliferate until confluency is around 80%.

3.2 Cardiac cECM Hydrogel Reconstitution

cECM hydrogel is processed as previously described [15]. The following steps must be performed under sterile conditions. From **step 2**, cECM must be kept on ice throughout all the wait steps.

1. Fill two-thirds of a 500 mL beaker with ice.
2. Take 4 mg aliquot of lyophilized cECM and keep it on ice (*see* **Note 1**).
3. Add 500 μL of sterile water to cECM to obtain concentration of 8 mg/μL (*see* **Note 2**).
4. Stir the mixture with the tip of P1000 and wait for 15 min.
5. Use a P1000 pipette to break the cECM particles by pipetting up and down making sure not to introduce bubbles and wait for 3–5 min.
6. Repeat **step 5** using a P200 pipette.
7. Centrifuge the tube in a minicentrifuge for 5 s to bring any bubbles to the surface and discard them.
8. Repeat **steps 6** and **7** until small particles are not visible.
9. Use a 25G needle syringe to carefully shear the entire aliquot without introducing bubbles (*see* **Note 3**).
10. Repeat **steps 6, 7,** and **9** as needed to ensure that the final solution is homogeneous and free of bubbles.

3.3 Cell Labeling with DiR Dye

This is an optional step that enables tracking of cells post injection using IVIS Spectrum in vivo imaging system. Other methods can be used for different imaging modalities such as echocardiography, computed tomography (CT), and magnetic resonance imaging (MRI). All steps must be performed under sterile conditions.

1. Add DiR dye to dimethyl sulfoxide (DMSO) to prepare stock solution at final concentration of 5 mg/mL (*see* **Note 4**).
2. Treat cells with trypsin and count them.
3. Centrifuge at 180 × *g* for 5 min to pellet desired number of cells and aspirate the supernatant (*see* **Note 5**).
4. Suspend one million cells per 1 mL of warm cell culture media.

5. Add 2.5 μL of prepared DiR solution per 1 mL of cell solution and mix gently using a pipette.
6. Place the tube in an incubator set at 37 °C and 5% CO_2 for around 20 min.
7. Centrifuge the tube at 180 × *g* for 5 min and aspirate all supernatant.
8. Resuspend in cell culture media and gently pipette to mix.
9. Repeat **steps** 7 and **8** for two more times.

3.4 Cell Encapsulation in cECM Hydrogel

The following steps must be performed under sterile conditions.

1. Treat cells with 5 mL trypsin and count them (*see* **Note 6**).
2. Centrifuge at 180 × *g* for 5 min to pellet desired number of cells and aspirate the supernatant (*see* **Note 5**).
3. Add 75 μL of cECM hydrogel per one million of cells.
4. Gently pipette up and down using a P200 pipette to mix thoroughly making sure not to introduce any bubbles.
5. Let the cells in cECM mixture sit at room temperature for 2–3 min.
6. Transfer the material into microcentrifuge tube using a pipette.

3.5 Ultrasound-guided Myocardial Injection

The protocol procedures must be performed under the approval of the *Institutional Animal Care and Use Committee* (IACUC). In-plane technique must be adopted with the needle tip and the heart visible in a wide field of view on the ultrasound monitor throughout the entire injection procedure under Subheading 3.5.4.

3.5.1 Preparation of Ultrasound Machine

1. Set up the Vevo2100 ultrasound machine with required cardiology parameters. The procedure will be performed in the B-mode.
2. Connect the MS-250 transducer at 21 MHz (*see* **Note 7**).
3. Create an electronic label of the study associated with the animal.
4. Set the temperature of the echo imaging platform, and ultrasound gel warmer to 37 °C.

3.5.2 Preparation of Rodent Model

1. Anesthetize the rat in an induction chamber by using 2–3% isoflurane mixed with 100% oxygen at a flow rate of 0.2–0.5 L/min.
2. Weigh the anesthetized rat by placing it in a plastic beaker on a weighing scale.

3. Place the rat on the echo imaging platform in a supine position with the rat's nose placed in a nose cone with 2–3% isoflurane flowing at the rate of 0.2–0.3 L/min.
4. Tape the upper and lower limbs of the rat to the imaging platform while securing the electrocardiogram and respiration electrodes.
5. Completely remove the fur from the upper thorax of the rat by using hair remover.
6. Apply lubricating eye ointment to both eyes of the rat to maintain moisture throughout the entire procedure.

3.5.3 Visualization of Heart and Needle Positioning

1. Apply a generous amount of ultrasound transmission gel to the upper thorax of the rat.
2. Advance the transducer scan head down into the gel to project a clear long axis, B-mode field of view on to the Vevo2100 monitor.
3. Rotate the transducer to the short-axis view to visualize both left ventricular and right ventricular chambers simultaneously while keeping the transducer in an upward 45° angle position to the rat's chest.
4. Withdraw 75 μL of prepared solution into a tuberculin syringe carefully from the microcentrifuge tube.
5. Tap the syringe to move any air bubbles to the top and discard them.
6. Place syringe with the needle's bevel facing up and secure the syringe by tightening down the clamp of the needle mount.
7. Make sure both the transducer and the needle are parallel to each other in the short-axis mode.
8. Advance the needle in a downward 20° angle towards the rat's thorax until the needle comes into the field of view on the Vevo2100 monitor. Refer to Fig. 2 for completed experimental setup prior to injection.

3.5.4 Injection of Cell-Laden Hydrogel

1. Confirm sedation with a firm tail pinch.
2. Proceed to advance the needle towards the rat by rotating the control knob of the needle mount clockwise.
3. Continue to advance the needle 2–3 mm deep inside the thoracic cavity while carefully observing the bevel of the needle on the monitor. Figure 3 shows monitor view of transthoracic needle penetration.
4. Once the desired injection site is reached, stop advancing the needle and slowly inject the therapeutic in the myocardial wall (*see* **Notes 8** and **9**).

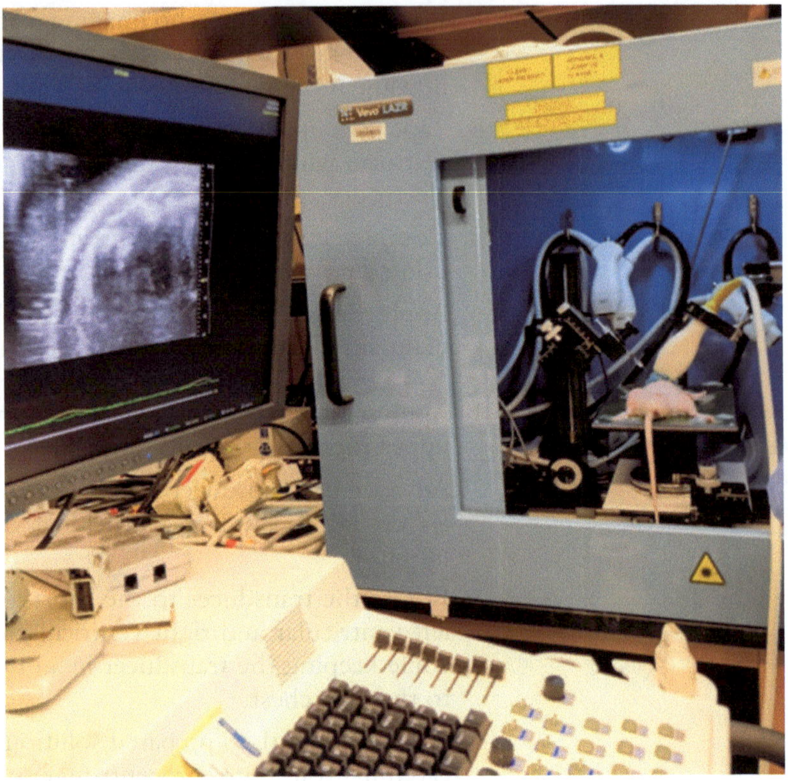

Fig. 2 Experimental setup for ultrasound-guided injection in a juvenile male athymic rat

5. Slowly withdraw the needle by rotating the control knob of the needle mount counterclockwise. Remove and discard the needle.
6. Keep the rat under echocardiographic observation for several minutes to ensure that there are no procedural complications.
7. Elevate the transducer off the rat's chest. Remove the gel from the thorax and the tape from the rat's limbs.
8. Place the rat back to a warm cage with food and water for further monitoring.

4 Notes

1. Lyophilized cECM can be stored in −80 °C freezer for long-term use.
2. Use sterile phosphate-buffered saline (PBS) solution instead of sterile water for rehydration if PBS salts were not added to the cECM material prior to lyophilization.

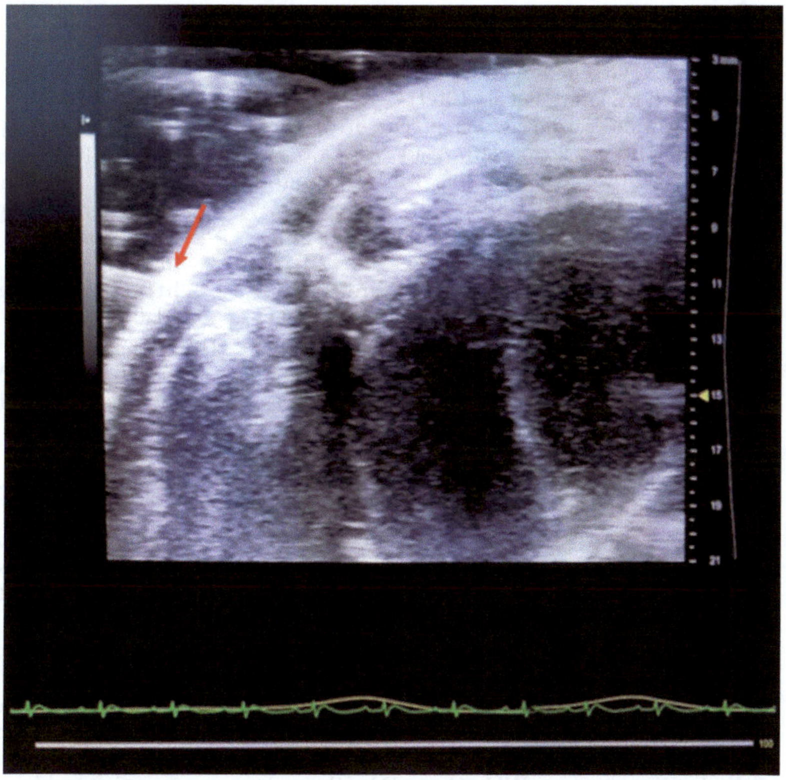

Fig. 3 Transthoracic echocardiogram of rodent heart showing advancement of needle (*arrow*) into right ventricular wall for myocardial injection

3. Minimal resistance during shearing is a sign of successful cECM reconstitution.
4. DiR stock solution can be stored in freezer at −20 °C for up to 6 months.
5. Maximum of one million cells can be injected per rat.
6. Skip this step if continuing from Subheading 3.4.
7. This parameter is ideal for rats.
8. Make sure the needle does not penetrate past the myocardial wall and into the cardiac chamber.
9. Administer in 2–3 small doses until the syringe is empty for best results.

Acknowledgments

This material is based upon work that is supported by a grant from the National Institutes of Health HL146147 to MED.

References

1. Tallini YN, Greene KS, Craven M, Spealman A, Breitbach M, Smith J et al (2009) C-Kit expression identifies cardiovascular precursors in the neonatal heart. Proc Natl Acad Sci U S A 106(6):1808–1813. https://doi.org/10.1073/pnas.0808920106
2. Agarwal U, Smith A, French KM, Boopathy AV, George A, Trac D et al (2016) Age-dependent effect of pediatric cardiac progenitor cells after juvenile heart failure. Stem Cells Transl Med 5(7):883–892. https://doi.org/10.5966/sctm.2015-0241
3. Trac D, Maxwell JT, Brown ME, Xu C, Davis ME (2018) Aggregation of child cardiac progenitor cells into spheres activates notch signaling and improves treatment of right ventricular heart failure. Circ Res 124(4):526–538. https://doi.org/10.1161/CIRCRESAHA.118.313845
4. Bejleri D, Streeter BW, Nachlas ALY, Brown ME, Gaetani R, Christman KL, Davis ME (2018) A bioprinted cardiac patch composed of cardiac-specific extracellular matrix and progenitor cells for heart repair. Adv Healthc Mater 1800672(7):1–13. https://doi.org/10.1002/adhm.201800672
5. Ye L, Zimmermann WH, Garry DJ, Zhang J (2013) Patching the heart: Cardiac repair from within and outside. Circ Res 113(7):922–932. https://doi.org/10.1161/CIRCRESAHA.113.300216
6. Feyen DAM, Gaetani R, Doevendans PA, Sluijter JPG (2016) Stem cell-based therapy: improving myocardial cell delivery. Adv Drug Deliv Rev 106:104–115. https://doi.org/10.1016/j.addr.2016.04.023
7. Singelyn JM, Sundaramurthy P, Johnson TD, Schup-magoffin PJ, Hu DP, Faulk DM et al (2012) Catheter-deliverable hydrogel derived from decellularized ventricular extracellular matrix increases endogenous cardiomyocytes and preserves cardiac function post-myocardial infarction. J Am Coll Cardiol 59(8). https://doi.org/10.1016/j.jacc.2011.10.888
8. Traverse JH, Henry TD, Dib N, Patel AN, Pepine C, Schaer GL et al (2019) First-in-man study of a cardiac extracellular matrix Hydrogel in early and late myocardial infarction patients. JACC: Basic Transl Sci 4(6): 659–669. https://doi.org/10.1016/j.jacbts.2019.07.012
9. Ungerleider JL, Johnson TD, Hernandez MJ, Elhag DI, Braden RL, Dzieciatkowska M et al (2016) Extracellular matrix hydrogel promotes tissue remodeling, arteriogenesis, and perfusion in a Rat Hindlimb Ischemia model. JACC: Basic Transl Sci 1(1–2). https://doi.org/10.1016/j.jacbts.2016.01.009
10. Hernandez MJ, Gaetani R, Pieters VM, Ng NW, Chang AE, Martin TR et al (2018) Decellularized extracellular matrix hydrogels as a delivery platform for microRNA and extracellular vesicle therapeutics. Adv Ther 1800032(1):1–9. https://doi.org/10.1002/adtp.201800032
11. Dequach JA, Mezzano V, Miglani A, Lange S, Keller GM, Christman KL (2010) Simple and high yielding method for preparing tissue specific extracellular matrix coatings for cell culture. PLoS One 5(9):1–11. https://doi.org/10.1371/journal.pone.0013039
12. Singelyn JM, Dequach JA, Seif-naraghi SB, Littlefield RB, Schup-magoffin PJ, Christman KL (2010) Naturally derived myocardial matrix as an injectable scaffold for cardiac tissue engineering. Biomaterials 30(29):5409–5416. https://doi.org/10.1016/j.biomaterials.2009.06.045
13. French KM, Boopathy AV, DeQuach JA, Chingozha L, Lu H, Christman KL, Davis ME (2013) A naturally-derived cardiac extracellular matrix enhances cardiac progenitor cell behavior in vitro. Acta Biomater 8(12): 4357–4364. https://doi.org/10.1016/j.actbio.2012.07.033.A
14. French KM, Davis ME (2014) Isolation and expansion of C-Kit-positive cardiac progenitor cells by magnetic cell sorting. In: Radisic M, Black LD III (eds) Cardiac tissue engineering: methods and protocols, vol 1181. Humana Press, New York, NY, pp 39–50. https://doi.org/10.1007/978-1-4939-1047-2
15. Ungerleider JL, Johnson TD, Rao N, Christman KL (2015) Fabrication and characterization of injectable hydrogels derived from decellularized skeletal and cardiac muscle. Methods:15–17. https://doi.org/10.1016/j.ymeth.2015.03.024

Chapter 19

Characterization of the Monocyte Response to Biomaterial Therapy for Cardiac Repair

Sarah McLaughlin, David Smyth, Emilio I. Alarcon, and Erik J. Suuronen

Abstract

Biomaterials are scaffolds designed to mimic the extracellular matrix and stimulate tissue repair. Biomaterial therapies have shown promise for improving wound healing in cardiac tissue after ischemic injury. An unintentional consequence of biomaterial delivery may be the stimulation of inflammation through recruitment of circulating monocytes into the tissue. Monocytes are a type of leukocyte (white blood cell) that play a critical role in pathogen recognition, phagocytosis of foreign material, and presentation of antigens to initiate an adaptive immune response. An increase in the pro-inflammatory subset of monocytes, marked by Ly6C antigen expression, in response to biomaterials can lead to rapid material degradation, ineffective treatment, and worsening of tissue injury. Flow cytometry is a leading method for screening the recruitment of monocytes to the heart in response to biomaterial injection. Here, we describe the isolation of leukocytes from the heart, blood, and spleen of mice treated with a biomaterial post-myocardial infarction and describe a flow cytometry protocol used to quantify the levels of major leukocyte subtypes, including $Ly6C^+$ inflammatory monocytes.

Key words Inflammatory $Ly6C^+$ monocytes, Cardiac tissue repair, Biomaterials, Flow cytometry, Tissue engineering

1 Introduction

Biomaterial therapies for cardiac repair were initially developed to provide a protective niche for therapeutic cells and to improve their survival in vivo [1–4]. However, several types of biomaterials such as injectable hydrogels and fibrous patches have also shown promise as stand-alone therapies to stimulate cardiac tissue repair post-ischemic injury [5–13]. Biomaterials are extracellular matrix (ECM)-inspired structures that can be either bioinert or bioactive [14, 15]. Bioactive biomaterials can interact with resident cells in the tissue where they are delivered [16]. Such interactions play an important role in the mechanism that leads to tissue repair and regeneration with biomaterial therapy [17, 18]. However, the interaction of some biomaterials with resident tissue macrophages

Kareen L. K. Coulombe and Lauren D. Black III (eds.), *Cardiac Tissue Engineering: Methods and Protocols*, Methods in Molecular Biology, vol. 2485, https://doi.org/10.1007/978-1-0716-2261-2_19,

may stimulate an immune response [19]. The subsequent chemokine production by the stimulated macrophages can lead to infiltration of the tissue with pro-inflammatory monocytes from the circulation, with additional monocytes often released from splenic reserves [20–22]. This may cause a foreign body response to the implanted biomaterial, which will undermine the therapeutic and translational potential of the material [23]. In addition, cardiac tissue injuries such as myocardial infarction (MI) have a multiphase immune response during wound healing including: an acute inflammatory phase (day 0–3 post-MI), a proliferative phase (day 3–7 post-MI), and a final maturation phase (beyond day 7 post-MI) [24, 25] (time points indicated are for mice). Careful timing of biomaterial delivery is required in order to avoid interfering with the clearance of cell debris during the inflammatory phase and to reduce inflammatory monocyte infiltration in the proliferative phase, which can lead to decreased wound healing [26, 27]. Thus, a method to assess acute monocytic infiltration in the heart would help screen for effective biomaterial therapy candidates for cardiac repair.

Multipanel flow cytometry is the most widely used technique used for immunophenotyping [28, 29]. Immunophenotyping is the process of detecting and quantifying specific cell types within a heterogeneous population of cells, which can be a tissue for example [30]. Single cell suspensions are isolated from tissues and incubated with fluorescently labeled antibodies raised against specific proteins that are used to distinguish between leukocyte lineages. Classical inflammatory and alternative (or resolving) subpopulations of monocytes, for example, are distinguished by their relative expression levels of the Ly6C antigen [31, 32]. Flow cytometers allow hydrodynamic focusing of cell suspensions, forcing individual cells to pass one or multiple lasers one at a time. Depending on the optical properties of the antibody-fluorophore conjugates used, a range of emitted photons specific to the bound antigen–antibody complexes can be simultaneously induced and detected [28]. This potentially allows multiple receptor expression levels to be detected for each cell in the suspension, which can be used to classify the cells into major leukocyte subsets [33–35].

The set of experiments outlined herein describe a method used to perform 7-color flow cytometry immunophenotyping of leukocytes to characterize the inflammatory response in cardiac tissue to biomaterial therapy. Also included are the methods for isolating leukocytes from the mouse heart, spleen, and blood. This procedure was successfully used in a previous study to assess the different immune response to two collagen biomaterials developed for cardiac repair [36]. The type I biomaterial therapy in this study was superior in improving cardiac function and did not stimulate an acute immune response at two days post-injection [36]. In contrast, the less effective type III biomaterial had an acute infiltration of Ly6C$^+$ monocytes into the heart and as a result was rapidly degraded by two days post-injection. Therefore, this flow cytometry procedure may be a valuable tool to screen biomaterials for an acute monocytic response that may help predict their potential for clinical translation.

2 Materials

2.1 Tissue Harvest and Blood Collection

1. Control and biomaterial treated female C57Bl/6 mice (8–10 weeks old). Mice in this study had myocardial infarction (MI) induced by left anterior descending (LAD) artery ligation 1 week prior to biomaterial injection into the myocardium.
2. CO_2 tank and cervical dislocation tool for mouse sacrifice.
3. Dissection scissors and tweezers.
4. Phosphate-buffered saline (PBS), calcium, and magnesium free (*see* **Note 1**).
5. 50 mM EDTA solution in PBS.
6. EDTA coated 1.5 mL microcentrifuge tubes (*see* **Note 2**).
7. 1 mL syringe with 25G, 3/8-inch needle coated with EDTA (*see* **Note 3**).
8. 70% ethanol.
9. 1.5 mL microcentrifuge tubes filled with PBS.
10. Digital microbalance.

2.2 Cell Isolation from Harvested Tissues

1. Cardiac digestion buffer: 10 μM HEPES, 5 μM sodium bicarbonate, 1 mM $MgCl_2$, 0.5 mM KCl, DNAase I 50 U/mL, collagenase II 188 U/mL, collagenase D 0.15 U/mL, hyaluronidase 10 U/mL in PBS (*see* **Note 4**).
2. 1.5 mL microcentrifuge tubes.
3. 50 mL and 15 mL conical tubes.
4. Dissection scissors and tweezers.
5. Red blood cell (RBC) lysis buffer: dilute 10× stock solution to 1× by adding 10 mL to 90 mL deionized water (*see* **Note 5**).
6. 2% fetal bovine serum (FBS) in PBS.
7. 10 mL syringes.
8. 70 μm cell strainers.
9. PBS.
10. Flow cytometry buffer: 2% FBS, 0.5 mM EDTA, 1.4 U/mL DNAase 1, 0.01% sodium azide in PBS (*see* **Note 6**).
11. 10 cm cell culture plates.
12. Temperature-controlled standard capacity centrifuge.

Table 1
Antibodies for labeling common markers of leukocyte subpopulations

Marker	Fluorophore	Dilution	Company	ID	Function
CD45	APC/Fire 750	1:160	Biolegend	103,154	Leukocyte marker
Ly6G	PerCP/Cy5.5	1:160	Biolegend	127,615	Granulocyte marker
CD11b	PE/Cy7	1:80	Biolegend	101,216	Monocyte marker
F480	AF647	1:160	Biolegend	123,122	Mouse macrophage marker
Ly6C	BV421	1:80	BD	562,727	Monocyte subset marker
CD3	PE	1:160	Biolegend	100,205	T cell marker
B220	AF488	1:80	Biolegend	103,228	Mouse B cell marker

This table lists the primary fluorophore-conjugated antibodies used in a panel to label mouse tissue leukocytes from the heart, blood, and spleen post-biomaterial delivery in cardiac tissue

13. Temperature-controlled shaker.
14. Rotary shaker.

2.3 Antibody Labeling of Cell Suspensions for Flow Cytometry

1. Trypan blue stain.
2. Hemocytometer.
3. Inverted brightfield microscope.
4. Flow cytometry buffer.
5. Temperature-controlled centrifuge.
6. Zombie Aqua™ (Biolegend) viability dye reconstituted in 100 μL dimethyl sulfoxide (DMSO) (*see* **Note 7**).
7. 5 mL round bottom flow cytometry tubes.
8. Antibodies for leukocyte subset panel (*see* Table 1).
9. Fluorophore-conjugated isotype antibodies for combined isotype control (*see* **Note 8**).
10. TruStainX™ Fc receptor blocking agent (Biolegend) (*see* **Note 9**).
11. 4% paraformaldehyde (PFA) in PBS.

2.4 Immunophenotyping Flow Cytometry and Gating of Leukocyte Subpopulations

1. Extra flow cytometry tubes for each sample.
2. 40 μm cell strainers.
3. Flow cytometer capable of detecting at least seven fluorescent probes. In the optimization of this protocol, a BD FACSAria III flow cytometer with a four-laser setup (405 nm, 488 nm, 561 nm, and 640 nm) with BD FACSDiva software v. 8.0.1 was used.
4. Software for flow cytometry data analysis (e.g., FlowJo).

3 Methods

Isolated cells must always be stored at 4 °C unless otherwise specified. Cell suspensions stained with fluorophore-conjugated antibodies must be protected from light. All animal work described was approved by the UOHI animal care committee. If using this protocol for florescence-activated cell sorting (FACS) of leukocyte subtypes, it is necessary to work under sterile conditions. For an overview of this protocol, see the schematic in Fig. 1.

3.1 Tissue Harvest and Blood Collection

1. Sterilize the necropsy bench, dissection tools with 70% ethanol.
2. Sacrifice the mouse by CO_2 and cervical dislocation.
3. Spray the mouse with 70% ethanol. Make a large incision down the middle of the chest and dissect the skin away from the ribs and intercostal muscles.
4. With a fine pair of scissors cut through the connective tissue and the ribs on either side of the heart. Remove the rib cage to expose the thoracic cavity (*see* **Note 10**).
5. Use an EDTA-coated syringe to collect blood from the heart through cardiac puncture. This yields approximately 500 μL blood (*see* **Note 11**). Add the blood to 100 μL of 50 mM EDTA in an EDTA-coated tube and place on ice.
6. Cut the left atrium of the heart and, using a 10 mL syringe filled with PBS, perfuse the heart until the tissue becomes blanched (~5 mL of PBS per mouse).
7. Remove the heart and the spleen, blot dry, and record their mass.
8. Place the heart and the spleen in microcentrifuge tubes filled with PBS. Leave on ice and proceed directly to cell isolation.

3.2 Cell Isolation from Harvested Tissues

3.2.1 Cell Isolation from the Heart

1. Use tweezers to remove the heart from the PBS and place it in a microcentrifuge tube containing 750 μL of cardiac digestion buffer cocktail.
2. Mince the heart into approximately 1 mm^2 pieces using fine dissection scissors in the digestion buffer. Clean the scissors between each sample.
3. Place tubes with minced hearts in digestion buffer in a temperature-controlled shaker for microcentrifuge tubes. Incubate at 37 °C while shaking at 900 rpm for 60 min. Every 20 min take each tube and tap the against the lab bench to keep debris from settling at the bottom of the tube (*see* **Note 12**).
4. Place the tubes on ice for 5 min to stop the enzymatic activity of the digestion buffer.

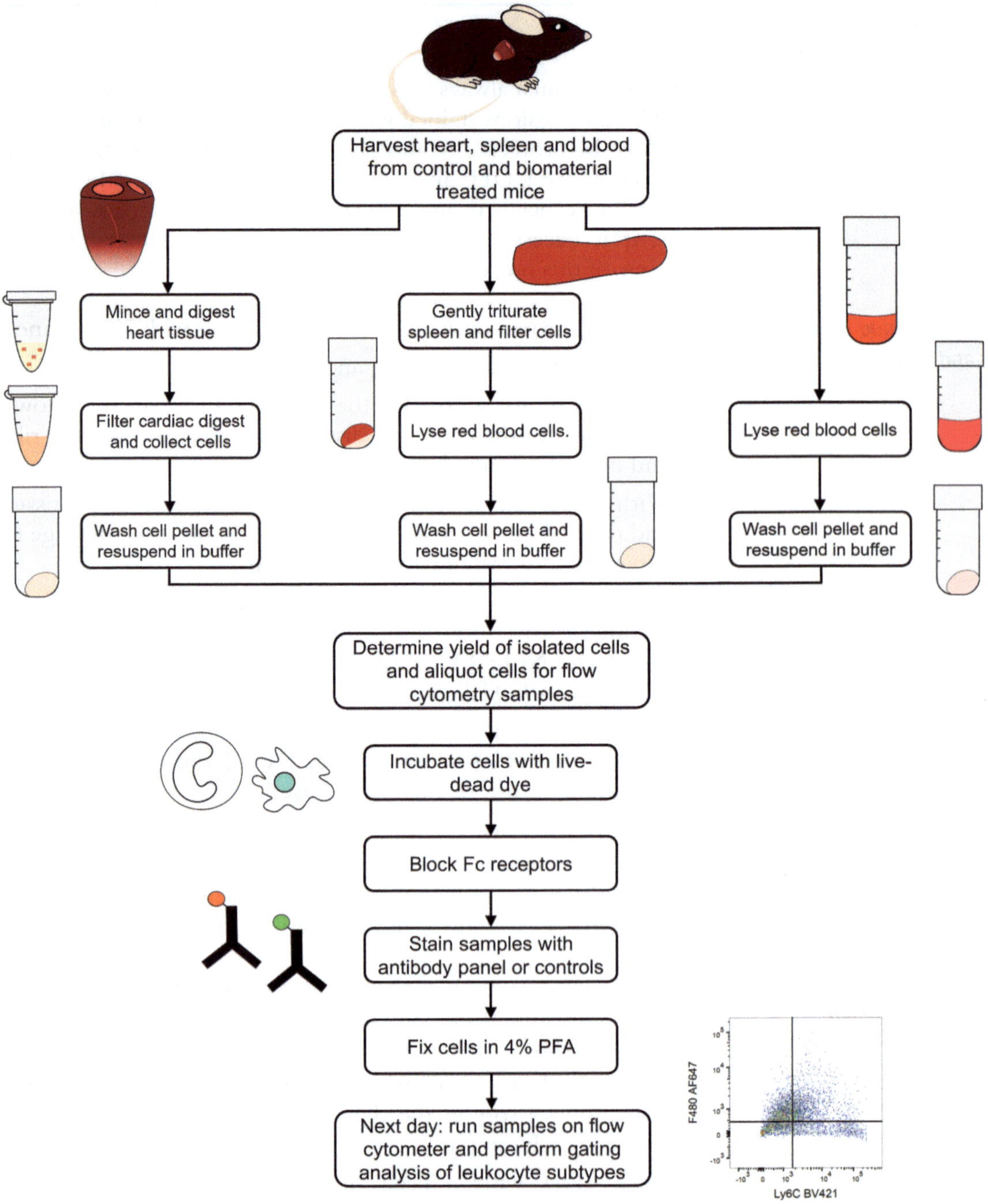

Fig. 1 Overview of the experimental protocol for the isolation and identification of leukocyte subpopulations from mice treated with a biomaterial for cardiac repair. Biomaterial therapy is delivered in a mouse model of myocardial infarction (MI) one-week after MI is induced by left anterior descending (LAD) artery ligation. Cells are isolated from the heart, spleen, and blood of mice at 2 days post-biomaterial delivery. The cell suspensions are then stained with an antibody panel to detect leukocyte subpopulations by flow cytometry

5. Pass the solution through a 70 μm cell strainer into a 50 mL conical tube. Pass 10 mL of PBS through the strainer into the conical tube to collect any additional cells from the debris.
6. Centrifuge the filtered cells at 350 × *g* for 10 min at 4 °C.
7. Resuspend each cell pellet in 300 μL of flow cytometry buffer.

3.2.2 Cell Isolation from Blood

1. Warm 10× RBC lysis buffer to room temperature before use. Dilute to 1× in deionized water.
2. Add 500 μL of blood to 10 mL 1× RBC lysis buffer. Shake vigorously to mix.
3. Incubate for 20 min at room temperature protected from light on a rotary shaker at high speed (*see* **Note 13**).
4. Centrifuge at 400 × *g* for 10 min at room temperature.
5. Discard the supernatant and wash the pellet in 2 mL of flow cytometry buffer.
6. Centrifuge at 400 × *g* for 5 min at 4 °C.
7. Resuspend each cell pellet in 250 μL of flow cytometry buffer. The single cell suspension should be no darker than light pink in color to indicate sufficient RBC lysis (*see* **Note 14**).

3.2.3 Cell Isolation from the Spleen

1. Mash the spleen in 2% FBS with the back of a 10 mL BD syringe plunger in a 10 cm cell culture dish. The red pulp of the tissue will separate from the white fibrous connective tissue of the spleen (*see* **Note 15**).
2. Pass the solution through a 70 μm strainer into a 50 mL conical tube and wash the strainer until the red pulp is pale. This takes about 30 mL of 2% FBS passed through the strainer into the conical tube.
3. Centrifuge at 400 × *g* for 10 min at 4 °C.
4. Add 5 mL of 1× RBC lysis buffer. Invert the tube vigorously 20× to mix and incubate at 4 °C on a rotary shaker for 5 min.
5. Centrifuge at 400 × *g* for 5 min at 4 °C. The pellet should be very pink or pale if the RBC lysis was successful.
6. Discard the supernatant and wash pellet with 2 mL flow cytometry buffer and centrifuge for 5 min at 400 × *g* and 4 °C.
7. Resuspend the pellet in 2 mL of flow cytometry buffer.

3.3 Antibody Labeling of Cells for Flow Cytometry

1. For each of the isolated cell suspensions, dilute a 1 μL aliquot into 9 μL of PBS. Add 1 μL of trypan blue dye to this solution and mix. To get the live cell concentration per sample, count the cells that exclude trypan blue using a hemocytometer and brightfield microscope. On average, an isolated cell suspension will yield 2.2 million, eight million, and 50 million cells for the mouse heart, blood, and spleen, respectively. At the

recommended cell resuspension volumes (as detailed in Subheading 3.2), this will amount to approximately 7 × 10^6 cells/mL, 26 × 10^6 cells/mL, and 25 × 10^6 cells/mL for the heart, blood, and spleen, respectively.

2. For antibody labeling, use 0.5–1 × 10^6 cells per sample. Dilute cell suspensions if necessary and transfer the required sample volume from each heart, blood, and spleen cell suspension into a flow tube (*see* **Note 16**).
3. A minimum of four additional cell samples of 0.5–1 × 10^6 cells per tissue will be required for flow cytometry controls. Each tissue requires an unstained control, an isotype control, and florescence minus one (FMO) controls for at least Ly6C-BV421 and F480-AF647. FMO controls can be prepared for each fluorophore-conjugated antibody. For the heart cell suspensions, pool the remaining sample volume in order to have enough cells to prepare the controls (*see* **Note 17**).
4. The final controls to prepare are the single stained samples (one corresponding to each antibody-fluorophore conjugate) for fluorescence compensation to control for spectral overlap between fluorophores in the flow cytometer. Use the leftover spleen cell suspension to prepare seven samples of 0.5–2 × 10^6 cells per sample (*see* **Note 18**).
5. Centrifuge the sample and control cell suspensions for 5 min at 350 × *g* and 4 °C. Aspirate the supernatant from pellet.
6. Dilute Zombie Aqua™ cell viability dye 1 in 500 in PBS. Each sample requires 100 μL of 1× Zombie Aqua™ dye solution per sample.
7. Add 100 μL of viability dye to each cell sample and isotype control and resuspend the pellets. Do not add Zombie Aqua™ viability dye to unstained controls or single stain controls except for the Zombie Aqua™ single stain control sample. Incubate for 20 min at room temperature protected from light (*see* **Note 19**).
8. Add 1 mL of flow cytometry buffer to wash the cells and centrifuge for 5 min at 350 × *g* and 4 °C. Aspirate the supernatant.
9. Prepare 1× TruStainX™ Fc receptor blocking solution by diluting it to 1:100 in flow cytometry buffer.
10. Add 100 μL Fc receptor blocking reagent to each sample and resuspend the cell pellet. Incubate for 15 min at room temperature protected from light.
11. Add 1 mL of flow cytometry buffer to wash the cells and centrifuge for 5 min at 350 × *g* and 4 °C. Aspirate the supernatant.

12. Prepare the antibody panel (*see* Table 1), single antibody solutions, and isotype control cocktail. Prepare a sufficient volume for 100 μL per cell sample (*see* **Note 20**).
13. Resuspend each experimental sample in 100 μL of the antibody panel. Add flow cytometry buffer to unstained controls. Add 100 μL single stain solutions to compensation controls and 100 μL isotype cocktail to isotype controls. Incubate for 30 min at room temperature protected from light.
14. Add 1 mL of flow cytometry buffer to wash the cells and centrifuge for 5 min at 350 × *g* and 4 °C. Aspirate the supernatant.
15. Resuspend the pellets in 500 μL PBS. Add 500 μL of 4% PFA and mix gently. Incubate for 25 min at 4 °C protected from light.
16. Add 2 mL of flow cytometry buffer and centrifuge for 5 min at 350 × *g* and 4 °C.
17. Aspirate the supernatant carefully to not disturb the pellet. Resuspend each sample in 500 μL flow cytometry buffer (*see* **Note 21**).
18. Leave the cell samples at 4 °C protected from light. The samples should be run on the flow cytometer the following day for best results. The samples should not be kept at 4 °C for more than a week prior to analysis.

3.4 Immunophenotyping Flow Cytometry and Gating of Leukocyte Subpopulations

1. Immediately prior to running samples on the flow cytometer, filter the solution through a 40 μm cell strainer into a new flow tube (*see* **Note 22**).
2. Optimize the photomultiplier tube (PMT) voltages (e.g., *see* Table 2) and then perform the fluorescence compensation with the flow cytometer software (e.g., *see* Table 3) using the unstained and single stain controls (*see* **Note 23**).
3. Run the samples and controls for each tissue type on the flow cytometer. Use a forward scatter and side scatter collection gate (*see* Fig. 2a) to collect at minimum 550,000, 30,000, and 100,000 gated events for the heart, blood, and spleen, respectively (*see* **Note 24**).
4. Clean the flow cytometry line with detergent for 5 min followed by water for 10 min.
5. For flow cytometry data analysis, set up gates using controls for each tissue type separately (*see* Fig. 2). The unstained control can be used to set the live gate, as dead cells will not exclude Zombie Aqua™ viability dye (*see* Fig. 2c).

Table 2
Example photomultiplier tube voltage settings for leukocyte antibody panel

PMT	Voltage (V)
FSC-A	347
FSC-H	347
SSC-A	470
PE-A	595
PE-Cy7-A	571
BV421-A	427
AmCyan-A	497
AF488-A	465
PerCP-A	543
AF647-A	528
APC-Cy7-A	506

This Table lists the photomultiplier tube (PMT) voltage settings optimized for the leukocyte antibody panel using a BD FACSAria III flow cytometer and FACSDiva software v. 8.0.1. The pulse characteristics are area (A) or height (H)

Table 3
Example compensation matrix setup for leukocyte antibody panel

Fluorophore	PE	PE-Cy7	BV421	AmCyan	AF-488	PerCP	AF-647	APC-Cy7
PE	100	0.413	0.017	0.106	0.498	6.232	0.013	0
PE-Cy7	3.558	100	0.126	0.339	0.249	0.876	0.255	4.542
BV421	0.181	0.036	100	9.655	0.192	0.33	0.175	0.015
AmCyan	0	0	7.145	100	0.204	0	0	0
AF488	0	0.007	0	1.65	100	2.421	0	0
PerCP	0.324	3.899	0.193	0.441	0.195	100	4.407	2.184
AF647	0	1.316	0.076	0	0	0.784	100	4.716
APC-Cy7	0.266	44.041	0.339	1.233	0.574	0.494	7.915	100

Compensation was calculated by the BD FACSDiva software v. 8.0.1 using the FACSAria III cytometer. The photomultiplier tube voltages were set as in Table 3. Percent spillover correction is listed for each channel. All signals are pulse area

6. Check to see that the combined isotype control for each tissue shows a similar level of fluorescence signal as the unstained control. This can indicate that the washing steps for the staining have been successful (*see* **Note 25**).

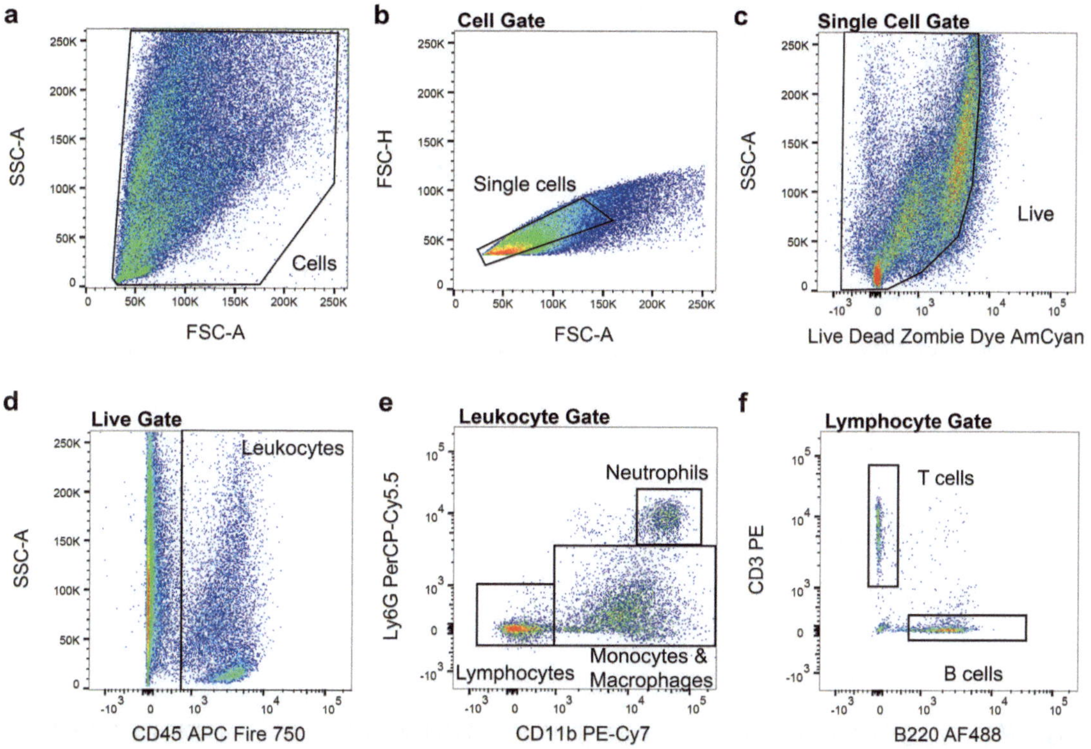

Fig. 2 Gating strategy for flow cytometry of mouse leukocytes after biomaterial injection for cardiac repair. This example is for a heart digest isolated 2 days post-biomaterial injection in a mouse model of MI. (**a**) A gate is placed around the cell population, so that the flow cytometer will analyze enough sample volume to record 550,000 events in this gate. (**b**) Cells are gated to remove doublets. (**c**) Single cells are gated on the live population based on the exclusion of live-dead dye. (**d**) Live CD45 expressing leukocytes. (**e**) Leukocyte subpopulations of neutrophils, monocytes, macrophages, and lymphocytes. (**f**) A lymphocyte gate is used to identify T cell and B cell subpopulations

7. Use the unstained or FMO control for each tissue to set the negative gates for CD45-APC/Fire 750, CD11b-PE/Cy7, Ly6G PerCP-Cy5.5, CD3-PE, and B220-AF488 using the live single cell population (*see* Fig. 2d–f).
8. Use the tissue-specific FMO controls to set the negative gates from F480-AF647 and Ly6C-BV421 in the monocyte- and macrophage-gated population (*see* Fig. 3). For the blood only, use the Ly6C marker for monocyte subpopulation gates as there are not F480^{+} macrophages in the blood (*see* Fig. 4) (*see* **Note 26**). For a comprehensive expression profile of all gated leukocyte subpopulations, *see* Table 4.

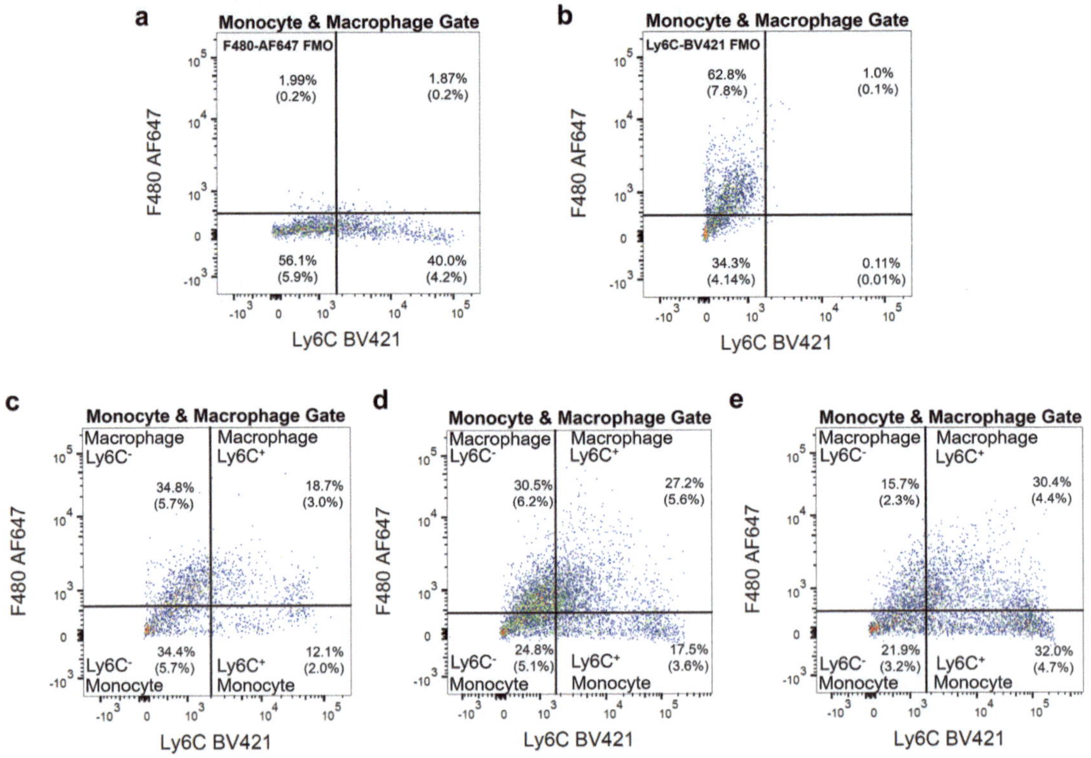

Fig. 3 Monocyte gating strategy of heart samples to detect the monocytic response to a biomaterial treatment. Mouse heart digests were isolated 2 days post-biomaterial delivery in a mouse model of MI. The monocyte and macrophage cell gate was separated into monocytes and macrophages based on F480 expression using an F480-AF647 FMO control (**a**). Each population was then split based on Ly6C expression using an Ly6C-BV421 FMO control (**b**). (**c**) A healthy heart sample shows less inflammation as compared to (**d**) a PBS injected MI heart. (**e**) A heart injected with a type III collagen biomaterial exhibits increased Ly6C^{+} monocyte infiltration as compared to a PBS-treated control heart. The percentage of monocyte and macrophage cell gate and percentage of live gate (in brackets) is displayed for each gated subpopulation

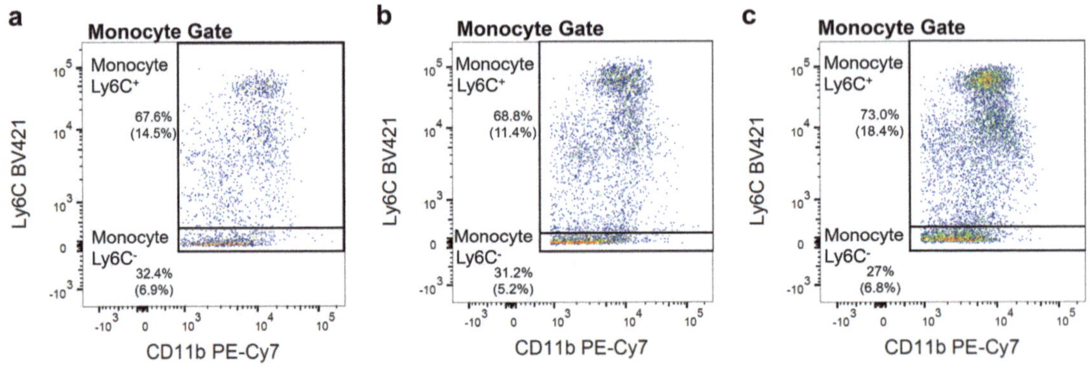

Fig. 4 The monocyte gating strategy of blood samples to detect the circulating monocytic response to biomaterial treatment of the heart. Mouse blood leukocytes were isolated 2 days post-biomaterial delivery in a mouse model of MI. The monocyte and macrophage cell gate was separated into Ly6C^{+} and Ly6C^{-} subpopulations. (**c**) The blood of a mouse treated with a type III collagen biomaterial has higher circulating levels of Ly6C^{+} monocytes as compared to (**a**) a healthy animal, and (**b**) a PBS-treated control MI mouse. Percentage of monocyte and macrophage cell gate and percentage of live (in brackets) is displayed for each gated subpopulation

Table 4
Expression profile of gated leukocyte subpopulations detected with this flow cytometry protocol

Cell type	Gated population
Leukocyte	$CD45^+$
Monocyte/Macrophage	$CD45^+CD11b^+Ly6G^-$
Neutrophil	$CD45^+CD11b^+Ly6G^+$
Lymphocyte	$CD45^+CD11b^-Ly6G^-$
Monocyte	$CD45^+CD11b^+Ly6G^-F480^-$
$Ly6C^+$ Monocyte	$CD45^+CD11b^+Ly6G^-F480^-Ly6C^+$
$Ly6C^-$ Monocyte	$CD45^+CD11b^+Ly6G^-F480^-Ly6C^-$
Macrophage	$CD45^+CD11b^+Ly6G^-F480^+$
$Ly6C^+$ Macrophage	$CD45^+CD11b^+Ly6G^-F480^+$ $Ly6C^+$
$Ly6C^-$ Macrophage	$CD45^+CD11b^+Ly6G^-F480^+$ $Ly6C^-$
T lymphocyte	$CD45^+CD11b^-Ly6G^-CD3^+B220^-$
B Lymphocyte	$CD45^+CD11b^-Ly6G^-CD3^-B220^+$

This Table lists the presence or absence of markers in gated leukocyte subpopulations. All cells were previously gated on single cell and live populations

9. Calculate the cell density for each leukocyte subpopulation. The percentage of the live cell gate that corresponds to each leukocyte subpopulation gate can be converted to cell density using the live isolated cell count and the mass of tissue and volume of blood (*see* **Note 27**). The conversion formula is as follows:

$$\text{Cell density} = \%\text{live cells} \times (\text{live cell count/mg tissue or mL blood})$$

4 Notes

1. PBS buffer that contains calcium and magnesium ions can cause cells to aggregate in suspension. Using calcium and magnesium-free PBS will help maintain a single cell suspension and reduce potential clogs in the flow cytometer line.
2. To coat microcentrifuge tubes with EDTA, fill 5 mL tubes with 50 mM EDTA and incubate overnight at 4 °C. Prior to use, aspirate the EDTA and then add exactly 100 μL of 50 mM EDTA to each tube to mix with the isolated blood. Alternatively, commercially available EDTA-coated blood isolation

tubes can be used. The 50 mM EDTA is prepared by diluting 0.5 M pH 8.0 stock solution to 1:10 in deionized water. To prepare the EDTA stock solution, dissolve 186 g of EDTA in 800 mL deionized water. While stirring, add ~20 g of NaOH pellets until the solution reaches a pH of 8.0. The EDTA will not fully dissolve until the pH is 8.0. Autoclave to sterilize if required.

3. To prepare the EDTA-coated syringes, fill a 1 mL syringe (with 25G needle) with 50 mM EDTA. Prior to blood collection via cardiac puncture, eject the EDTA out of the syringe. This will ensure the blood does not coagulate during collection. The 25G needle should be 3/8 inches in length for cardiac puncture from mouse hearts as longer needles lead to puncturing through the heart ventricle wall, which are shallow. This will lead to blood leakage from the ventricles and less blood drawn.
4. Prepare cardiac digestion buffer using sterile technique and sterile filtered (0.22 μm filter) solutions. To prepare 10 mL of cardiac digestion buffer (enough for 15 hearts at 750 μL per heart), mix on ice: 200 μL 0.25 M sodium bicarbonate, 200 μL 0.5 M HEPES buffer pH 7.5, 10 μL $MgCl_2$ 1 M stock, 5 μL 1 M KCL stock, and 17.8 μL DNAase 1 from a 0.28 U/mL stock solution. Dissolve 15 mg Collagenase Type II (188 U/mL) and 10 mg Collagenase D (0.15 U/mL) in 1 mL of cold PBS, and add to previous mixture. Add 200 μL hyaluronidase 489 U/mL stock and fill the buffer up to 10 mL with sterile PBS. Prepare 1 mL aliquots of the cardiac digestion buffer and freeze at −20 °C. If stored as described, the cardiac digestion buffer can be used for up to 6 months after preparation. Avoid freeze-thaw.
5. An alternative RBC lysis buffer can be prepared in the lab using 155 mM ammonium chloride, 0.1 mM EDTA, and 12 mM sodium bicarbonate. For 100 mL of RBC solution, dissolve 0.83 g of ammonium chloride in 80 mL deionized water, add 20 μL of 0.5 M EDTA stock solution, and 4.8 mL of 0.25 M sodium bicarbonate. Top up the volume to 100 mL using deionized water.
6. To prepare 500 mL of flow cytometry buffer: add 10 mL of FBS, 0.25 mL of 0.5 M EDTA stock, 5 μL of 1.4 U/mL (10 mg/mL) DNAase 1, 5 mL of 1% sodium azide stock, and fill up to 500 mL with PBS. When handling sodium azide, operate in a well-ventilated area with personal protective equipment and protect the stock solution from light.
7. To prepare Zombie Aqua™ stock solution, dissolve lyophilized powder in 100 μL of DMSO and make aliquots of 20 μL to freeze at −20 °C. Do not freeze-thaw. The dye solution should be used within 1 month of dissolving in solution or else the

fluorescence signal will be faded. Another cell viability dye can be used, but ensure that the fluorescence from the dye is in the AmCyan filter; otherwise, the fluorophores of the antibodies will need to be adapted.

8. Isotype antibodies have the same immunoglobulin (Ig) subtype as the detection antibody and are conjugated to the same fluorophores. For example, Ly6C-BV421 is a rat IgM antibody, and therefore the isotype control is a rat IgM-BV421 nonspecific antibody. This negative control can estimate the efficiency of the washing steps in the protocol and the Fc nonspecific binding of antibodies by comparing it to the unstained sample [37]. An isotype cocktail contains all fluorescently conjugated isotypes of antibodies in the experimental panel, as well as the cell viability dye, at the same concentration as the experimental antibody solution.

9. Any alternative Fc blocking solution for mouse samples can be used. This solution blocks Fc receptors on leukocytes that can bind to Fc domains on experimental antibodies rather than their specific variable domains, yielding a false-positive signal.

10. Be careful not to pierce the heart when cutting through the rib cage. Use fine dissection scissors to pierce the diaphragm and cut along the most lateral portion of the ribs on the right-hand side of the mouse. This will allow you to avoid the heart and have a clearer visualization of the heart when dissecting the left side of the ribs.

11. For cardiac puncture, ensure that the heart is clearly dissected away from the ribs and is lying flat. In MI mice, the heart can be stuck to the chest wall due to fibrotic tissue, so take care to not pierce the heart with scissors as you dissect. Move the plunger of the syringe up slightly (10 μL mark) prior to inserting the needle into the heart. The mouse ventricles are very shallow, insert the needle bevel gently, and begin to slowly pull the syringe plunger to aspirate blood. If the flow of blood into the needle slows, adjust the depth of the needle (pull towards operator) or insert into a different position in the heart. Lacerating the heart by pushing the needle through the ventricle wall will cause blood loss. An alternative method is to collect blood from an anesthetized animal, this can yield a greater blood volume drawn due to the live heartbeat. When doing the cardiac puncture in a mouse post-sacrifice, ensure to draw the blood within 5 min of death as the heart will still be beating moderately, allowing for easier blood collection.

12. The cardiac tissue must be constantly agitated during digestion, as insufficient agitation will result in poor cell yield and cell death. A successful cardiac digest will be a cloudy yellow-

brown solution with only a few non-digested fibrous tissue pieces.

13. Insufficient agitation of the blood sample will yield poor RBC lysis.

14. Resuspended cell pellets from blood post-RBC lysis should be no darker than pale pink to clear in color. If RBC lysis is incomplete, there is a risk for the samples to clog the flow cytometer line leading to difficulties in running the analysis.

15. The fibrous plug from the spleen can be incubated for 15–30 min in a collagenase D cocktail (in PBS) to fully extract embedded cells. This can be important for optimal isolation of dendritic cells and some macrophages subpopulations.

16. For flow cytometry samples, cell density lower than 250,000 cells/sample will require long run times on the machine and possibly have too low a sample volume to collect enough cellular events for leukocyte subpopulation analysis. This is especially true for cardiac digests where the leukocyte population is only 20% of the live cell population isolated. However, cell concentrations greater than 3×10^6 cells/sample can exceed the labeling capacity of the antibody solution and yield false negatives.

17. FMO controls include all antibodies and dyes of the experimental samples except for the antibody being controlled for. FMO samples control for spectral overlap in antibody panels [38]. This provides a negative control for gating, especially for weaker fluorescent signals or rare cell subpopulations. FMO controls can be used for all antibodies in a panel, but an additional mouse may need to be sacrificed to have enough cells for all seven control samples per tissue. Strong positive population gates that separate well from negative populations can be set using the unstained controls.

18. The spleen is used for single stain compensation controls for two reasons: there is a high yield of cells from the spleen that are predominantly leukocytes and the spleen contains strong positive populations for all leukocyte subsets that are detected by the antibody panel. Strong, clear positive signals from each antibody in the panel are required for compensation. Also, the cells should be the same type (mouse leukocytes) as those being analyzed in the experimental samples.

19. Zombie viability dye is required for isotype controls as dead cells can give off higher fluorescent background that is not associated with the background staining of live cells in the sample.

20. The antibody panel described in this method can be used as a base panel for leukocyte flow cytometry. Table 1 lists the

antibodies used herein and their suppliers; however, these can be substituted with alternative antibodies and products from other companies, by following the methods described. For example, some flow cytometers can run 10 or more fluorophores at once so at least three different fluorophore-conjugated antibodies can be added to this panel. Or alternatively, if the objective was to study macrophage subpopulations in the heart, then CD3 and B220 can be removed from the current panel while two macrophage subpopulation markers can be added conjugated to PE and AF488 (CD206 and CD38 or MHC-II and CCR2) [39–42]. There are enough cells isolated from a mouse heart to have two cell samples per mouse. These samples can be labeled with two alternative versions of the basic leukocyte panel. This multiple panel analysis increases the amount of information you can get from one mouse trial.

21. Final sample volumes lower than 400 μL can lead to air being drawn into the flow cytometer.
22. Cells need to be strained prior to loading in the flow cytometer to reduce doublets and to prevent clog formation in the flow cytometer lines.
23. Compensation should be performed using the software provided with the flow cytometer rather than manually [43]. Be sure to use the same voltage settings in the FSC and SSC channels as would be required for the experimental sample. FSC and SSC are measures of cell size and granularity, respectively. To perform proper compensation, the positive gate in the single stain controls must be from the same fluorophore and be of similar or greater intensity as the positive cell population in the experimental sample. For antigens with low expression, use an antibody with the same conjugated fluorophore, but that recognizes a more highly expressed antigen. For F480-AF647, a single stained control using spleen cells labeled with CD3e-AF647 provided a cleaner positive signal that enhanced compensation for the leukocyte antibody panel.
24. The cell gate placed on the FSC vs. SSC plot is used to instruct the flow cytometer to analyze enough sample volume to detect a certain number of cellular events. The event numbers per tissue type indicated in the methods resulted in sufficient events detected in each gated leukocyte subpopulation for analysis.
25. In this protocol, isotype controls showed similar intensities for each fluorophore as compared to the unstained control. There was no difference in the negative gates set using either control. All markers, except for F4/80 and Ly6C, that were controlled

using FMOs, showed clear, strong positive populations that were well-separated from the negative unstained cells.

26. The monocyte population could be further divided into low, medium, and high Ly6C expressing cells.

27. The cell density conversion is necessary, as the flow cytometer does not record the sample volume taken up to analyze a given cell event number. An alternative way for absolute cell counting is to add a known concentration of commercially available counting beads to the flow cytometry sample.

Acknowledgments

This work was supported by the Canadian Institutes of Health Research and the Natural Sciences and Engineering Research Council (EJS and EIA), a UOHI start-up grant (EIA), and a CIHR Frederick Banting and Charles Best Canada Graduate Scholarship (SM).

References

1. Suuronen EJ, Venoit JP, Wong S et al (2006) Tissue-engineered injectable collagen-based matrices for improved cell delivery and vascularization of ischemic tissue using CD133+ progenitors expanded from the peripheral blood. Circulation 114:I-138–I-144
2. Davis ME, Motion JPM, Narmoneva DA et al (2005) Injectable self-assembling peptide nanofibers create intramyocardial microenvironments for endothelial cells. Circulation 111:442–450
3. Padin-Iruegas ME, Misao Y, Davis ME et al (2009) Cardiac progenitor cells and biotinylated insulin-like growth factor-1 nanofibers improve endogenous and exogenous myocardial regeneration after infarction. Circulation 120:876–887
4. Lin Y-D, Yeh M-L, Yang Y-J et al (2010) Intramyocardial peptide nanofiber injection improves postinfarction ventricular remodeling and efficacy of bone marrow cell therapy in pigs. Circulation 122:S132–S141
5. Christman KL, Fok HH, Sievers RE et al (2004) Fibrin glue alone and skeletal myoblasts in a fibrin scaffold preserve cardiac function after myocardial infarction. Tissue Eng 10: 403–409
6. Dorsey SM, McGarvey JR, Wang H et al (2015) MRI evaluation of injectable hyaluronic acid-based hydrogel therapy to limit ventricular remodeling after myocardial infarction. Biomaterials 69:65–75
7. Seif-Naraghi SB, Singelyn JM, Salvatore MA et al (2013) Safety and efficacy of an injectable extracellular matrix hydrogel for treating myocardial infarction. Sci Transl Med. 5:173ra25-173ra25
8. Leor J, Tuvia S, Guetta V et al (2009) Intracoronary injection of in situ forming alginate hydrogel reverses left ventricular remodeling after myocardial infarction in swine. J Am Coll Cardiol 54:1014–1023
9. Ahmadi A, McNeill B, Vulesevic B et al (2014) The role of integrin α2 in cell and matrix therapy that improves perfusion, viability and function of infarcted myocardium. Biomaterials 35: 4749–4758
10. Purcell BP, Lobb D, Charati MB et al (2014) Injectable and bioresponsive hydrogels for on-demand matrix metalloproteinase inhibition. Nat Mater 13:653–661
11. Hosoyama K, Ahumada M, McTiernan CD et al (2018) Nanoengineered electroconductive collagen-based cardiac patch for infarcted myocardium repair. ACS Appl Mater Interfaces 10:44668–44677
12. Fujimoto KL, Tobita K, Merryman WD et al (2007) An elastic, biodegradable cardiac patch induces contractile smooth muscle and improves cardiac remodeling and function in

subacute myocardial infarction. J Am Coll Cardiol 49:2292–2300

13. Dai W, Wold LE, Dow JS, Kloner RA (2005) Thickening of the infarcted wall by collagen injection improves left ventricular function in rats. J Am Coll Cardiol 46:714–719
14. Christman KL, Lee RJ (2006) Biomaterials for the treatment of myocardial infarction. J Am Coll Cardiol 48:907–913
15. Rane AA, Christman KL (2011) Biomaterials for the treatment of myocardial infarction. J Am Coll Cardiol 58:2615–2629
16. Sadtler K, Wolf MT, Ganguly S et al (2019) Divergent immune responses to synthetic and biological scaffolds. Biomaterials 192:405–415
17. Yanez M, Blanchette J, Jabbarzadeh E (2018) Modulation of inflammatory response to implanted biomaterials using natural compounds. Curr Pharm Des 23:6347–6357
18. Morris AH, Stamer DK, Kyriakides TR (2017) The host response to naturally-derived extracellular matrix biomaterials. Semin Immunol 29:72–91
19. Anderson JM, Rodriguez A, Chang DT (2008) Foreign body reaction to biomaterials. Semin Immunol 20:86–100
20. Vasconcelos DP, Águas AP, Barbosa MA et al (2019) The inflammasome in host response to biomaterials: bridging inflammation and tissue regeneration. Acta Biomater 83:1–12
21. Tang L, Eaton JW (1995) Inflammatory responses to biomaterials. Am J Clin Pathol 103:466–471
22. Ogle ME, Segar CE, Sridhar S, Botchwey EA (2016) Monocytes and macrophages in tissue repair: implications for immunoregenerative biomaterial design. Exp Biol Med 241:1084–1097
23. Mariani E, Lisignoli G, Borzì RM, Pulsatelli L (2019) Biomaterials: foreign bodies or tuners for the immune response? Int J Mol Sci 20:636
24. Frangogiannis N (2002) The inflammatory response in myocardial infarction. Cardiovasc Res 53:31–47
25. Yang F, Liu Y-H, Yang X-P et al (2002) Myocardial infarction and cardiac remodelling in mice. Exp Physiol 87:547–555
26. Blackburn NJR, Sofrenovic T, Kuraitis D et al (2015) Timing underpins the benefits associated with injectable collagen biomaterial therapy for the treatment of myocardial infarction. Biomaterials 39:182–192
27. Horckmans M, Ring L, Duchene J et al (2017) Neutrophils orchestrate post-myocardial infarction healing by polarizing macrophages towards a reparative phenotype. Eur Heart J 38:187–197
28. Adan A, Alizada G, Kiraz Y et al (2017) Flow cytometry: basic principles and applications. Crit Rev Biotechnol 37:163–176
29. Laerum OD, Farsund T (1981) Clinical application of flow cytometry: A review. Cytometry 2:1–13
30. Baumgarth N, Roederer M (2000) A practical approach to multicolor flow cytometry for immunophenotyping. J Immunol Methods 243:77–97
31. Lagasse E, Weissman IL (1996) Flow cytometric identification of murine neutrophils and monocytes. J Immunol Methods 197: 139–150
32. Nahrendorf M, Swirski FK, Aikawa E et al (2007) The healing myocardium sequentially mobilizes two monocyte subsets with divergent and complementary functions. J Exp Med 204:3037–3047
33. Swirski FK, Nahrendorf M, Etzrodt M et al (2009) Identification of splenic reservoir monocytes and their deployment to inflammatory sites. Science 325:612–616
34. Heidt T, Courties G, Dutta P et al (2014) Differential contribution of monocytes to heart macrophages in steady-state and after myocardial infarction. Circ Res 115:284–295
35. Cheng HS, Besla R, Li A et al (2017) Paradoxical suppression of atherosclerosis in the absence of microRNA-146a. Circ Res 121: 354–367
36. McLaughlin S, McNeill B, Podrebarac J et al (2019) Injectable human recombinant collagen matrices limit adverse remodeling and improve cardiac function after myocardial infarction. Nat Commun 10:4866
37. Keeney M, Gratama JW, Chin-Yee IH, Sutherland DR (1998) Isotype controls in the analysis of lymphocytes and CD34+ stem and progenitor cells by flow cytometry--time to let go! Cytometry 34:280–283
38. Maecker HT, Trotter J (2006) Flow cytometry controls, instrument setup, and the determination of positivity. Cytometry A 69A:1037–1042
39. Epelman S, Lavine KJ, Beaudin AE et al (2014) Embryonic and adult-derived resident cardiac macrophages are maintained through distinct

mechanisms at steady state and during inflammation. Immunity 40:91–104

40. Lavine KJ, Epelman S, Uchida K et al (2014) Distinct macrophage lineages contribute to disparate patterns of cardiac recovery and remodeling in the neonatal and adult heart. Proc Nat Acad Sci U S A 111:16029–16034
41. Shiraishi M, Shintani Y, Shintani Y et al (2016) Alternatively activated macrophages determine repair of the infarcted adult murine heart. J Clin Invest 126:2151–2166
42. Jablonski KA, Amici SA, Webb LM et al (2015) Novel markers to delineate murine M1 and M2 macrophages. PLoS One 10:e0145342
43. Roederer M (2001) Spectral compensation for flow cytometry: visualization artifacts, limitations, and caveats. Cytometry 45:194–205

Chapter 20

Right Ventricular Outflow Tract Surgical Resection in Young, Large Animal Model for the Study of Alternative Cardiovascular Patches

Elizabeth C. Porter, Whitney L. Stoppel, Raymond K. Kudej, and Lauren D. Black III

Abstract

Tetralogy of Fallot (ToF) is a severe congenital heart defect (CHD) that requires surgical reconstruction soon after birth. Reconstructive surgery involves the implantation of synthetic cardiovascular patches to widen the right ventricular outflow tract (RVOT) and repair defects in the septal wall. However, synthetic patches can cause complications for these patients later in life as they do not integrate or adapt in the tissue of a growing patient; a limitation that could be solved with the development of a patch fabricated from a degradable biomaterial. Unfortunately, the lack of appropriate pre-clinical models has hindered the development of novel patch materials. Currently, most studies use rodent models to study the efficacy of new patch materials; however, large animal models are necessary to develop realistically sized patches in a clinically relevant growing heart where gradients in diffusion and length scales for cell migration are more similar to the human. Here, we describe a novel method by which a Satinsky vascular clamp is used to isolate RVOT muscle for resection followed by implantation of a cardiovascular patch in an appropriately young, rapidly growing porcine model.

Key words Tetralogy of Fallot, Right ventricular outflow tract, Surgical model, Porcine

1 Introduction

Congenital heart defects (CHD) are the most common type of birth defect, affecting nearly 1 in 100 live births [1–3]. Tetralogy of Fallot (ToF) is one of the most severe CHDs and estimated to occur at a rate of about 1 in 2500 live births per year [4]. The current clinical treatment is surgical widening of the right ventricular outflow tract (RVOT) and repair of the ventricular septal defect utilizing a synthetic GORE® ACUSEAL polytetrafluorethylene (PTFE) or a Dacron™ polyethylene terephthalate cardiovascular patch [5–8]. However, these synthetic patches can result in a variety of sequalae for young patients that persist throughout adulthood.

Kareen L. K. Coulombe and Lauren D. Black III (eds.), *Cardiac Tissue Engineering: Methods and Protocols*, Methods in Molecular Biology, vol. 2485, https://doi.org/10.1007/978-1-0716-2261-2_20,

The most problematic of these complications is the formation of a fibrotic capsule around the patch and fibrosis of the proximal tissue [9–11]. It is well understood that severe fibrosis limits the contractile ability of the heart in patients of all ages. To compensate, the heart pumps with greater force and more rapidly to supply the body with sufficient oxygenated blood. This is especially problematic in children because over time this results in right ventricular dilation and hypertrophy. Ultimately, this poses long-term health risks and reduces patient prognosis [12–14]. Secondly, recent advances in biomaterial design and development aimed to engineer cardiac tissues for the repair of damaged cardiac muscle have shown that material structure and architecture influences tissue level force generation and individual cardiomyocyte function [15, 16]. The synthetic patches commonly used in the clinic do not mimic the native alignment or electrical properties of the cardiac muscle limiting true integration of the patch and native tissue, likely contributing to the arrhythmias commonly found in postoperative adult patients who underwent RVOT reconstructive surgery as children [12, 17, 18]. Finally, as ToF is a severe CHD, most patients require reconstructive surgery in the first few years of their lives; however, as GORE® ACUSEAL or Dacron™ patches are synthetic materials, they do not "grow" with patients nor do they encourage patient tissue growth. This can necessitate reoperation to remove the old patch and implant a new patch to further widen the RVOT as the patient's heart grows. The combination of these complications and secondary interventions increases the mortality and morbidity associated with the condition [11, 19–21]. Therefore, there exists a clinical need to develop a patch or tissue graft that prevents severe fibrosis, recapitulates the native properties of the heart, promotes tissue growth, and limits the need for reintervention.

Interest in identifying such alternative cardiovascular patch or graft materials is underway [9, 10, 22, 23]. Fujimoto and colleagues showed that implantation of an elastomeric, biodegradable polyester urethane urea scaffold implanted over a surgical defect in the RVOT of adult rats resulted in more capillary formation and fibroblast cell infiltration than control PTFE patches [9]. Wainwright and colleagues showed decellularized porcine cardiac extracellular matrix patches implanted into the RVOT of rats were associated with better overall tissue remodeling and heart function than control Dacron patches [10]. However, most of these studies have been conducted using small animal models which presents problems for determining realistic alternative patch materials. Rodents have relatively short life spans, reaching adulthood by 8–12 weeks of age [24]. Pigs, on the other hand, are considered juvenile until about 1–1.5 years of age [25], allowing experimental timelines that stay within the bounds of clinically relevant young, growing hearts. Additionally, translating the experimental patch from a rodent model to a human is not as straightforward as allometric scaling of the patch size. For one, while rodent hearts and human hearts have anatomical similarities, the long-term effects on implanted patches cause by significant differences between the two species' heart rates and pressure loads are concerning and worthy of investigation [26–28]. Furthermore, for an implanted patch to truly integrate with the host tissue, it must be properly perfused and be infiltrated with the host's cells. This is feasible in a small patch implanted in a rodent's heart via diffusion and cell migration; however, diffusion and migration to the center of the patch will be limited in physically larger patches needed for human patients. The use of large-animal models for evaluation of alternative patch materials may avoid inherent problems with the translation from small animal models.

Outlined in this chapter are detailed methods for a surgical approach to mimic the widening of the RVOT by implantation of a cardiovascular patch in a young, large porcine animal model. We developed a novel method by which a Satinsky vascular clamp is used to isolate RVOT muscle for resection and implantation of a cardiovascular patch in approximately 2-month-old male pigs. To validate this model, the procedure describes the placement of a standard GORE® ACUSEAL patch. Future work could use this same procedure to investigate the efficacy of newly developed alternative patch materials that prevent severe fibrosis, recapitulates the native properties of the heart, promotes tissue growth, and limits the need for reintervention.

2 Materials

All materials (equipment, tools, and consumable supplies) that come in contact with the animal during surgery should be sterile or cleaned and autoclaved before the surgery.

1. Male pig (approximately 2 months or 20 kg).
2. General surgical equipment: Surgical table, surgical lights, heating pad, mechanical ventilator, isoflurane anesthesia system, cauterizer, electric razor.
3. General surgical tools: #10 and #15 scalpels, fine forceps, fine scissors, regular forceps, regular scissors, needle holders.
4. Specialized surgical tools: Rib spreader, Satinksy vascular clamp.
5. General surgical supplies: Drapes, towels, towel clamps, bowls, gauze, cotton-tipped applicators, ophthalmic ointment, chlorhexidine, isopropyl alcohol, saline, skin glue.
6. Intubation materials: Nylon rope, laryngoscope, endotracheal tube, cuff syringe, sterile lubricant.

7. Needles: 18 g × 1″, 20 g × 1″, 20 g × 1.5″, 22 g × 1″ hypodermic needles.
8. Syringes: 1 cc–10 cc hypodermic syringes.
9. Catheters: 20 g × 1.25″ SurFlash® intravenous catheter with injection cap.
10. Sutures: 5-0 polypropylene non-absorbable monofilament, poliglecaprone 25 absorbable monofilament, 2-0 and 3-0 nylon non-absorbable monofilament, 1 and 2-0 polydioxanone absorbable monofilament.
11. Cardiovascular patch, such as GORE® ACUSEAL Cardiovascular Patch (or alternative patch for investigation).
12. Thoracotomy tube: 20Fr trocar catheter.
13. Immunosuppressant: PO Mycophenolate (500 mg).
14. Analgesics: Transdermal Fentanyl patch (50 mcg/h), intravenous and oral Meloxicam (0.3 mg/kg), perineural Bupivacaine (1.5 mg/kg).
15. Antibiotics: Intramuscular Zoetis Excede® (5 mg/kg) or ceftiofur (1 mL/20 kg).
16. Anesthetics: Intramuscular TKX mixture (0.08 mL/kg) for anesthetic induction (TKX mixture is a combination of telazol (8 mg/kg), ketamine (4 mg/kg), and xylazine (4 mg/kg)), 1–4% isoflurane by inhalation via endotracheal tube for maintenance anesthesia.

3 Methods

All experiments must be performed in compliance with institutional, local, and federal regulation. For our studies, all experiments were approved by Tufts University Institutional Animal Care and Use Committee. 20 kg pigs were obtained from the Cummings School Swine Unit and singly housed in 25 sq. ft. hog panel enclosures before and after surgery. Pigs were fed twice daily at 8 am and 3 pm and had continuous access to water.

3.1 Animal and Surgical Suite Preparation

1. Approximately 72 h before surgery, begin immunosuppression regime of Mycophenolate (500 mg) twice daily by PO. Following the procedure, reduce administration to once daily and continue until the end of experiment.
2. Approximately 24 h before surgery, place a transdermal fentanyl patch (50mcg/h) on the hind leg of the pig to provide continuous analgesia relief for 72 h.
3. Approximately 8 h prior to surgery, fast pigs. However, ensure continuous access to water.

4. The day before surgery, begin preparations of the surgical suite by cleaning and disinfecting all surfaces, including the floors.
5. The morning of the surgery, but before anesthetizing the pig, confirm all tools and supplies, as well as extra sets/packs, listed in Subheading 2 section are on hand. This includes general and specialized sterile surgical tools, sterile supplies, sterile needles, sterile syringes, sterile catheters, and sterile suture packs.
6. Additionally, prior to anesthetizing the pig, turn on and check anesthesia machines and oxygen and scavenge system for issues and proper function.

3.2 Animal Preoperative Care

1. On the day of surgery, administer an intramuscular injection of Excede (5 mg/kg) or ceftiofur (1 mL/20 kg) as a long-acting antibiotic.
2. On the day of surgery, administer a TKX mixture (0.08 mL/kg) via intramuscular injection into the neck of the pig as an induction agent for general anesthesia. The TKX mixture is a combination of telazol (8 mg/kg), ketamine (4 mg/kg), and xylazine (4 mg/kg).
3. Confirm anesthesia by checking for jaw tone and abolition of pharyngeal reflexes by touching the roof of the mouth and around the back of the throat for resistance with a laryngoscope.
4. Intubate pig (*see* **Note 1**): Place a nylon rope in the mouth to hold the head up and the upper jaw open. With a clean piece of gauze, extend the tongue off to the side of the incisors to not cause trauma to the tongue. Place the laryngoscope into the mouth gently pressing up on the soft pallet releasing the epiglottis. The laryngoscope blade will hold the epiglottis down and the arytenoids will be easily visualized. Next, place the endotracheal tube into the trachea.
5. Confirm proper intubation by feeling breaths coming out of tube. Successful intubation will also be confirmed using the capnograph on the anesthesia monitor once connected.
6. Place pig on a heating pad on a surgical table in right lateral recumbency to expose surgical site. Mildly extend the left front leg forward and restrain to the surgery table (*see* **Note 2**).
7. Connect endotracheal tube to a mechanical ventilator and set at a rate of 12–18 breaths per min with 10–15 mL/kg tidal volume for inspiration. Heart rate, respiratory rate, SpO_2%, temperature, $ETCO_2$%, and noninvasive blood pressure are monitored continuously throughout the procedure. $ETCO_2$ will determine if the animal is being properly ventilated.
8. At this point, administer isoflurane (1–4%) via inhalation and continue to administer to maintain anesthesia.

9. Administer eye lubrication.
10. Place an IV catheter in the ear vein. Administer 0.9% saline or lactated Ringer's solution at a rate of 10–20 mL/kg/h and continue to administer throughout the procedure.
11. Administer Meloxicam (0.3 mg/kg) intravenously for additional analgesia relief.
12. Aseptically prepare the left thorax for surgery by shaving the entire left thoracic wall, from thoracic inlet to 13th rib and sternum to spine. The surgical field is then prepped by alternate cleansing three times with chlorhexidine scrub/solution and 70% isopropyl alcohol.
13. Make final preparations of surgical suite: In an aseptic manner, drape clean ¾ "drape sheets on all sterile tables. Aseptically, place the sterile tools and supplies listed in Subheading 2 section on the covered tables.

3.3 Surgical Technique to Mimic Surgical Outcome of RVOT Widening

1. Using a #10 scalpel, incise the skin over the left fourth intercostal space to expose the latissimus dorsi muscle.
2. Incise and retract the pectoral muscle to expose the intercostal and serratus anterior muscles.
3. At this point, administer a perineural injection of bupivacaine (1.5 mg/kg) as a high intercostal nerve block immediately caudal to ribs 3–6. This will provide preventative local analgesia at the thoracotomy site.
4. Using a fine scissor, perform a left fourth intercostal thoracotomy.
5. Using a rib spreader, separate the fourth and fifth rib to fully expose the heart.
6. Elevate the pericardium using fine forceps and scissors to incise over the right ventricular outflow tract (RVOT) and proximal pulmonary artery, up to but not including the phrenic nerve. Push the pericardium aside using a cotton-tipped applicator to expose the right ventricular muscle and outflow tract (*see* **Note 3**).
7. Using a Satinksy vascular clamp, clamp the muscle of the RVOT (*see* **Note 4**) (Fig. 1a, b).
8. Using a #15 scalpel, incise and then resect (*see* **Note 5**) the clamped ventricular muscle creating an elliptical shaped transmural defect with a diameter that is at least 2 cm in length and 1 cm in width (Fig. 1c, d).

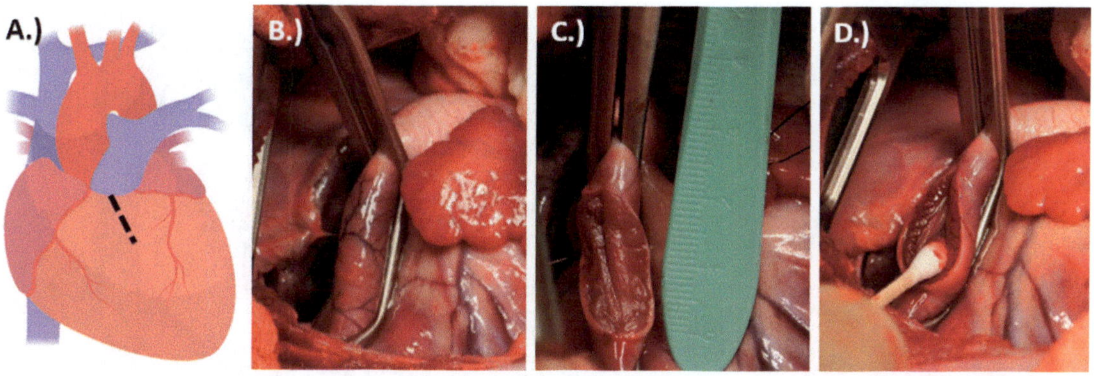

Fig. 1 Surgical technique to mimic surgical outcome of Right Ventricular Outflow Tract (RVOT) widening. (**a**) Pictorial representation of approximate location to clamp RVOT and make incision. (**b**) Isolate RVOT tissue with a Satinsky vascular clamp. (**c**) Resect approximately 2 cm of tissue from the RVOT. (**d**) Use a sterile cotton-tip applicator to separate muscle layers and confirm creation of a transmural defect

3.4 Surgical Technique for Implantation of a Cardiovascular Patch

1. Using fine scissors, cut a cardiovascular patch into an elliptical shape slightly larger than the defect (Fig. 2a, b).
2. Place a 5-0 polypropylene (non-absorbable) or poliglecaprone 25 (rapidly absorbable) suture through one end of the patch and secure one complete side of the patch over the defect using a simple continuous suture pattern. Tie-off the suture at the other end of the elliptical patch (*see* **Note 6**) (Fig. 2c, d).
3. Repeat suturing on the second side of the patch (Fig. 2e).
4. Release the vascular clamp and confirm clotting (*see* **Note 7**) (Fig. 2f).
5. Close the pericardium with 3-0 polydioxanone suture using a simple continuous pattern.

3.5 Post-Implantation and Postoperative Care

1. Using a scalpel make a small skin incision over the mid-seventh intercostal space for thoracostomy tube placement.
2. Insert a thoracostomy tube and prepare a loose stitch around the tube with 2-0 or 3-0 nylon suture.
3. Approximate the ribs with 1 polydioxanone using multiple simple interrupted sutures.
4. Approximate the thoracic muscle layers with 1 and 2-0 polydioxanone suture using a simple continuous pattern.
5. Close the subcutaneous and dermal layers with 2-0 polydioxanone suture using a simple continuous pattern.
6. Close the skin with 2-0 or 3-0 nylon suture using a simple continuous pattern.
7. Using a 60 cc syringe and 3-way stopcock attached to the thoracostomy tube, evacuate the thorax of air and fluid. Repeat evacuation to remove residual fluid and air and confirm maintenance of negative pressure.

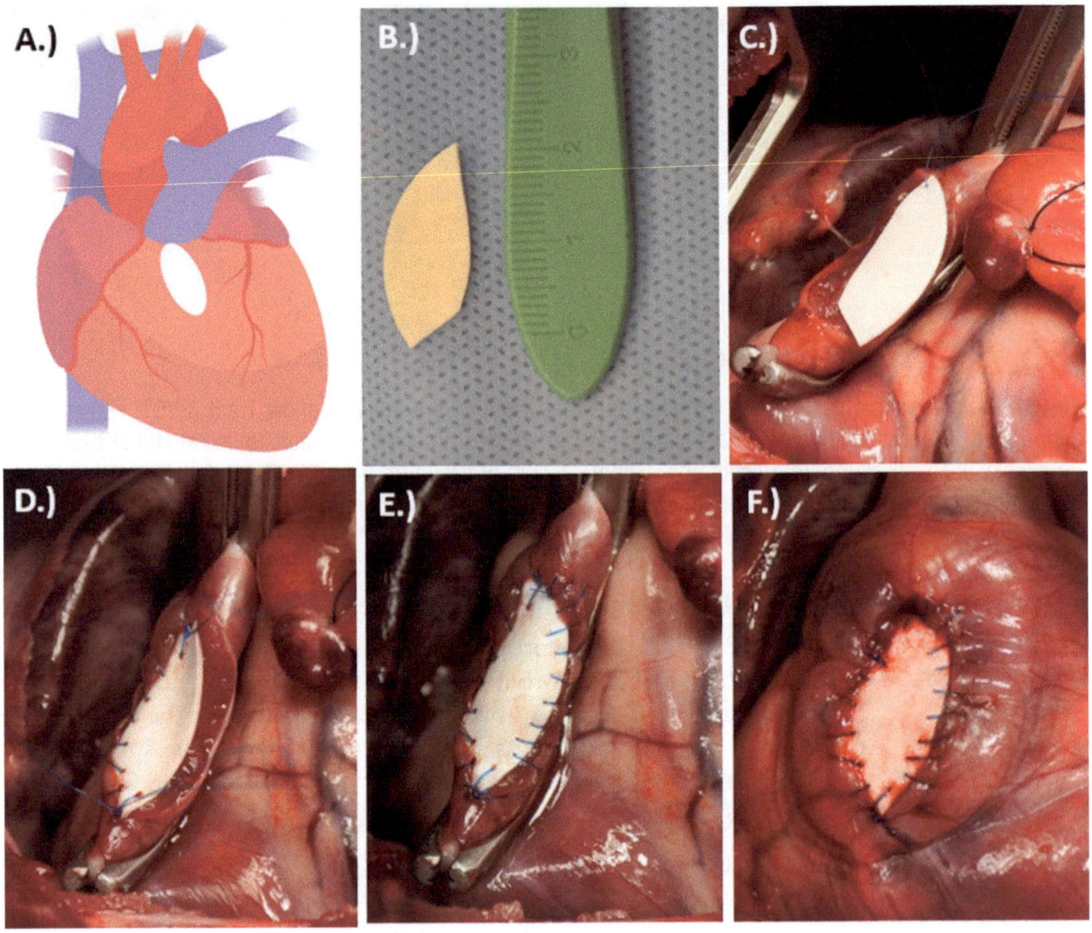

Fig. 2 Surgical technique for implantation of a cardiovascular patch. (**a**) Pictorial representation of approximate location to implant cardiovascular patch. (**b**) Cut a sterile cardiovascular patch into an elliptical shape slightly larger than the defect created using sterile surgical scissors. (**c**) Place a 5-0 polypropylene or poliglecaprone 25 suture through one end of patch. (**d**) Secure one complete side of the patch using a continuous over and over suture pattern and tie off the suture at the other end using a square knot. (**e**) Repeat suturing to secure the second side of the patch. (**f**) Release the vascular clamp and confirm clotting

8. Simultaneously, remove the thoracostomy tube and tighten the previously placed loose stitch to prevent leakage of air back into the chest.
9. Double-knot the suture.
10. To help prevent environmental contamination, apply skin glue to the thoracotomy incision and the incision made for the thoracostomy tube.
11. At this point, discontinue administration of isoflurane.
12. Slowly wean the pig from mechanical ventilation.
13. Once pharyngeal and laryngeal reflexes are restored, extubate the pig.

14. Following the procedure, reduce administration of mycophenolate (500 mg) to once daily and continue until the end of experiment.
15. Following the procedure, administer oral meloxicam (0.3 mg/kg) once daily for 3 days post-surgery for additional analgesic relief.

4 Notes

1. Proper intubation requires two persons to complete: one person assists with restraint and positioning of the pig, while the second performs the intubation. The first person places a nylon rope in the mouth and utilizes it to hold the head up and the upper jaw open. The other hand extends the tongue using a clean piece of gauze making sure the tongue is off to the side of the incisors as to not cause trauma to the tongue. The second person can then operate the laryngoscope and place the endotracheal tube.
2. Positioning the pig by extending and restraining the left front leg to the surgery table will help facilitate the surgical approach to the fourth intercostal space.
3. If the pericardium does not remain pushed aside, it can be temporarily "tacked" aside. This can be accomplished by temporary stay sutures placed between pericardium tissue and an adjacent rib.
4. The mass of ventricular muscle clamped between the Satinsky clamps is not trivial. Enough muscle should be isolated to allow for the creation of a transmural defect such that if released the hole would be at least 2 cm in length and 1 cm wide. However, collecting too much muscle between the clamp can have negative effects on the heart rhythm as well as disrupt blood flow, resulting in cardiac arrest in the worst-case scenario.
5. After incision, the tissue is resected to allow for expansion of the outflow tract as well as uniform adherence of the patch to the margins of the defect.
6. Using two sets of hands is the most efficient and effective way to implant the patch. One surgeon uses a cotton-tipped applicator to fully open the defect, while the second surgeon performs the suturing. For suturing, use a small, taper point needle and sutures at 2–3 mm depth and 2–3 mm apart.
7. After release of the clamp, expect some leaking around the edge of the patch. However, if securely implanted the blood will clot around the patch. Wait for bleeding to stop before continuing to the next step of closing the pericardium.

Acknowledgments

This work was supported by the Department of Defense CDMRP (Award Number W81XWH-16-1-0304). The authors would also like to thank Ms. Kimberly Gaspie and the LAMS Staff at the Cummings School of Veterinary Medicine for their expertise in anesthesia, surgical procedures, and animal care support.

References

1. Botto LD, Correa A, Erickson JD (2001) Racial and temporal variations in the prevalence of heart defects. Pediatrics 107(3):E32. https://doi.org/10.1542/peds.107.3.e32
2. Jenkins KJ, Correa A, Feinstein JA, Botto L, Britt AE, Daniels SR, Elixson M, Warnes CA, Webb CL, American Heart Association Council on Cardiovascular Disease in the Y (2007) Noninherited risk factors and congenital cardiovascular defects: current knowledge: a scientific statement from the American Heart Association Council on Cardiovascular Disease in the Young: endorsed by the American Academy of Pediatrics. Circulation 115(23): 2995–3014. https://doi.org/10.1161/CIRCULATIONAHA.106.183216
3. Benjamin EJ, Blaha MJ, Chiuve SE, Cushman M, Das SR, Deo R, de Ferranti SD, Floyd J, Fornage M, Gillespie C, Isasi CR, Jimenez MC, Jordan LC, Judd SE, Lackland D, Lichtman JH, Lisabeth L, Liu S, Longenecker CT, Mackey RH, Matsushita K, Mozaffarian D, Mussolino ME, Nasir K, Neumar RW, Palaniappan L, Pandey DK, Thiagarajan RR, Reeves MJ, Ritchey M, Rodriguez CJ, Roth GA, Rosamond WD, Sasson C, Towfighi A, Tsao CW, Turner MB, Virani SS, Voeks JH, Willey JZ, Wilkins JT, Wu JH, Alger HM, Wong SS, Muntner P, American Heart Association Statistics C, Stroke Statistics S (2017) Heart disease and stroke statistics-2017 update: a report from the American Heart Association. Circulation 135(10): e146–e603. https://doi.org/10.1161/CIR.0000000000000485
4. Parker SE, Mai CT, Canfield MA, Rickard R, Wang Y, Meyer RE, Anderson P, Mason CA, Collins JS, Kirby RS, Correa A, National Birth Defects Prevention N (2010) Updated National Birth Prevalence estimates for selected birth defects in the United States, 2004–2006. Birth Defects Res A Clin Mol Teratol 88(12):1008–1016. https://doi.org/10.1002/bdra.20735
5. Turrentine MW, McCarthy RP, Vijay P, McConnell KW, Brown JW (2002) PTFE monocusp valve reconstruction of the right ventricular outflow tract. Ann Thorac Surg 73(3):871–879.; discussion 879–880. https://doi.org/10.1016/S0003-4975(01)03441-5
6. Brown JW, Ruzmetov M, Vijay P, Rodefeld MD, Turrentine MW (2007) Right ventricular outflow tract reconstruction with a polytetrafluoroethylene monocusp valve: a twelve-year experience. J Thorac Cardiovasc Surg 133(5): 1336–1343. https://doi.org/10.1016/j.jtcvs.2006.12.045
7. Hauser M, Eicken A, Kuehn A, Hess J, Fratz S, Ewert P, Kaemmerer H (2013) Managing the right ventricular outflow tract for pulmonary regurgitation after tetralogy of Fallot repair. Heart Asia 5(1):106–111. https://doi.org/10.1136/heartasia-2013-010319
8. Sasson L, Houri S, Raucher Sternfeld A, Cohen I, Lenczner O, Bove EL, Kapusta L, Tamir A (2013) Right ventricular outflow tract strategies for repair of tetralogy of Fallot: effect of monocusp valve reconstruction. Eur J Cardio-Thorac Surg 43(4):743–751. https://doi.org/10.1093/ejcts/ezs479
9. Fujimoto KL, Guan J, Oshima H, Sakai T, Wagner WR (2007) In vivo evaluation of a porous, elastic, biodegradable patch for reconstructive cardiac procedures. Ann Thorac Surg 83(2):648–654. https://doi.org/10.1016/j.athoracsur.2006.06.085
10. Wainwright JM, Hashizume R, Fujimoto KL, Remlinger NT, Pesyna C, Wagner WR, Tobita K, Gilbert TW, Badylak SF (2012) Right ventricular outflow tract repair with a cardiac biologic scaffold. Cells Tissues Organs 195(1–2):159–170. https://doi.org/10.1159/000331400
11. Sandeep B, Huang X, Xu F, Su P, Wang T, Sun X (2019) Etiology of right ventricular restrictive physiology early after repair of tetralogy of Fallot in pediatric patients. J Cardiothorac Surg 14(1):84. https://doi.org/10.1186/s13019-019-0909-8

12. Perloff JK, Warnes CA (2001) Challenges posed by adults with repaired congenital heart disease. Circulation 103(21):2637–2643. https://doi.org/10.1161/01.cir.103.21.2637
13. Yoo BW, Kim JO, Kim YJ, Choi JY, Park HK, Park YH, Sul JH (2012) Impact of pressure load caused by right ventricular outflow tract obstruction on right ventricular volume overload in patients with repaired tetralogy of Fallot. J Thorac Cardiovasc Surg 143(6): 1299–1304. https://doi.org/10.1016/j.jtcvs.2011.12.033
14. Gatzoulis MA, Elliott JT, Guru V, Siu SC, Warsi MA, Webb GD, Williams WG, Liu P, McLaughlin PR (2000) Right and left ventricular systolic function late after repair of tetralogy of Fallot. Am J Cardiol 86(12): 1352–1357. https://doi.org/10.1016/s0002-9149(00)01241-8
15. Zhao Y, Korolj A, Feric N, Radisic M (2016) Human pluripotent stem cell-derived cardiomyocyte based models for cardiotoxicity and drug discovery. Expert Opin Drug Saf 15(11):1455–1458. https://doi.org/10.1080/14740338.2016.1223624
16. Kolewe ME, Park H, Gray C, Ye X, Langer R, Freed LE (2013) 3D structural patterns in scalable, elastomeric scaffolds guide engineered tissue architecture. Adv Mater 25(32): 4459–4465. https://doi.org/10.1002/adma.201301016
17. Gatzoulis MA, Balaji S, Webber SA, Siu SC, Hokanson JS, Poile C, Rosenthal M, Nakazawa M, Moller JH, Gillette PC, Webb GD, Redington AN (2000) Risk factors for arrhythmia and sudden cardiac death late after repair of tetralogy of Fallot: a multicentre study. Lancet 356(9234):975–981. https://doi.org/10.1016/s0140-6736(00)02714-8
18. Khairy P, Aboulhosn J, Gurvitz MZ, Opotowsky AR, Mongeon FP, Kay J, Valente AM, Earing MG, Lui G, Gersony DR, Cook S, Ting JG, Nickolaus MJ, Webb G, Landzberg MJ, Broberg CS, Alliance for Adult Research in Congenital C (2010) Arrhythmia burden in adults with surgically repaired tetralogy of Fallot: a multi-institutional study. Circulation 122(9):868–875. https://doi.org/10.1161/CIRCULATIONAHA.109.928481
19. Nollert G, Fischlein T, Bouterwek S, Böhmer C, Klinner W, Reichart B (1997) Long-term survival in patients with repair of tetralogy of Fallot: 36-year follow-up of 490 survivors of the first year after surgical repair. J Am Coll Cardiol 30(5):1374–1383. https://doi.org/10.1016/s0735-1097(97)00318-5
20. Therrien J, Marx GR, Gatzoulis MA (2002) Late problems in tetralogy of Fallot—recognition, management, and prevention. Cardiol Clin 20(3):395–404. https://doi.org/10.1016/s0733-8651(02)00010-3
21. Geva T (2012) Tetralogy of Fallot repair: ready for a new paradigm. J Thorac Cardiovasc Surg 143(6):1305–1306. https://doi.org/10.1016/j.jtcvs.2012.01.076
22. Sakai T, Li RK, Weisel RD, Mickle DA, Kim ET, Jia ZQ, Yau TM (2001) The fate of a tissue-engineered cardiac graft in the right ventricular outflow tract of the rat. J Thorac Cardiovasc Surg 121(5):932–942. https://doi.org/10.1067/mtc.2001.113600
23. Ozawa T, Mickle DAG, Weisel RD, Matsubayashi K, Fujii T, Fedak PWM, Koyama N, Ikada Y, Li RK (2004) Tissue-engineered grafts matured in the right ventricular outflow tract. Cell Transplant 13(2): 169–177. https://doi.org/10.3727/000000004773301852
24. Jackson SJ, Andrews N, Ball D, Bellantuono I, Gray J, Hachoumi L, Holmes A, Latcham J, Petrie A, Potter P, Rice A, Ritchie A, Stewart M, Strepka C, Yeoman M, Chapman K (2017) Does age matter? The impact of rodent age on study outcomes. Lab Anim 51(2):160–169. https://doi.org/10.1177/0023677216653984
25. Reiland S (1978) Growth and skeletal development of the pig. Acta Radiol Suppl 358:15–22
26. Papadimitriou D, Xanthos T, Dontas I, Lelovas P, Perrea D (2008) The use of mice and rats as animal models for cardiopulmonary resuscitation research. Lab Anim 42(3): 265–276. https://doi.org/10.1258/la.2007.006035
27. Hamlin RL, Altschuld RA (2011) Extrapolation from mouse to man. Circ Cardiovasc Imaging 4(1):2–4. https://doi.org/10.1161/CIRCIMAGING.110.961979
28. Milani-Nejad N, Janssen PM (2014) Small and large animal models in cardiac contraction research: advantages and disadvantages. Pharmacol Ther 141(3):235–249. https://doi.org/10.1016/j.pharmthera.2013.10.007

Index

Kareen L. K. Coulombe and Lauren D. Black III (eds.), *Cardiac Tissue Engineering: Methods and Protocols*, Methods in Molecular Biology, vol. 2485, https://doi.org/10.1007/978-1-0716-2261-2,

U

V

MIX
Papier aus verantwortungsvollen Quellen
Paper from responsible sources
FSC® C105338

If you have any concerns about our products,
you can contact us on
ProductSafety@springernature.com

In case Publisher is established outside the EU,
the EU authorized representative is:
Springer Nature Customer Service Center GmbH
Europaplatz 3, 69115 Heidelberg, Germany

Printed by Libri Plureos GmbH
in Hamburg, Germany